Real Analytic and Algebraic Geometry

Real Analytic and Algebraic Geometry

Proceedings of the International Conference,
Trento (Italy), September 21–25, 1992

Edited by

Fabrizio Broglia
Margherita Galbiati
Alberto Tognoli

Walter de Gruyter · Berlin · New York 1995

Editors

Fabrizio Broglia
Dipartimento di
Matematica
Università di Pisa
I-56127 Pisa, Italy

Margherita Galbiati
Dipartimento di
Matematica
Università di Pisa
I-56127 Pisa, Italy

Alberto Tognoli
Dipartimento di
Matematica
Università di Trento
I-38050 Povo (Trento), Italy

1991 Mathematics Subject Classification: 14-06; 14G30, 21B20, 32CXX, 58A07, 68C2C
Keywords: Real algebraic varieties, ordered fields, computational geometry, Nash functions, semialgebraic sets, analytic and subanalytic sets

⊗ Printed on acid-free paper which falls within the guidelines of the ANSI to ensure permanence and durability.

Library of Congress Cataloging-in-Publication Data

Real analytic and algebraic geometry : proceedings of the international conference, held in Trento, September 21–25, 1992 / edited by Fabrizio Broglia, Margherita Galbiati, Alberto Tognoli.
p. cm.
The conference was organized by the Centro internazionale per la ricerca matematica in Trento.
ISBN 3-11-013778-X
1. Geometry, Analytic – Congresses. 2. Geometry, Algebraic – Congresses. I. Broglia, Fabrizio, 1948– . II. Galbiati, M. (Margherita), 1946– . III. Tognoli, Alberto. IV. Centro internazionale per la ricerca matematica (Trento, Italy)
QA551.R38 1995
516.3'5–dc20 94-42179
CIP

Die Deutsche Bibliothek – Cataloging-in-Publication Data

Real analytic and algebraic geometry: proceedings of the international conference, held in Trento, September 21–25, 1992 / ed. by Fabrizio Broglia ... – Berlin ; New York : de Gruyter, 1995
ISBN 3-11-013778-X
NE: Broglia, Fabrizio [Hrsg.]

Printed in Germany. Printing: Gerike GmbH, Berlin. Binding: Lüderitz & Bauer GmbH, Berlin.
Cover design: Thomas Bonnie, Hamburg.

Introduction

The meeting "Real Analytic and Algebraic Geometry", held in Trento from September 21 to 25, 1992, was organized by the Centro Internazionale per la Ricerca Matematica (C.I.R.M.) of the Istituto Trentino di Cultura (I.T.C) in Trento, and partially sponsored by the Department of Mathematics of the University of Trento.

It was the second conference in Trento (the first was in 1988) in the series of the yearly meetings of the research group on Real Geometry, characterized by the variety of problems, aspects and methods (topological, analytical, algebraic), and at the same time by the unitary interests and language, with a special attention to the applications, this year mainly in robotics and vision.

This meeting was however the first time we met but our friend Mario Raimondo was not with us; his absence cast a shadow on our feelings and on the meeting.

All of us, we wish to dedicate to his memory this volume.

Fabrizio Broglia
Margherita Galbiati
Alberto Tognoli

The Editors would like to emphasize that this volume contains the papers presented at the meeting. All papers have been refereed.

The Editors would like to thank the authors, the referees, and Mr. Augusto Micheletti, for the always competent help in organizing the meeting and preparing the manuscript.

Table of Contents

Algebraic versus analytic basicness

*Carlos Andradas*and Jesús M. Ruiz**

Abstract. We study in this paper when a semialgebraic set which is basic semianalytic is basic semialgebraic, that is, when local analytic basicness implies global polynomial basicness. The main notion involved is that of *fan*, which we present in a purely geometric form. The final solution to the problem is given in terms of blowing-ups of fans. This contrasts with prior presentations of these matters, using valuations and quadratic forms. The reason why we can avoid all of that is the *ultrafilter theorem*. This "non-algebraic" approach can be pursued as far as we do not enter into proofs. These really require algebra, even hard commutative algebra as M. Artin's approximation theory. In fact, proofs are not given in this "geometric" paper, which we consider a necessary complement to the very technical and quite long "algebraic" monography where full details will be published.

1991 Mathematics Subject Classification: 14P10, 14P15.

Introduction. In this paper we deal with the problem of deciding under which circumstances a semialgebraic set which is basic semianalytic is also basic semialgebraic. The main result is Theorem II*, which is stated in paragraph 10. However, the paper contains no proof. Indeed, the proof of Theorem II* is very technical, long and complicated, and cannot be even sketched in a few pages. To quote the theories behind the scenes, first of all, we use the real spectrum, which makes the bridge between geometry and quadratic forms. Then fans are the crucial aspect of the abstract theory of quadratic forms over spaces of orderings, which in turn are a special type of real spectra. To deal with fans in the real spectrum setting, we use heavily henselization and more generally, ind-etale limits. One important ingredient to compare fans before and after henselization is M. Artin's approximation theory. Finally, we call in valuation theory, which is the main tool to define trivializations.

Quite surprisingly, all this stuff can be hidden if one does not enter into proofs. The reason for this masquarade to be possible is the so-called *ultrafilter theorem*, which permits defining orderings (and therefore also fans) as ultrafilters of constructible sets, so that statements and their meaning can be put in a suggestive "geometric" form. We believe that this hidding exercise is highly instructive, till the point that we conceive this note as a very convenient complement to the

*Partially supported by DGICYT, PB 89-0379-C02-02

"algebraic" proof of the theorem, [AR]. We also hope that this presentation will motivate some readers to study that paper and, more important, the theory of spaces of orderings and its geometric applications.

1. Real algebraic sets, regular functions and analytic functions. A *real algebraic set* is a subset $X \subset \mathbb{R}^n$ that can be described as the zero set of finitely many polynomials, that is:

$$X = \{x \in \mathbb{R}^n \mid f_1(x) = \cdots = f_s(x) = 0\},$$

with $f_1, \ldots, f_s \in \mathbb{R}[\mathrm{x}_1, \ldots, \mathrm{x}_n]$. Real algebraic sets can be endowed with different sheaves of functions, and here we will concentrate on the following two.

(a) A *regular function* on an open set U of $\mathbb{R}^n$ is a function $f : U \to \mathbb{R}$ that can be written as the quotient of two polynomials with a denominator that never vanishes in U. A *regular function on an open set W of X* is a function $f : X \to \mathbb{R}$ that extends to a regular function on some open neighborhood U of W in $\mathbb{R}^n$. All regular functions on W form a ring, which will be denoted by $\mathcal{R}(W)$. The family of these rings $\mathcal{R}(W)$ define the *sheaf of regular functions of X*, denoted by $\mathcal{R}_X$.

(b) An *analytic function* on an open set U of $\mathbb{R}^n$ is a differentiable function $f : U \to \mathbb{R}$ whose Taylor expansions converge all to the given function f. An *analytic function on an open set W of X* is a function $f : X \to \mathbb{R}$ that extends to an analytic function on some open neighborhood U of W in $\mathbb{R}^n$. All analytic functions on W form a ring, denoted by $\mathcal{O}(W)$. The collection of these $\mathcal{O}(W)$ define the *sheaf of analytic functions of X*, denoted by $\mathcal{O}_X$.

From now on, we fix a real algebraic set $X \subset \mathbb{R}^n$.

2. Semialgebraic and semianalytic sets. Once we are given a ring of real functions, we can introduce the so-called *constructible sets*, that is, the sets that can be described by means of sign conditions on finitely many functions of the given ring. In our case we find two different types of constructible sets, according to the two types of functions introduced above. At the algebraic level we have:

(a) A *semialgebraic set* is a subset $S \subset X$ such that

$$S = \bigcup_{i=1}^{p} \{x \in X \mid f_{i1}(x) > 0, \ldots, f_{is_i}(x) > 0, g_i(x) = 0\}$$

for some regular functions $f_{ij}, g_i \in \mathcal{R}(X)$.

For the analytic category we adopt exactly the same approach taken by Łojasiewicz in [Ł]:

(b) A *semianalytic set* is a subset $S \subset X$ such that for each point $a \in X$ there is an open neighborhood W with

$$S \cap W = \bigcup_{i=1}^{p} \{x \in W \mid f_{i1}(x) > 0, \ldots, f_{is_i}(x) > 0, g_i(x) = 0\}$$

for some analytic functions $f_{ij}, g_i \in \mathcal{O}(W)$.

We stress the two-fold difference between semialgebraic and semianalytic sets: first we pass from regular to analytic functions, then from a global to a local formula.

3. Complexity of semialgebraic and semianalytic sets. Clearly, the sign conditions that describe a fixed constructible set $S \subset X$ are far from unique, and depending on our choice, we will understand better the nature of S. For instance, if $X = \mathbb{R}^2$ and

$$S = \{x > 0, y > 0\} = \{x \geq 0, y \geq 0, x + y > 0\},$$

the first description tells us that S is open. This type of remark is summarized in the fact that every constructible set has a description that gives the closure by relaxing all strict inequalities involved, [BCR]. The construction of such a description is based on the fundamental fact that constructible sets can be very nicely stratified. In particular, this allows many arguments by induction on the dimension, and explains many special topological properties ([BCR]). In addition, in recent years, dimension has been connected, somehow surprisingly, to the notion of *complexity*. Here by *complexity of a constructible set* we understand the minimal number of functions needed to describe it by sign conditions. Thus, the question of deciding whether a point x belongs to the constructible set

$$S = \bigcup_{i=1}^{t} \{x \in X \mid f_{i1}(x) > 0, \ldots, f_{is_i}(x) > 0, g_i(x) = 0\}$$

can be represented by a decision tree of width t and depth s, the maximum of s_i's. The most important result concerning complexity is that for any constructible subset S of X, both, *the width and depth of this tree can be bounded by a number that depends only on the dimension d of X*. More precisely, s, that is, the number of strict inequalities, can be always taken $\leq d$, and there are examples which really need d ([B], [S], [ABR]). These results lead to many interesting questions, among which we select here a special one concerning the comparison of the algebraic and analytic categories. In fact, every semialgebraic set $S \subset X$ produces a semianalytic germ S_a at each point $a \in X$, and we can compare the complexities of S and S_a. This comparison is in general rather intricate and not much is known yet. Here we restrict ourselves to the study of a more concrete question which will be made precise after the next paragraph.

4. Basic sets. For the sake of simplicity, and to avoid technical details that could hide the essential ideas involved, we will henceforth assume that our real algebraic set X is *irreducible*, that is, that any proper algebraic subset $Y \subset X$ has dimension $< d = \dim(X)$.

We want to study constructible sets of minimal width complexity, that is, those that can be described without unions. To deal with a less restrictive condition we introduce the following definitions:

(a) A semialgebraic set S is *basic semialgebraic* if there are regular functions $f_1, \dots, f_s, h \in \mathcal{R}(X)$ such that

$$S \setminus \{x \in X \mid h(x) = 0\} = \{x \in X \mid f_1(x) > 0, \dots, f_s(x) > 0\} \setminus \{x \in X \mid h(x) = 0\},$$

and $\dim(\{x \in X \mid h(x) = 0\}) < d$.

(b) A semianalytic set S is *basic semianalytic* if for each point $a \in X$ there are an open neighborhood W and analytic functions $f_1, \dots, f_s, h \in \mathcal{O}(W)$ such that

$$S \setminus \{x \in W \mid h(x) = 0\} = \{x \in W \mid f_1(x) > 0, \dots, f_s(x) > 0\} \setminus \{x \in W \mid h(x) = 0\},$$

and $\dim(\{x \in W \mid h(x) = 0\}) < d$.

The above definitions have *generic nature*, that is, the sets involved coincide up to some piece of dimension smaller that d. In fact the above notions are ussually defined as *generic* basicness. However we will drop the term generic, because there will be no other type of basicness in this discussion. There is a nice geometric characterization of basic semialgebraic (resp. basic semianalytic) sets due to Bröcker, [B], which can be expressed in terms of ultrafilters and which we will recall later. Coming to the comparison between the algebraic and the analytic categories, it is obvious that a basic semialgebraic set is also basic semianalytic, but it is easy to find counterexamples for the converse assertion. However the search for such counterexamples gives rise to a limited series of obstructions that soon become repetitive. This makes reasonable to expect a consistent solution to the next question:
PROBLEM: Characterize geometrically basic semianalytic sets that are basic semialgebraic.

5. Examples. To get some feeling about the problem, the reader can check directly the following examples.

a) *Łojasiewicz's example: a semialgebraic set which is not basic semianalytic.*

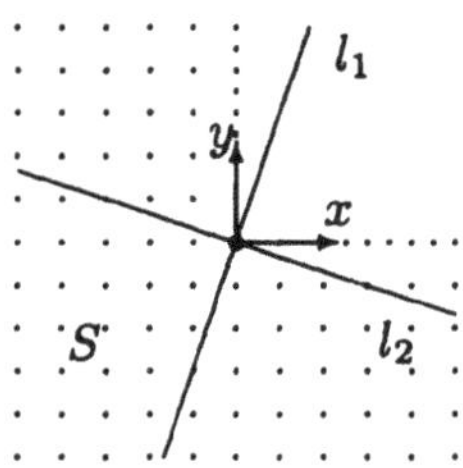

$$S = \{x < 0\} \cup \{y < 0\}$$

Very roughly said, the changes of sign through the origin O of a function f restricted to a line l_i are ruled by the order of the Taylor expansion of f at O. If there is a change along l_i, then the order must be odd, and if there is no change along l_i the order must be even. This remark explains the example. Similar examples are the following:

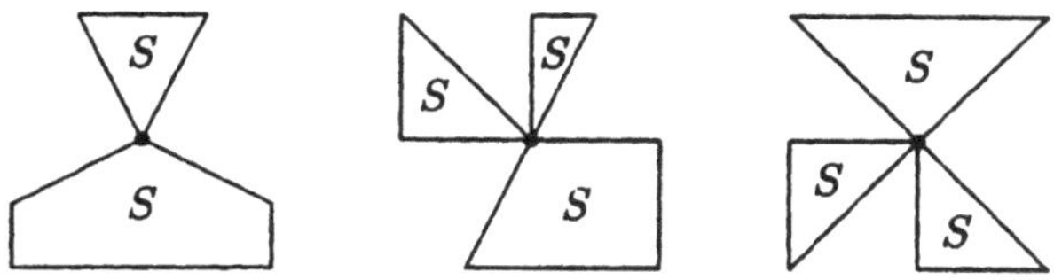

b) *A basic semianalytic set which is not basic semialgebraic.*

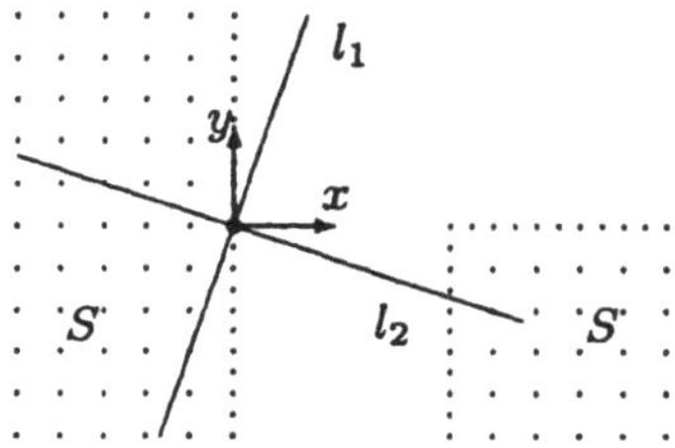

$$S = \{x < 0\} \cup \{x > 1, y < 0\}$$

Obviously S is basic semianalytic. However working as in the preceding example, using the degree of a polynomial f instead of its order at the origin, and analyzing the changes of sign through infinity of the restriction of f to l_i we easily get that S is not basic semialgebraic. Notice that working with infinity in the projective plane $(x_0 : x_1 : x_2)$, we find again an obstruction of Łojasiewicz's type (and therefore analytic) which explains this example:

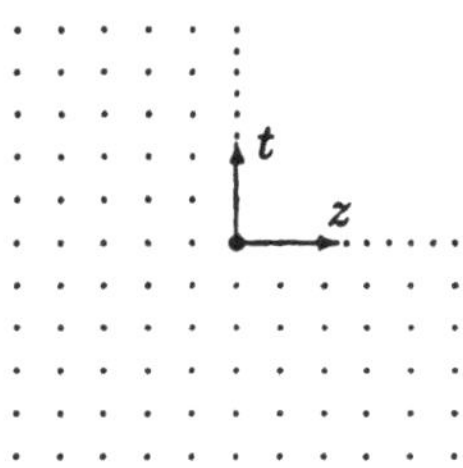

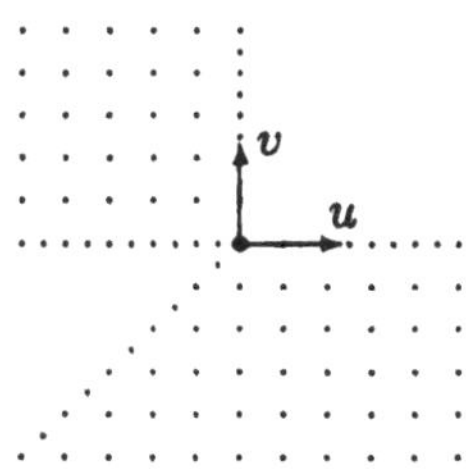

$S = \{z < 0\} \cup \{z < 1, t < 0\}$ $\qquad$ $S = \{uv < 0\} \cup \{v < u < 0\}$

$$\begin{cases} x = x_1/x_0 \\ y = x_2/x_0 \end{cases} \qquad \begin{cases} z = x_0/x_1 \\ t = x_2/x_1 \end{cases} \qquad \begin{cases} u = x_0/x_2 \\ v = x_1/x_2 \end{cases}$$

However one should not think that the obstructions are only at infinite. In fact if we restrict an open ball $B_r = \{x^2 + y^2 < r\}$ with $r > 1$, $S \cap B_r$ is not basic semialgebraic either. The argument to see this involves changes of sign along the line $y = 0$, and is similar to the one explained in example d) below. Nevertheless, notice that if we blow down this line, we obtain again Łojasiewicz's obstruction:

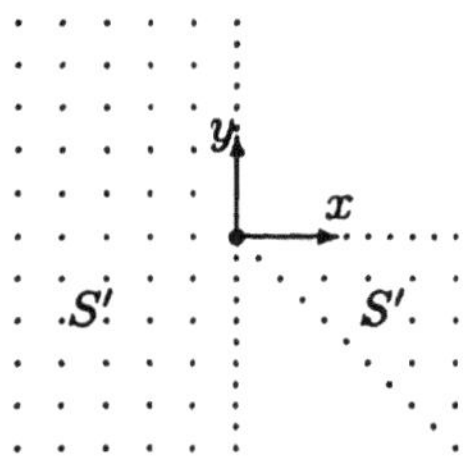

c) *An example without analytic obstructions, either finite or infinite.*
Consider the semialgebraic set $S = S_1^+ \cup S_1^- \cup S_2^+$ described in the picture:

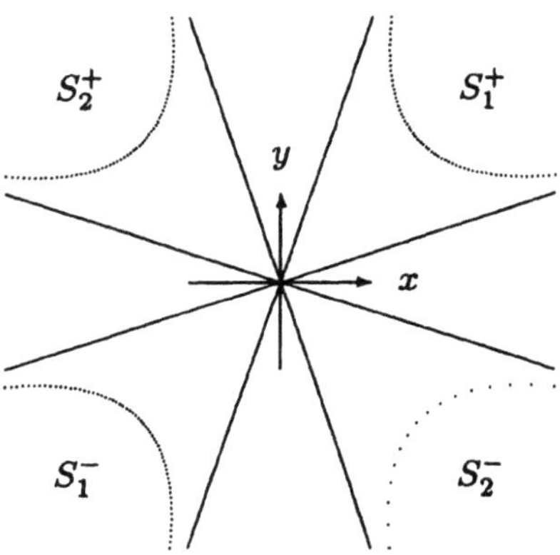

$S_1 : 3x^2 + y^2 - 4xy + 1 < 0$

$S_2 : 3x^2 + y^2 + 4xy + 1 < 0$

$+ \equiv y > 0, \ - \equiv y < 0$

$S = S_1^+ \cup S_1^- \cup S_2^+$

S is basic semianalytic, but it is not basic semialgebraic as one can easily check by arguing with lines passing through the origin as we did in the previous example. Moreover, if we study S at infinity, we encounter the following pictures:

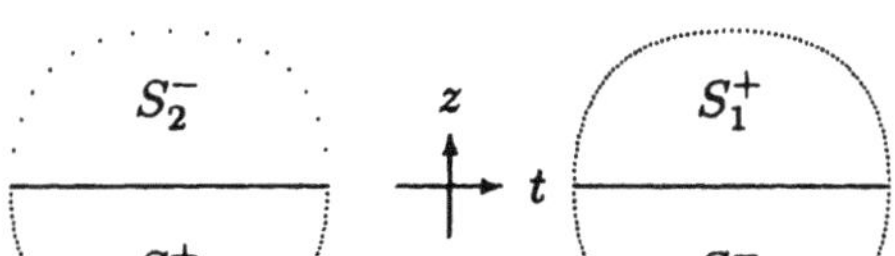

$$\begin{cases} z = 1/x \\ t = y/x \end{cases}$$

$S_1 : z^2 + t^2 - 4t + 3 < 0, \ S_2 : z^2 + t^2 + 4t + 3 < 0$

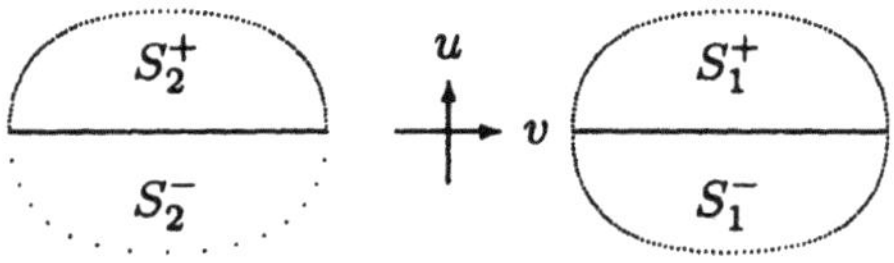

$$\begin{cases} u = 1/y \\ v = x/y \end{cases}$$

$S_1 : u^2 + 3v^2 - 4v + 1 < 0, \ S : u^2 + 3v^2 + 4v + 1 < 0$

Hence, S is also basic semianalytic at infinity. What happens is that the obstruction is globally located in the line at infinite. In fact, if we collapse this line to a single point, or in other words, if we consider the Alexandroff compactification instead of the projective closure, we again encounter at the infinite point Lojasiewicz's obstruction to semianalytic basicness.

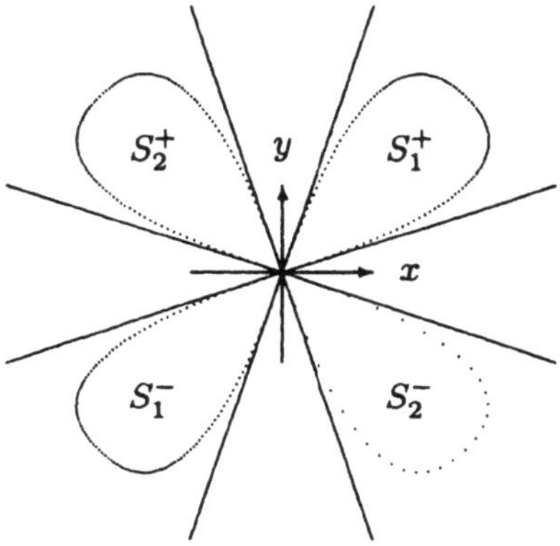

d) *More delicate obstructions.*
Consider now the following example:

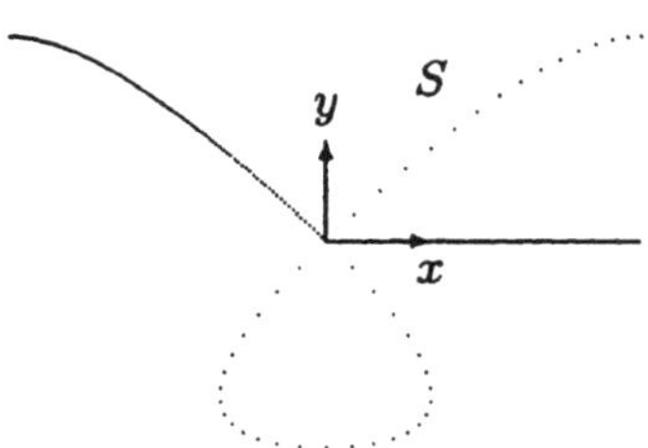

$$S = \{x^2 - y^2 - y^3 < 0, y > 0\} \cup \{x > 0, y > 0\}$$

The set S is basic semianalytic (near the origin it can be described as $x > 0, x + y\sqrt{1+y} > 0$), but not basic semialgebraic. The idea is to look at changes of sign, this time through the nodal curve: they must propagate along the curve and therefore they also appear in the part contained in S, which explains the example.

Again looking at infinity we find Łojasiewicz's obstruction at some point:

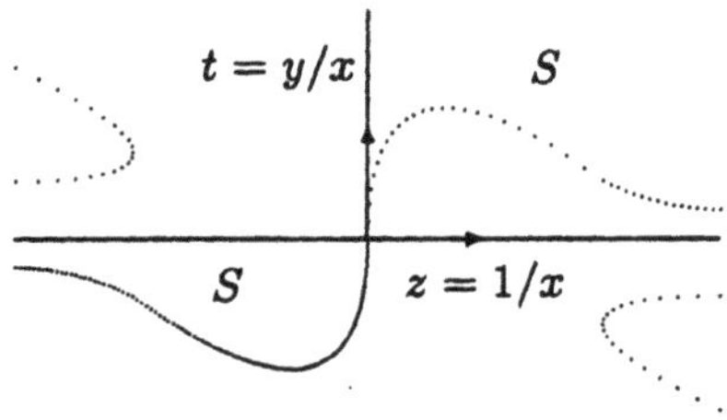

$$S = \{z - zt^2 - t^3 > 0, z < 0, t < 0\} \cup \{z > 0, t > 0\}$$

S is not basic semianalytic

but not always:

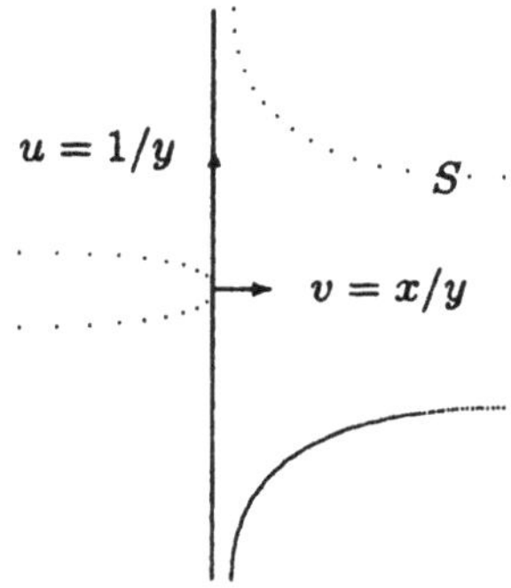

$$S = \{uv^2 - u - 1 < 0, v > 0\} \cup \{u > 0, v > 0\}$$

S is basic semianalytic, not basic semialgebraic

As mentioned in example b) one should not think that the only obstructions to semialgebraic basicness come from infinity. Indeed, the restriction of S to any open ball around the origin is still not basic semialgebraic, as the argument given at the beginning of the example shows. In this case we can obtain Łojasiewicz's obstruction by collapsing the curve $x^2 - y^2 - y^3 = 0$ to one point, but we do not picture it here.

e) *Two examples that look quite similar but behave differently.*

In the following two examples we consider the subset S of X defined by the equations $S = \{y > 0, z > 0\} \cup \{y > 0, x > 0\}$.

i) *A semialgebraic set which is not basic semianalytic.*

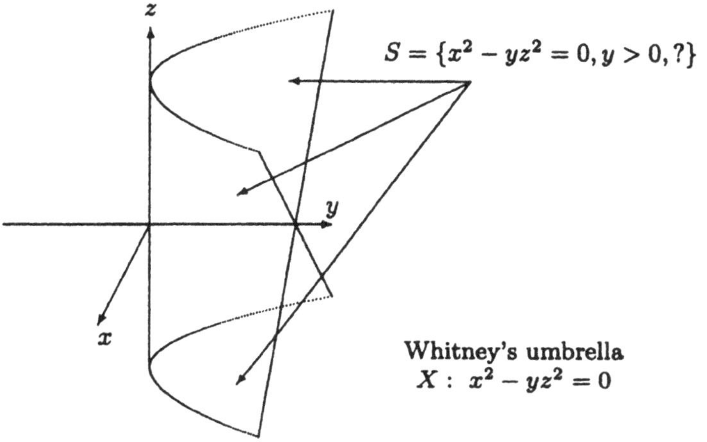

One can check, although it is not immediate, that S is not basic semianalytic.

ii) *A semialgebraic set that looks worse than the last example, but is better: it is basic semianalytic.*

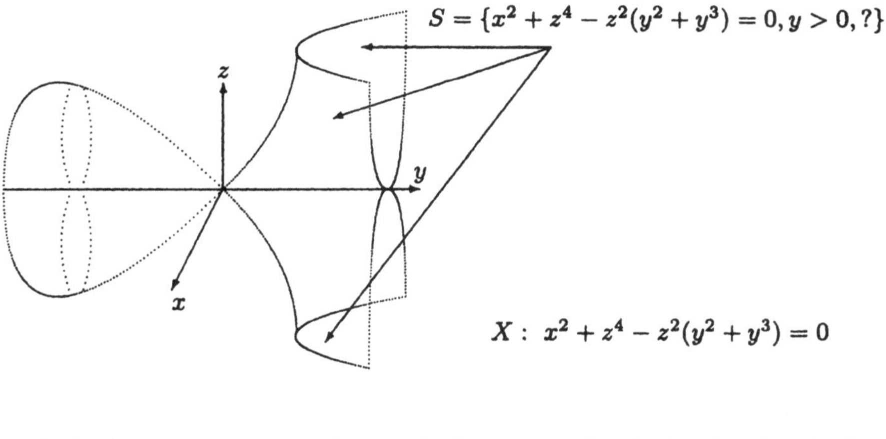

In both cases we want to know whether or not S is basic, that is, whether or not there exists a function "?" that separates one sheet from the other three in our algebraic set X. It is not difficult to see that there is not such an *algebraic* "?", and what we claim is that an *analytic* "?" does not exist in the first case, but does exist in the second. Even more, we want to understand why this is so; at a first glance, separation could look more likely on Whitney's umbrella, since sheets intersect (or so they seem) more transversally.

6. Algebraic and analytic fans. Let us denote by $\mathcal{B}$ the Boole algebra of all semialgebraic sets of X and by Ω the *Stone space* of $\mathcal{B}$; we recall that the points of this space are the *ultrafilters* of $\mathcal{B}$, and that it is a compact totally disconnected space. We will distinguish a special subspace $\Sigma \subset \Omega$. First, we define the *dimension* $\dim(\phi)$ of an ultrafilter $\phi \in \Omega$ to be the smallest dimension of the sets belonging to ϕ. Clearly, $\dim(\phi) \leq d = \dim(X)$, and we denote by Σ the subspace of all ultrafilters of maximal dimension d; an element of Σ is called *ordering* and denoted by σ. All this belongs to the theory of the real spectrum ([BCR]).

To say more explicitely when an ultrafilter is an ordering, we consider the collection $\varphi \subset \mathcal{B}$ of all the sets $\{x \in X \mid h(x) \neq 0\}$, where h is a regular function not vanishing on X. One checks easily that the elements of Σ are exactly the ultrafilters that contain φ.

Let $\sigma \in \Sigma$ be an ordering and $f \in \mathcal{R}(X)$ a regular function. If

$$\{x \in X \mid f(x) > 0\} \in \sigma,$$

we say that f is *positive* in σ, and if

$$\{x \in X \mid f(x) < 0\} \in \sigma,$$

that f is *negative* in σ. Note that it always holds exactly one of these two assertions, since σ is an ultrafilter that contains the set

$$\{f \neq 0\} = \{f > 0\} \cup \{f < 0\}.$$

In other words, f is positive (resp. negative) in σ if f is positive (resp. negative) on a collection of generators D of σ.

Now, an *algebraic fan* is a collection F of four orderings

$$\sigma_1, \sigma_2, \sigma_3, \sigma_4 \in \Sigma$$

such that every regular function which is positive in three of them is also positive in the fourth. This can be expressed by saying that there is no function that *separates* one ordering from the others. Another formulation is that the *product* of the first three orderings gives the fourth, with the obvious meaning of product.

The notion of an *analytic fan* is introduced similarly. The only technical difference is that one starts by fixing a point $a \in X$ and then considers the Stone space of the algebra of semianalytic set germs at a.

It is with the help of the notion of fan that Bröcker's characterization of basicness can be stated. We have:

Theorem I. [B] *A semialgebraic (resp. semianalytic) set S is basic semialgebraic (resp. basic semianalytic) if and only if there is no algebraic (resp. analytic) fan F such that S belongs to exactly three ultrafilters of F.*

7. Examples. We will give here a procedure to draw fans, and to distinguish between algebraic and analytic ones. What we really draw is a sample of the typical set generating the ultrafilters of the fan.

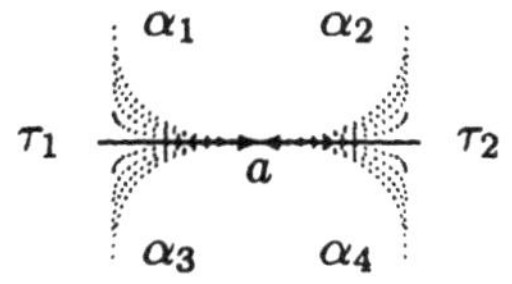

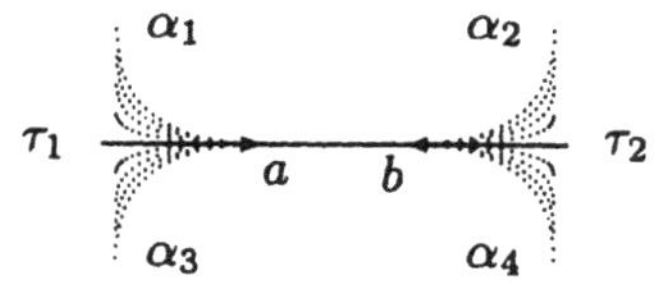

These pictures are done as follows. First we pick some curve (a straight line in these examples). Then we choose two points on it (possibly the same), and two branches τ_1, τ_2 of the curve ending at the two points. Finally, each branch separates the surface (the plane in this example) into two sides, and thus we get four orderings. We claim that they form a fan if the curve is irreducible. The reason is that any function separating, say, α_1 from α_3 must change sign along τ_1, and so it vanishes on the whole curve (by irreducibility); furthermore this vanishing must have odd order to give a change of sign. Consequently, the change of sign propagates along the curve to reach τ_2, and so the function also separates α_2 from α_4. This argument works equally well for regular and for analytic functions, provided that in the analytic case there is only one point, and the curve is *analytically* irreducible at that point; notice for instance that the curve of example d) in paragraph 5 is algebraically irreducible, but analytically reducible at the origin. This idea is also used in example e). Note also that irreducibility does not imply conectedness in the algebraic case. Thus in the picture, the left hand side example is an analytic fan, while the right hand side example is a non-analytic algebraic fan.

Next we point out the fans that explain the examples of paragraph 5. We have shadowed in black the ultrafilters that contain the semialgebraic set S under consideration, and left blank the one, σ, that does not, so giving the obstruction to basicness. Notice that $S \notin \sigma$ means that σ contains the complement of S, so that a typical generator of σ (which is what we actually draw) does not meet S.

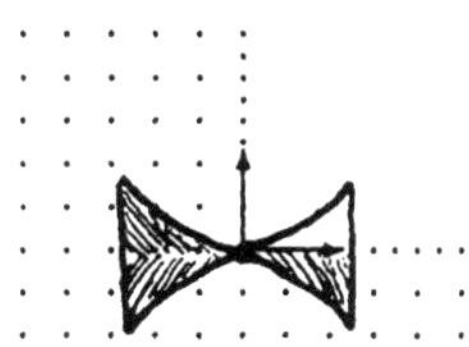

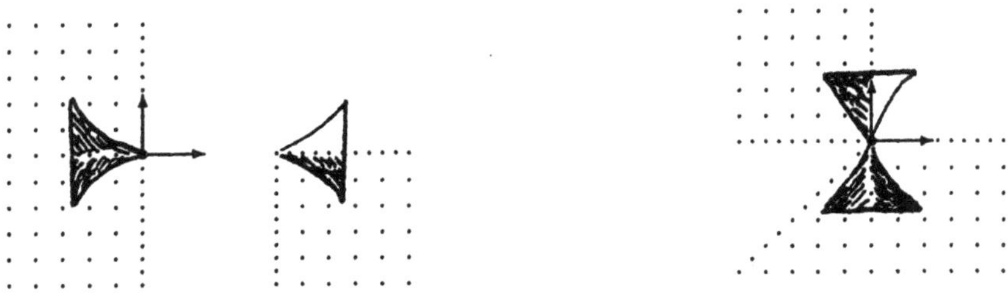

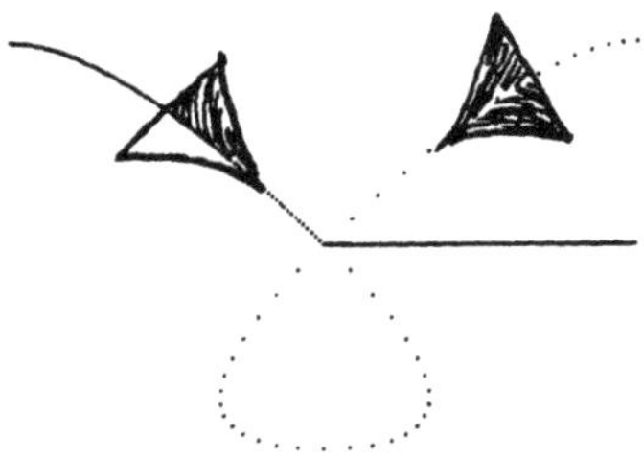

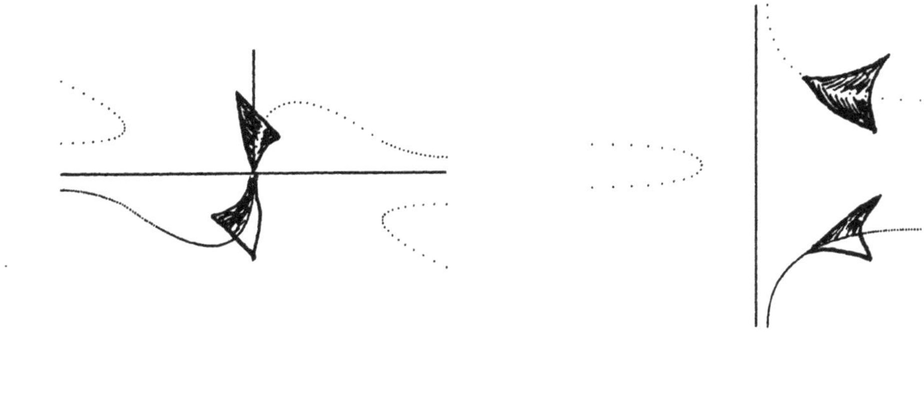

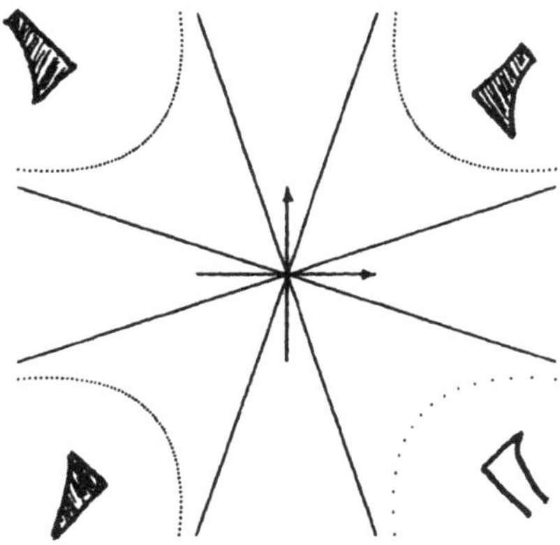

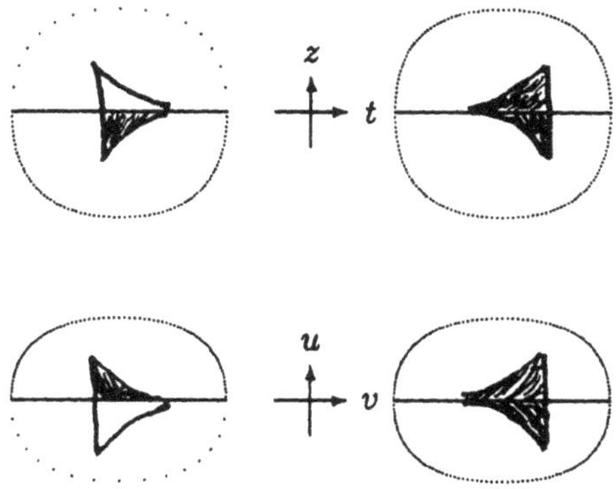

Notice that in the picture on the left, the fan is not "attached" to any curve at finite distance, but rather to the infinite line.

Finally, here there are the fans that correspond to example e) in 5.

An analytic fan

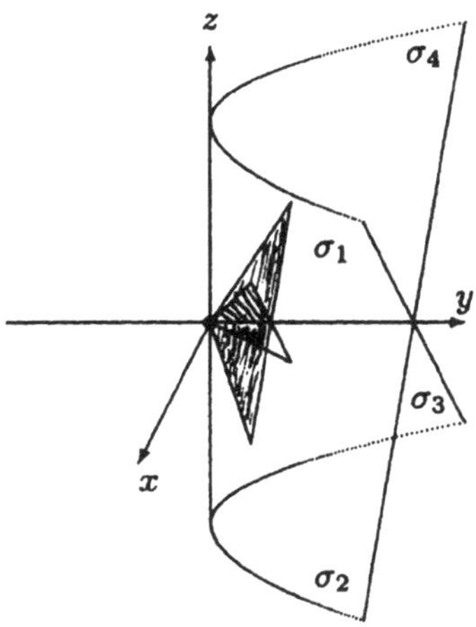

An algebraic fan which is not analytic

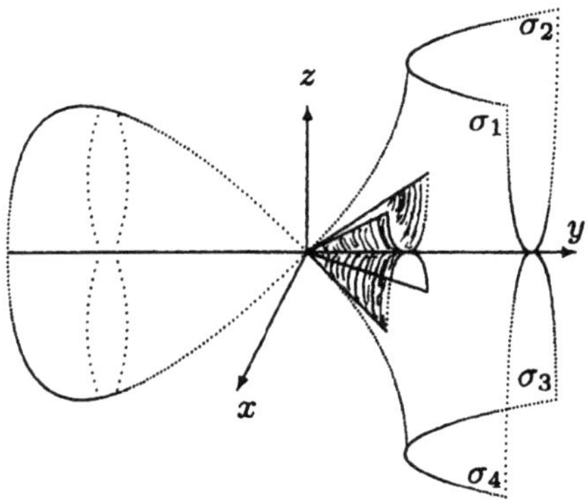

So far, we cannot explain completely the two last examples. This is in fact our main goal here: to understand examples like these.

8. Algebraic versus analytic fans. The preceding statements locate in fans the obstructions to basicness: if a semialgebraic set S is not basic semialgebraic, there is an algebraic fan $F = \{\sigma_1, \sigma_2, \sigma_3, \sigma_4\}$ such that S belongs to three of the σ_i's and does not to the fourth. But then, if S is basic semianalytic, F cannot be *analytic*. This means, according to our terminology, that there must be a point $a \in X$ and an analytic function f on a neighborhood of a such that

$$f|D_1 > 0,\ f|D_2 > 0,\ f|D_3 > 0,\ f|D_4 < 0,$$

where $D_i \in \sigma_i$, $D_i \subset W$, $i = 1,2,3,4$. Following our conventions, we will say that the analytic function f *separates* F. Thus, our PROBLEM takes the following form:
PROBLEM*. Characterize geometrically the algebraic fans that are analytic, that is, that cannot be separated by analytic functions.

To solve this we have to take a closer look at the properties of fans.

9. Trivialization of algebraic fans and limits. Let $Y \subset \mathbb{R}^m$ be another irreducible real algebraic set, and $f : Y \to X \subset \mathbb{R}^n$ a regular mapping, that is, a mapping whose components $f_i : Y \to X \subset \mathbb{R}^n \to \mathbb{R}$ are regular functions. We say that $f : Y \to X$ is a *birational diffeomorphism* if there are regular functions $h \in \mathcal{R}(X), g \in \mathcal{R}(Y)$ such that f restricts to a diffeomorphism from $Y \setminus \{y \in Y \mid g(y) = 0\}$ onto $X \setminus \{x \in X \mid h(x) = 0\}$ whose inverse is also a regular mapping. Let $f : Y \to X$ be a birational diffeomorphism and $F = \{\sigma_1, \sigma_2, \sigma_3, \sigma_4\}$ an algebraic fan of X. Since the set $\{x \in X \mid h(x) \neq 0\}$ belongs to all the ultrafilters σ_i of F, the inverse images $f^{-1}(\sigma_i)$ are well defined ultrafilters of Y of maximal dimension, and one easily checks that $f^{-1}(F) = \{f^{-1}(\sigma_1), f^{-1}(\sigma_2), f^{-1}(\sigma_3), f^{-1}(\sigma_4)\}$ is an algebraic fan of Y. In other words, f induces a bijection between the space of orderings of X and that of Y, and this bijection preserves fans.

In this situation, we say that f is *compatible* with F if there is an irreducible real algebraic set $Z \subset Y$ such that for every $\sigma_i \in F$ the *specialization*

$$\tau_i = f^{-1}(\sigma_i)|Z = \{f^{-1}(D) \cap Z \neq \emptyset \mid D \in \sigma_i\}$$

is an ultrafilter of maximal dimension of Z; this Z is called a *center* of F in Y. Thus, f is compatible with F if F has a center in Y, or more properly, the fan $f^{-1}(F)$ has a center in Y. In fact, there are very few centers: there are finitely many of them, and they are totally ordered by inclusion (this corresponds to the very rigid specialization behaviour of our Stone space). Moreover, if X is compact centers do always exist. Therefore, replacing X by its projective closure, for instance, the notion of compatibility is vacuous. However, in the affine situation it can happen that no center exists, or in other words, that centers are *at infinity*. For instance, this is the case of the fan which explains example *c)* of paragraph **5**. Since we do not assume X to be projective, we need the notion of compatibility, meaning that $f^{-1}(F)$ is centered *at finite distance* in Y.

Consider now a second compatible birational diffeomorphism $f' : Y' \to X$. We say that f' is *posterior* to f if there is a third birational diffeomorphism $h : Y' \to Y$ such that $f' = f \circ h$ and $Z \supset h(Z')$, $\dim(Z) = \dim(h(Z'))$ for some center Z' of F in Y'.

For instance, let $X = \mathbb{R}^2$, where we fix coordinates x, y, let Y be the affine chart of the blowing-up of the origin in X defined by the coordinates x', y', where $x' = x$ and $y' = y/x$, and let $f : Y \to X$ be the canonical mapping $(x', y') \mapsto (x', x'y')$. Finally let F be the fan

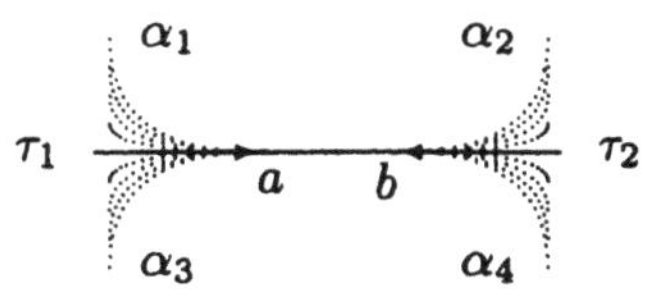

where the line represents the exceptional divisor Z (with equation $x' = 0$) in Y. Then it is immediate to check that the only center of F in Y is Z, and that the only center of F in X is the origin. Then it is obvious that f is posterior to Id and that for every f' posterior to f, F has a center Z' of dimension 1.

Finally, a *trivialization* of F is a birational diffeomorphism f compatible with F such that for any other "posterior enough" f' it holds (up to permutation)

$$\tau_1' = \tau_3', \; \tau_2' = \tau_4'.$$

If that is the case, there is a maximal center Z for f, which is the one always chosen. For that Z we have $\tau_1 = \tau_3$, $\tau_2 = \tau_4$, and the tuple $(f, Y, Z, \tau_1, \tau_2)$ is uniquely determined by f.

The reason to consider all posterior enough f''s is that for a given f two or more orderings of F can collapse to the same ordering of the center Z, so faking triviality. However, this feint will disappear for a posterior enough f'. This happens for instance in the two last fans pictured in the previous paragraph: their centers are the y-axis and F collapses to a single ordering in it (namely, the one defined by the rightward half-axis at the origin). However, we will see that for posterior f's this single ordering splits into two different ones.

Coming back to our notion, suppose that $(f, Y, Z, \tau_1, \tau_2)$ is a trivialization of F. Then we have the *limit sets*

$$\mathrm{Lim}(\tau_i) = \bigcap_{D \in \tau_i} \overline{D}, \ i = 1, 2,$$

and say that f *has a unique limit* if these two limit sets are not empty and both consist of the same point $a \in Z$. Notice also that our order relation is given by continuous mappings, and consequently, if $f' > f$ and f' has a unique limit, the same is true for f. Finally, a trivialization f is said to be *finite* if the restriction $f|Z : Z \to X$ is proper with finite fibres.

The birational diffeomorphisms compatible with F form a tree, which has trivializations and finite trivializations as a distinguished branch and subbranch respectively. In fact, we have:

> *The set of all trivializations of F form a directed system with respect to the order relation defined above.*

The roots of this tree to $f = \mathrm{Id}$. A typical picture that illustrates the tree of compatible diffeomorphisms and the branch of trivializations could be the following:

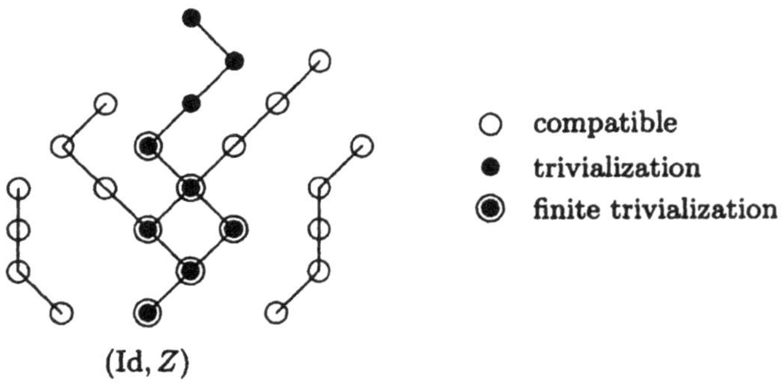

10. Solution to PROBLEM II*. Once the suitable notions are available, our analyticity problem can be solved *by computing limits*:

Theorem II*. *An algebraic fan is analytic if and only if all its finite trivializations have a unique limit.*

11. Examples. Let us apply our result to the last two examples of paragraph 5. Consider first the example with the Withney's umbrella:

An analytic fan

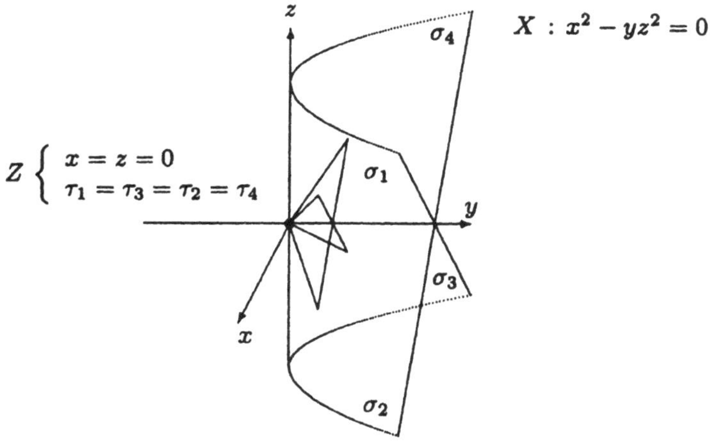

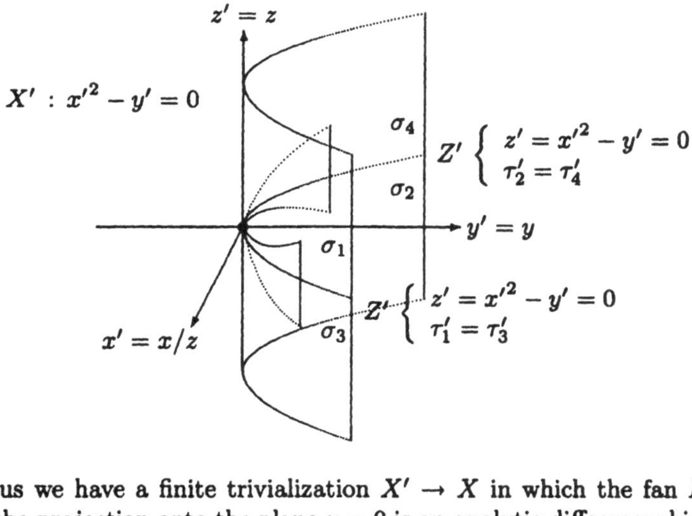

Thus we have a finite trivialization $X' \to X$ in which the fan F is analytic, since the projection onto the plane $y = 0$ is an analytic diffeomorphism that maps our fan to a typical analytic one. Now, by the theorem, F is analytic in X if all finite trivializations have a unique limit. But these trivializations form a directed branch, and so it is enough to check that property for a cofinal subbranch. Since F is analytic in X', the cofinal subbranch that starts at X' does the job. This shows that the given fan is analytic and therefore the considered semialgebraic set S is not basic semianalytic.

Consider now the other example and let us see that the situation is quite different despite the appearances:

An algebraic fan which is not analytic

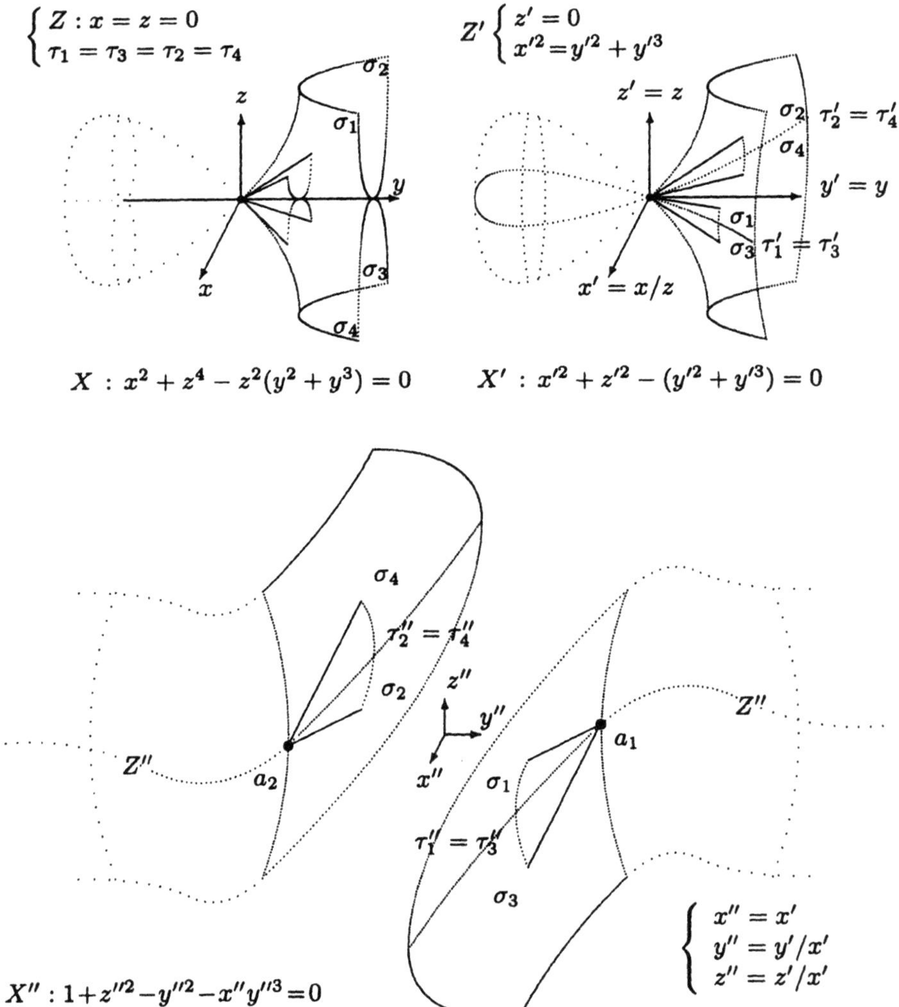

As pictured above, we obtain two finite trivializations $X'' \to X' \to X$, and in X'' there is not a unique limit: $\mathrm{Lim}(\tau_1'') = a_1 \neq a_2 = \mathrm{Lim}(\tau_2'')$. Hence the fan is not analytic which explains why the considered semialgebraic set is, in this case, basic semianalytic. Note also that in X' we find the nodal curve that already appeared in a planar example where algebraic and analytic basicness differed (see 5 d)).

12. Further scope of the procedure. Here we have restricted ourselves to four-element fans. However, the full theory extends to fans with an arbitrary number of elements and allows us to treat more delicate quantitative aspects. Mainly, once we know that a constructible set is basic, we want to know *how many* inequalities are necessary to describe it. For all of this we refer to [B] and [ABR]. Then we come to the comparison question, and the fact is that **Theorem II*** works exactly the same for fans with an arbitrary number of elements. Full proofs and a detailed description of the background involved, as well as several concrete applications will appear in the monograph [AR]. Finally, the ideas used here can also be applied to compare other types of functions; this has been done for global Nash and global analytic functions in [RS].

References

[ABR] C. Andradas, L. Bröcker, J.M. Ruiz, Constructible sets in real geometry. To appear.

[AR] C. Andradas, J.M. Ruiz, Algebraic and analytic geometry of fans. In: Mem. Amer. Math. Soc.

[BCR] J. Bochnak, M. Coste, M.-F. Roy, Géométrie algébrique réelle. Ergeb. Math. Grenzgeb. (3) **12**, Springer-Verlag, Berlin-Heidelberg-New York, 1987.

[B] L. Bröcker, On basic semialgebraic sets. Exposition. Math. **9** (1991), 289–334.

[Ł] S. Łojasiewicz, Ensembles semi-analytiques. Prépublication I.H.E.S., 1964.

[RS] J.M. Ruiz, M. Shiota, On global Nash functions. Ann. Sci. École Norm. Sup. (4) **27** (1994), 103–124.

[S] C. Scheiderer, Stability index of real varieties. Invent. Math. **97** (1989), 467–483.

Notes on the space of real places of a formally real field

Eberhard Becker and Danielle Gondard

Introduction

The intentions behind these notes are threefold. First we want to review some of the basic results about the space of real places by appealing consequently to the more recent concepts of the real holomorphy ring and the real spectrum. That this might be a worthwile approach grew out of discussions with various colleagues over the years, notably with L. Bröcker and M. Knebusch. Secondly we want to add a few new results. Thirdly we like to turn more attention to the space of real places as an apparently interesting and difficult invariant of a formally real field.

Let K be a formally real field. We assume familiarity with the theory of orders, real places and the valuation theory of formally real fields. Recommended references are the books by Prestel [P], Knebusch–Scheiderer [K–S] and Lam [L 1].

If K is a formally real field we consider its real holomorphy ring $H(K)$, its space of real places $M(K)$, its space of orders and the Witt ring of K. The various relations between these objects are the content of the first two sections. As said above the properties of the real holomorphy ring and the theory of the real spectrum are used as the central ingredients. This approach is also well displayed in the book of Knebusch and Scheiderer. The third section is devoted to real algebraic geometry. The main result is concerned with the fibers of the center map where we slightly extend an unpublished result of L. Bröcker. A description of the fibers via real spectra is also presented. In the last section previous results of H.W. Schülting are revisited. It turns out that the space of connected components $\pi_0(M(K))$ is homeomorphic to the maximal real spectrum of the Witt ring of $H(K)$. This closely resembles the fact that $M(K)$ itself is homeomorphic to Max Sper $H(K)$.

If V is any valuation ring we denote its maximal ideal by $\mathcal{M}_V$, V is called residually real if its residue field is formally real.

1. The real holomorphy ring and the space of real places

The real holomorphy ring $H(K)$ of a formally real field K is defined by

$$H(K) = \mathbb{Q}\left[\frac{1}{1+q} \mid q \in \sum K^2\right].$$

From [Be 2], [K–S], [L 1], [Schü 1] it is known that $H(K)$ is a Prüfer ring with quotient field K. It can be defined in various ways and has interesting properties. A sample is given in the next theorem. If P is any order of K it gives rise to the valuation ring

$$A(P) = \{a \in K \mid n \pm a \in P \text{ for some } n \in \mathbb{N}\}$$

with the maximal ideal

$$I(P) = \{a \in K \mid \frac{1}{n} \pm a \in P \text{ for every } n \in \mathbb{N}\}.$$

Theorem 1.1.

i) $H(K) = \{a \in K \mid \exists n \in \mathbb{N} : n \pm a \in \sum K^2\}$,

ii) *for any valuation ring V of K we have:*

$$H(K) \subseteq V \Leftrightarrow V \textit{ is residually real,}$$

iii) $H(K) = \cap V$, *V ranging over all residually real valuation rings,*

iv) $H(K) = \cap A(P)$, *P ranging over all orders P of K,*

v) $H(K) \cap \sum K^2 = \sum H^2$.

The statement in v) above is proved in the remark on p. 24 in [Be 2].

The real holomorphy ring is well behaved relative to places between formally real fields. To formulate this property adequately we are now going to introduce the *category of fields* $\mathcal{F}$ where we use the places as morphisms. So, $\mathcal{F}$ has the constituents:

objects of $\mathcal{F}$: fields,
$\mathrm{Mor}(K, L) = \{\lambda : K \to L \cup \infty \mid \lambda \text{ place}\}$.

To define the composition of morphisms in $\mathcal{F}$ let K, L and F be fields and $K \xrightarrow{\lambda} L \cup \infty$, $L \xrightarrow{\mu} F \cup \infty$ be places with valuation rings V_λ and V_μ respectively. The ring $\lambda^{-1}(V_\mu)$ is a valuation ring of K as can be readily verified. On $\lambda^{-1}(V_\mu)$ we have the ring homomorphism $\mu \circ \lambda$. Since $\lambda^{-1}(V_\mu)$ is a valuation ring we can interpret $\mu \circ \lambda : \lambda^{-1}(V_\mu) \to F$ as a place from K into $F \cup \infty$. This place is, by definiton, the composition of λ and μ and will be denoted by $\mu \circ \lambda$. Its valuation ring $V_{\mu \circ \lambda}$ satisfies

$$V_{\mu\circ\lambda} = \{x \in K \mid \lambda(x) \neq \infty, \mu(\lambda(x)) \neq \infty\}.$$

This description of $V_{\mu\circ\lambda}$ is useful in the proof of the associativity of the composition of places. The identity id : $K \to K$ serves as the identity morphism in $\mathrm{Mor}(K, K)$. Note that two fields are isomorphic in $\mathcal{F}$ if and only if they are isomorphic as rings.

Thus, we have introduced the category $\mathcal{F}$ of fields with places as morphisms. However, we are mainly interested in the full subcategory of formally real fields. By the way, if L is formally real and $\mathrm{Mor}(K, L) \neq \emptyset$ then K is formally real as well.

Proposition 1.2. *Let K and L be formally real fields and $\lambda : K \to L \cup \infty$ be a place. Then*

$$\lambda(H(K)) \subseteq H(L).$$

Proof. Since L is formally real we derive $H = H(K) \subseteq V_\lambda$ from 1.1, ii). Hence, $\lambda(H) \subseteq L$. To show $\lambda(H) \subseteq H(L)$ we use that H is generated by the elements $x_q = \frac{1}{1+q}$ where $q \in \sum K^2$. If $q \notin V_\lambda$ then $\lambda(x_q) = 0$. Otherwise $q = \sum x_i^2 \in V_\lambda$ which implies $x_i \in V_\lambda$ for every i since V_λ is residually real. Consequently, $\lambda(1+q) = 1 + \sum \lambda(x_i)^2 \neq 0$ and $\lambda(x_q) = 1/1 + \sum \lambda(x_i)^2 \in H(L)$. □

Proposition 1.3. *Let K and L be formally real fields. Then the map*

$$\mathrm{res} : \left\{ \begin{array}{ccc} \mathrm{Mor}(K, L) & \to & \mathrm{Hom}(H(K), H(L)) \\ \lambda & \mapsto & \lambda \mid_{H(K)} \end{array} \right.$$

is a bijection.

Proof. Let $\lambda \in \mathrm{Mor}(K, L)$. From 1.2 we know that $\lambda \mid_{H(K)} \in \mathrm{Hom}(H(K), H(L))$. To prove injectivity of res we show that λ is determined by $\varphi := \lambda \mid_{H(K)}$. As $H = H(K)$ is a Prüfer ring we get $V := V_\lambda = H_\wp$ where $\wp = \mathcal{M}_V \cap H = \ker \lambda \cap H = \ker \varphi$. Thus, every $x \in V$ can be written as $x = \frac{r}{s}$ where $r, s \in H$, $s \notin \wp$, i.e. $\lambda(s) = \varphi(s) \neq 0$. We obtain $\lambda(sx) = \lambda(r)$ and $\lambda(x) = \frac{\lambda(r)}{\lambda(s)} = \varphi(r)\varphi(s)^{-1}$.

To prove surjectivity of res consider any homomorphism $\varphi : H(K) \to H(L)$. φ extends to a homomorphism

$$\overline{\varphi} : H(K)_\wp \to L, \frac{r}{s} \mapsto \varphi(r)\varphi(s)^{-1}$$

where $\wp = \ker \varphi$, $r, s \in H(K)$, $s \notin \wp$. $H(K)_\wp$ is a valuation ring since $H(K)$ is a Prüfer ring. Therefore $\overline{\varphi}$ can be understood as a place from $K = \mathrm{quot}(H(K))$ to L. Obviously, $\overline{\varphi} \mid_{H(K)} = \varphi$. □

Corollary 1.4. *The assignment $K \mapsto H(K)$ is a full and faithful functor from the category of formally real fields into the category of rings.*

We will apply the last result to the case $L = \mathbb{R}$ to get that

$$\mathrm{Mor}(K, \mathbb{R}) = \{\lambda : K \to \mathbb{R} \cup \infty\} \text{ is in bijection with } \mathrm{Hom}(H(K), \mathbb{R}).$$

Next we impose a topology on $\mathrm{Hom}(H(K), \mathbb{R})$. The set $\mathbb{R}^{H(K)}$ is endowed with the product topology and $\mathrm{Hom}(H(K), \mathbb{R}) \subseteq \mathbb{R}^{H(K)}$ inherits the subspace topology.

Proposition 1.5. Hom$(H(K), \mathbb{R})$ *is a compact space.*

Proof. Set $H = H(K)$. By routine arguments we see that Hom$(H, \mathbb{R})$ is closed in $\mathbb{R}^H$. Therefore, the claim will follow once we can show that Hom$(H, \mathbb{R})$ is contained in a compact subspace of $\mathbb{R}^H$. To deduce this pick any $a \in H$. We claim

$$\{\varphi(a) \mid \varphi \in \operatorname{Hom}(H, \mathbb{R})\} \subseteq [n_a, -n_a]$$

for some $n_a \in \mathbb{N}$. In fact, from 1.1, i) and 1.1, v) we infer $n_a \pm a \in \sum H^2$ for some $n_a \in \mathbb{N}$. Applying φ we obtain $n_a \pm \varphi(a) \geq 0$, hence $\varphi(a) \in [n_a, -n_a]$. Consequently, Hom$(H(K), \mathbb{R}) \subseteq \prod_{a \in H} [-n_a, n_a]$. Since the latter space is compact the proof is complete. □

This last result allows to define a compact topology on Mor$(K, \mathbb{R}) = \{\lambda : K \to \mathbb{R} \cup \infty\}$ by using the bijection Mor$(K, \mathbb{R}) \to$ Hom$(H(K), \mathbb{R})$ to transfer the topology from Hom$(H(K), \mathbb{R})$ to Mor$(K, \mathbb{R})$. Following a traditional notation we introduce the

Definition 1.6. If K is a formally real field then the set of real places Mor$(K, \mathbb{R})$ endowed with the topology described above is denoted by $M(K)$ and called the space of real places of K.

To see first examples we study the space $M(K)$ of a totally Archimedean field K. By definition, this means that every order P is Archimedean, i.e. $A(P) = K$. In view of 1.1, iv), K is totally Archimedean iff $H(K) = K$. Hence,

Proposition 1.7. *If K is a totally Archimedean field then $M(K) =$* Hom$(K, \mathbb{R})$ *endowed with the topology inherited, as a subspace, from $\mathbb{R}^K$.*

Totally Archimedean fields exist in abundance. Clearly $\mathbb{Q}, \mathbb{R}$ are examples. More generally, if K is any subfield of $\mathbb{R}$ it carries an Archimedean order P. It is well known that every algebraic ordered extension of an Archimedean ordered field is Archimedean ordered as well. Hence, if we consider any uniquely ordered algebraic extension L of (K, P) where P is Archimedean then L is totally Archimedean as well. So far, only fields with just one order have been exhibited. However, the argument above stating that algebraic extensions are relatively Archimedean yields that

> every formally real algebraic extension of a totally
> Archimedean field is again totally Archimedean.

In particular, every real algebraic number field is totally Archimedean. Given any finite cardinality k there are number fields with exactly k real embeddings $K \to \mathbb{R}$. Consequently, every finite Hausdorff space occurs as a space $M(K)$. Which compact spaces occur as a space of real places seems to be an open, difficult and attractive problem. Therefore, we would like to list it explicitly:

Problem: Determine the compact spaces which occur as the spaces $M(K)$, K a formally real field.

Note that the corresponding question for the space $X(K)$ of orders (for the definition see next section or the literature) has been completely answered by T. Craven [C] in the following way:

- every space $X(K)$ is compact and totally disconnected,
- every compact, totally disconnected space is isomorphic to some space $X(K)$, K a formally real field.

The spaces $M(K)$, K totally Archimedean, are compact and totally disconnected since $M(K) \simeq X(K)$, see corollary 2.15 of the next section. However, there occur other types of spaces among the spaces $M(K)$'s. Samples will be given in section 3 where we will be concerned with infinite compact spaces having only finitely many connected components and admitting a presentation as a projective limit of compact real manifolds.

2. The real spectrum of the real holomorphy ring and the mapping $X(K) \to M(K)$

It is the essence of this section that the real holomorphy ring carries a great deal of information about the "real" aspects of a formally real field. We will use the real spectrum of this ring to treat uniformily the space of orders and real places. That this is a worthwile approach has been known for a while and is also well displayed in the book of Knebusch and Scheiderer, cf. [K–S], Kap. III, §12.

For any (commutative) ring A let Sper A denote its real spectrum, consult e.g. [BCR], chap. 7, [L 2], [Be 4], [K–S], Kap. III. The elements α of $\operatorname{Sper} A$ are either treated as prime cones of positivity or as pairs $\alpha = (\wp, \overline{\alpha})$, $\wp \in \operatorname{Spec} A$, $\overline{\alpha}$ an order of the residue field $k(\wp) = \operatorname{quot}(A/\wp)$, according to what is more convenient in a given situation.

If K is a formally real field we still write $X(K) = \operatorname{Sper} K$ to keep the traditional notation.

Note that $\operatorname{Sper} A$ is a quasi-compact space, in general not Hausdorff. However, $X(K)$ is a Hausdorff space for every field K. Moreover, $X(K)$ is a compact totally disconnected space.

Definition 2.1. An element $\alpha \in \operatorname{Sper} A$ is called minimal (resp. maximal) if there is no $\beta \in \operatorname{Sper} A$ such that

$$\beta \subseteq \alpha, \beta \neq \alpha \text{ (resp. } \alpha \subseteq \beta, \alpha \neq \beta).$$

We set

$$\operatorname{Min}\operatorname{Sper} A = \{\alpha \in \operatorname{Sper} A \mid \alpha \text{ minimal}\},$$
$$\operatorname{Max}\operatorname{Sper} A = \{\alpha \in \operatorname{Sper} A \mid \alpha \text{ maximal}\}$$

and endow these sets with the subspace topology. We also have

$$\operatorname{Max}\operatorname{Sper} A = \{\alpha \in \operatorname{Sper} A \mid \{\alpha\} \text{ closed}\},$$

Max Sper A is a compact space, [L 2], prop. 4.7, [BCR], prop. 7.1.24. From [L 2], prop. 4.6 one deduces that Min Sper A is a Hausdorff space. Every $\alpha \in$ Sper A is contained in a unique maximal element $\alpha^{\max}$ called its maximal specialization. The specialization map

$$\operatorname{sp} : \operatorname{Sper} A \to \operatorname{Max}\operatorname{Sper} A,\ \alpha \mapsto \alpha^{\max}$$

is continuous, and Max Sper A carries the quotient topology, cf. [L 2], 4.10, also [BCR], prop. 7.1.24. Consequently, the restriction

$$\operatorname{sp} : \operatorname{Min}\operatorname{Sper} A \to \operatorname{Max}\operatorname{Sper} A$$

is a continuous surjective map. If Min Sper A is compact then Max Sper A carries the quotient topology of Min Sper A.

It is our final goal to study the spaces Min Sper $H(K)$ and Max Sper $H(K)$, K any formally real field. As a matter of fact, they can be identified with $X(K)$ and $M(K)$ respectively, cf. [K–S], Kap. III, §12. These identifications turn out to be consequences of more general results we are now going to derive.

Let A be any integral domain with a formally real quotient field K. Then the inclusion

$$i : A \to K$$

induces a continuous map

$$\operatorname{Sper} i : X(K) = \operatorname{Sper} K \to \operatorname{Sper} A,\ P \mapsto P \cap A.$$

We see that the image of Sper i is contained in Min Sper A.

Proposition 2.2. Sper i *is a homeomorphism of* $X(K)$ *onto its image which is a closed subspace of* Min Sper A.

Proof. Sper i is injective since $\frac{a}{b} \in P$ iff $ab \in P \cap A$ where $a, b \in A$, $b \neq 0$. Since $X(K)$ is compact and Sper i is continuous the image is a quasicompact subspace of the Hausdorff space Min Sper A, hence compact as well. Any continuous bijection between compact spaces is a homeomorphism. □

The next theorem provides a characterization of integral domains A satisfying Sper $K \xrightarrow{\sim}$ Min Sper K via Sper i. Note that the surjectivity of Sper i already implies, by the last proposition, that Sper i is in fact a homeomorphism.

Theorem 2.3. *If A is an integral domain with a formally real quotient field K then the following statements are equivalent:*

i) $\operatorname{Sper} i : X(K) \to \operatorname{Min}\operatorname{Sper} A$ *is a homeomorphism,*

ii) *for every* $\alpha = (\wp, \overline{\alpha}) \in \operatorname{Sper} A$ *the canonical morphism* $A \to k(\wp)$ *extends to a place*

$$\lambda : K \to L \cup \infty$$

where L is a real closed extension of $(k(\wp), \overline{\alpha})$.

Proof. i) ⇒ ii) Pick $\alpha \in \operatorname{Sper} A$. Then $\beta \subseteq \alpha$ for some $\beta \in \operatorname{Min}\operatorname{Sper} A$. By assumption, $\beta = P \cap A$ for some $P \in X(K)$. Let R be a real closure of (K, P). From $P \cap A \subseteq \alpha$ we deduce the existence of a place $\lambda : R \to L \cup \infty$ satisfying

1) L is a real closed extension of $(K(\wp), \overline{\alpha})$,

2) λ extends the morphism $A \to A/\wp \hookrightarrow k(\wp)$,

see [Be 4], prop. 2.9 e.g. Restricting λ to K we get a place as desired.
ii) ⇒ i) By assumption we are facing the following situation:

$$\begin{array}{ccc} A & \to A/\wp \hookrightarrow & k(\wp) \\ | & & | \\ K & \xrightarrow{\lambda} & L \cup \infty \end{array}$$

where L is a real closed extension of $(k(\wp), \overline{\alpha})$. The order L^2 of L can be pulled back to K to get an order $P \in X(K)$ satisfying $\lambda(P \cap V_\lambda) \subseteq L^2$, see [L 1], sect. 3 e.g. Pick any $a \in P \cap A$. Then $\lambda(a) \in k(\wp) \cap L^2 = \overline{\alpha}$ yielding $a \in \alpha$. Thus, we have shown that every α is the specialization of an order of type $P \cap A$. This shows $\operatorname{Sper} i$ to be surjective. □

Corollary 2.4.

i) *If for every real prime ideal $\wp$ of A the canonical morphism $A \to k(\wp)$ extends to a place $\lambda : K \to k(\wp) \cup \infty$ then* $X(K) \xrightarrow{\sim} \operatorname{Min}\operatorname{Sper} A$ *via* $\operatorname{Sper} i$,

ii) *the hypothesis of* i) *is satisfied in each of the following cases:*

   a) *A is a Prüfer ring,*

   b) *A is a regular ring (i.e. all localizations are regular local rings).*

iii) $\operatorname{Sper} i : X(K) \xrightarrow{\sim} \operatorname{Min}\operatorname{Sper} H(K)$ *for every formally real field K.*

Proof. iii) follows from ii), a). The claim ii), a) is obvious since $A_\wp$ is a valuation ring with residue field equal to $k(\wp)$. In the case of regular local rings with maximal ideal $\mathcal{M}$ it is well known that the canonical epimorphism $\mathcal{O} \to \mathcal{O}/\mathcal{M}$ extends to a place $\lambda : K \to \mathcal{O}/\mathcal{M} \cup \infty$, $K = \operatorname{quot}(\mathcal{O})$, see [Be 2], lemma 1.4 e.g.. i) directly follows from 2.3. □

Real algebraic geometry is a rich source for regular integral domains. In fact, take any smooth affine integral variety V defined over $\mathbb{R}$ having at least one real point. Then the coordinate ring $\mathbb{R}[V]$ is a regular integral domain with a formally real quotient field $\mathbb{R}(V)$. E.g. $V = \mathbb{A}^n$ yields the polynomial ring $A = \mathbb{R}[X_1, \ldots, X_n]$.

In order to exhibit examples with Sper i not surjective we state the following necessary criterion.

Corollary 2.5. *Retain the assumptions of* 2.3. *If* Sper $i : X(K) \to \operatorname{Min}\operatorname{Sper} A$ *is surjective then every real prime ideal* $\wp$ *of* A *is the center of a residually real valuation ring* V *of* K, *i.e.*

$$A \subseteq V, \quad \wp = A \cap \mathcal{M}_V.$$

Proof. Any real prime ideal $\wp$ has, by definition, a formally real residue field. Choosing any order $\overline{\alpha}$ on $k(\wp)$ we therefore have defined an element $\alpha = (\wp, \overline{\alpha}) \in$ Sper A. By 2.3 the morphism $A \to A/\wp$ extends to a place $\lambda : K \to L \cup \infty$, L some real closed field. Hence, $A \subseteq V_\lambda$, $\wp = A \cap \mathcal{M}_V$, and V_λ is residually real. □

Consider the real curve $\Gamma : y^2 = x^2(x-1)$ over $\mathbb{R}$. Its coordinate ring

$$\mathbb{R}[\Gamma] = \mathbb{R}[X,Y]/_{(Y^2 - X^2(X-1))}$$

has a formally real quotient field. Let $\mathcal{M}$ be the maximal ideal of the point $x = (0,0)$ and assume $\mathcal{M}$ to be the center of a valuation ring V. We get $\left(\frac{y}{x}\right)^2 = x - 1 \in \mathbb{R}[\Gamma]$, hence $\frac{y}{x} \in V$ and $-1 = t^2$ in the residue field F of V. Thus, F is not formally real. A more detailed analysis shows

$$\operatorname{Min}\operatorname{Sper}\mathbb{R}[\Gamma] = X(\mathbb{R}(\Gamma)) \cup \{(\mathcal{M}, \mathbb{R}^2)\}.$$

After the previous study of Min Sper A we are now turning to the determination of Max Sper A where we allow *any commutative ring A*. We start by considering the map

$$\begin{cases} \operatorname{Hom}(A, \mathbb{R}) & \to \quad \operatorname{Sper} A \\ \quad \varphi & \mapsto \quad \alpha_\varphi := \varphi^{-1}(\mathbb{R}^2). \end{cases}$$

In other formulation: $\alpha_\varphi = (\ker \varphi, \overline{\alpha})$ where $\overline{\alpha} = \mathbb{R}^2 \cap \operatorname{quot}(\varphi(A))$ is an Archimedean order.

Claim:

i) $\alpha_\varphi \in \operatorname{Max}\operatorname{Sper} A$,

ii) α_φ determines φ, i.e. the map is injective.

To prove i) one may apply [K–S], Kor. 5, p. 134. Alternatively one may argue as follows. Let R be a real closure of $\operatorname{quot}(\varphi(A), \overline{\alpha})$. If $\alpha_\varphi \subseteq \beta$ then there exists a place $\lambda : R \to L \cup \infty$ into a real closed extension of $(k(\wp), \overline{\beta})$, where $\beta = (\wp, \overline{\beta})$, which extends the canonical epimorphism $A/_{\ker\varphi} \to A/_{\wp}$, cf. [Be 4], prop. 2.9 e.g. However, R is Archimedean ordered. Hence, $R = V_\lambda$, telling that $\lambda : R \hookrightarrow L$ is an embedding, and $\wp = \ker\varphi$ which in turn implies $\alpha_\varphi = \beta$. To derive ii) we decompose $\varphi : A \to \mathbb{R}$ as follows

$$\begin{array}{ccc} A & \xrightarrow{\varphi} & (\mathbb{R}, \mathbb{R}^2) \\ \pi \downarrow & & \uparrow \overline{\varphi}, \\ A/\wp & \underset{i}{\hookrightarrow} & (k(\wp), \overline{\alpha}) \end{array} \qquad \pi \text{ the canonical epimorphism.}$$

Note that $\overline{\varphi}$ is order-preserving. Since $\overline{\alpha}$ is Archimedean this implies $\overline{\varphi}$ to be uniquely determined by $\overline{\alpha}$. Moreover, π and i are uniquely determined by $\wp$. Altogether, φ is determined by $(\wp, \overline{\alpha})$. □

We endow $\mathrm{Hom}(A, \mathbb{R})$ with the subspace topology inherited from the product topology on $\mathbb{R}^A$.

Proposition 2.6. *The map*

$$j : \mathrm{Hom}(A, \mathbb{R}) \to \mathrm{Max\,Sper}\, A, \varphi \mapsto \alpha_\varphi$$

is a topological embedding.

Proof. Injectivity has been proved already. A subbasis in $\mathrm{Max\,Sper}\, A$ is given by the sets $D(a) = \{\alpha \in \mathrm{Max\,Sper}\, A \mid a(\alpha) > 0\}$, a ranging over A. We obtain $j^{-1}(D(a)) = \{\varphi \mid \varphi(a) > 0\}$ which is clearly open in $\mathrm{Hom}(A, \mathbb{R})$. Thus, j is continuous. To show that j is open it is sufficient to demonstrate that $U := j(\{\varphi \mid r < \varphi(a) < s\})$ is open in the image of j for any choice of $a \in A$ and $r, s \in \mathbb{Q}$. The condition $r < \varphi(a) < s$ can be rephrased as $\varphi(a - r) > 0$, $\varphi(s - a) > 0$. Thus, $U = D(a - r) \cap D(s - a) \cap \mathrm{im}(j)$, U is open. □

Wishing to characterize rings admitting a bijection between $\mathrm{Hom}(A, \mathbb{R})$ and $\mathrm{Max\,Sper}\, A$ we are led to consider the notion of "elements being finite on $\mathrm{Sper}\, A$". This concept, in greater generality, is well developed in [K–S], Kap. III, §11. It provides the basis for the definition and study of the so-called *real holomorphy rings of rings*, a topic not to be pursued here, but see the forthcoming paper [B–P].

Definition 2.7. $a \in A$ is *finite* on $X \subseteq \mathrm{Sper}\, A$ if there exists $n \in \mathbb{N}$ such that for every $\alpha \in X$:

$$(n + a)(\alpha) \geq 0 \text{ and } (n - a)(\alpha) \geq 0.$$

Lemma 2.8. *The following statements are equivalent:*

i) *a is finite on* $\mathrm{Sper}\, A$,

ii) *a is finite on* $\mathrm{Max\,Sper}\, A$,

iii) *a is finite on* $\mathrm{Min\,Sper}\, A$.

Proof. i) $\Rightarrow$ iii) and iii) $\Rightarrow$ ii) are trivial, ii) $\Rightarrow$ i) Let $(n \pm a)(\alpha) \geq 0$ for every $\alpha \in \operatorname{Max Sper} A$. Then $((n+1) \pm a)(\alpha) > 0$ for every $\alpha \in \operatorname{Max Sper} A$. Since every $\beta \in \operatorname{Sper} A$ specializes to some $\alpha \in \operatorname{Max Sper} A$ we deduce $((n+1) \pm a)(\beta) > 0$ for every $\beta \in \operatorname{Sper} A$. □

Theorem 2.9. *The following statements are equivalent:*

i) $j : \operatorname{Hom}(A, \mathbb{R}) \to \operatorname{Max Sper} A$ *is homeomorphism,*

ii) *every* $a \in A$ *is finite on* $\operatorname{Sper} A$.

Proof. i) $\Rightarrow$ ii) Being homeomorphic to the compact space $\operatorname{Max Sper} A$ the space $\operatorname{Hom}(A, \mathbb{R})$ is compact as well. The evaluation map $\varphi \mapsto \varphi(a)$ is continuous on $\operatorname{Hom}(A, \mathbb{R})$. By compactness, it is bounded by some $n \in \mathbb{N}$, i.e. $\varphi(n \pm a) \geq 0$ for every $\varphi \in \operatorname{Hom}(A, \mathbb{R})$. The surjectivity of j implies that $(n \pm a)(\alpha) \geq 0$ on $\operatorname{Max Sper} A$. Hence, by 2.8, a is finite on $\operatorname{Sper} A$. ii) $\Rightarrow$ i) Consider $\alpha = (\wp, \overline{\alpha}) \in \operatorname{Sper} A$ and let $\pi : A \to A_{\wp} \hookrightarrow k(\wp)$ be the canonical map. From $(n \pm a)(\alpha) \geq 0$ for a suitable $n \in \mathbb{N}$ we infer $\pi(a) \in A(\overline{\alpha})$. Thus $\pi(A) \subseteq A(\overline{\alpha})$. The residue field of $A(\overline{\alpha})$ is a subfield of $\mathbb{R}$. Let $\lambda : A(\overline{\alpha}) \to \mathbb{R}$ be the corresponding place. Then $\lambda \circ \pi$ gives rise to the specialization $\alpha_{\lambda \circ \pi}$ of α. Consequently, if $\alpha \in \operatorname{Max Sper} A$ then $\alpha = \alpha_{\varphi}$ for some $\varphi \in \operatorname{Hom}(A, \mathbb{R})$. In view of 2.6 the proof is now complete. □

The result above will be applied to an integral domain A. Assume

$$\operatorname{Hom}(A, \mathbb{R}) \underset{j}{\overset{\sim}{\to}} \operatorname{Max Sper} A$$

then every a is finite on $\operatorname{Sper} A$, in particular on $\operatorname{Sper} K = X(K) \subseteq \operatorname{Sper} A$ where $K = \operatorname{quot}(A)$. This means $n \pm a \in \cap P$, $P \in X(K)$, implying $n \pm a \in \sum K^2$ for every $a \in A$. Therefore, in our case, $A \subseteq H(K)$.

Proposition 2.10. *Let A be an integral domain with a formally real quotient field K. If*

$$A \subseteq H(K), \ \operatorname{Min Sper} A = X(K)$$

then

$$\operatorname{Hom}(A, \mathbb{R}) \textit{ is homeomorphic to } \operatorname{Max Sper} A$$

(via j).

Proof. We have to show that every $a \in A$ is finite on $\operatorname{Sper} A$. From $A \subseteq H(K)$ we deduce $n \pm a \in \sum K^2$ for some $n \in \mathbb{N}$. This means $(n \pm a)(P \cap A) \geq 0$ for every $P \in X(K)$. Using $X(K) = \operatorname{Min Sper} A$ and 2.8 we derive that a is finite on $\operatorname{Sper} A$. □

All the results above will now be applied to the real holomorphy ring $H(K)$ of a formally real field K. The following diagram arises:

$$\begin{array}{ccccc} X(K) & & \xrightarrow[\sim]{\text{Sper}\, i} & & \text{Min Sper}\, H(K) \\ \downarrow \Lambda & & /// & & \downarrow \text{sp} \\ M(K) & \xrightarrow[\sim]{\text{res}} & \text{Hom}(H(K), \mathbb{R}) & \xrightarrow[\sim]{j} & \text{Max Sper}\, H(K) \end{array}$$

where Λ is defined by requiring the diagram to be commutative. Since $\text{Min Sper}\, H(K)$ is compact we see that $\text{Max Sper}\, H(K)$ carries the quotient topology relative to the specialization map sp. Thus Λ is continuous and surjective imposing the quotient topology on $M(K)$. It remains to make the assignment $P \mapsto \lambda_P = \Lambda(P)$ more explicit. Following the mappings in the diagram above then λ_P is the unique real place $\lambda \in M(K)$ satisfying

2.11. $\lambda(P \cap H) \geq 0.$

This property implies

$$V_\lambda = A(P) \text{ and } \lambda(P \cap V_\lambda) \geq 0.$$

To see this first note that $V_\lambda = H_{\ker \lambda}$, $A(P) = H_{I(P) \cap H}$.Thus, as a first step, we prove $\ker \lambda = I(P) \cap H$. Assume $a \in H$, $\lambda(a) = 0$, but $a \notin I(P)$, say $\frac{1}{n} - a \notin P$. Then $a - \frac{1}{n} \in P \cap H$ yielding the contradiction $-\frac{1}{n} = \lambda(a - \frac{1}{n}) \geq 0$. Therefore $\ker \lambda \subseteq I(P) \cap H$. Conversely, if $a \in I(P) \cap H$ then, for every $n \in \mathbb{N}$, $\frac{1}{n} - a \in P \cap H$, $\frac{1}{n} + a \in P \cap H$. Applying λ and using 2.11 the bound $| \lambda(a) | \leq \frac{1}{n}$ results for every n. Obviously, $\lambda(a) = 0$, and $\ker \lambda = I(P) \cap H$ is proved. Pick any $x = \frac{a}{b} \in P \cap V_\lambda$ where $a, b \in H$, $\lambda(b) \neq 0$. By passing to $x = \frac{ab}{b^2}$ if necessary we may assume $b \in P \cap H$ in addition. Then $a = xb \in P \cap H$ yielding $\lambda(a) \geq 0$ and $\lambda(x) = \frac{\lambda(a)}{\lambda(b)} \geq 0$. □

If $P \in X(K)$ is given then the real place $\lambda_P = \Lambda(P)$ characterized by 2.11 is called the *canonical place*. Our definition is equivalent to anyone met in the literature. This follows from our consideration in combination with theorem 2.11 in [L 1], p. 19.

The next theorem summarizes all what has been derived so far in the case of the real holomorphy ring $H(K)$ of a formally real field K.

Theorem 2.12. *In the diagram*

$$\begin{array}{ccc} X(K) & \xrightarrow{\text{Sper}\, i} & \text{Min Sper}\, H(K) \\ \downarrow \Lambda & & \downarrow \text{sp} \\ M(K) & \xrightarrow{j \circ \text{res}} & \text{Max Sper}\, H(K) \end{array}$$

the horizontal maps are homeomorphisms, the vertical ones are continuous and surjective where Λ is defined by 2.11 *and* sp *is the specialization map. All spaces are compact and $M(K)$ and* Max Sper $H(K)$ *carry the quotient topologies relative to Λ and* sp *respectively.*

The rest of this section will be devoted to the study of the fibers of $\Lambda : X(K) \to M(K)$. The next results i) - iv) are all known, e.g. [P], [L 1] and are listed here just for the sake of completeness. Given $\lambda \in M(K)$ then 2.11 states

2.13. $\lambda_P = \lambda \Leftrightarrow \lambda(P \cap H) \geq 0.$

Fix λ and consider the valuation ring V_λ. Set

$$T_\lambda = K^2 \cdot \{\epsilon \in V_\lambda \mid \lambda(\epsilon) > 0\}.$$

Then

i) T_λ is a fan, i.e. a preordering satisfying $T_\lambda + T_\lambda a \subseteq T_\lambda \cup T_\lambda a$ whenever $a \notin -T_\lambda$,

ii) $\lambda_P = \lambda \Leftrightarrow T_\lambda \subseteq P$ (use 2.13),

iii) $\Lambda^{-1}(\{\lambda\}) = X_{T_\lambda} := \{P \in X(K) \mid T_\lambda \subseteq P\} \simeq \left(K^*/(T_\lambda^* \cup -T_\lambda^*)\right)^\wedge \simeq (\Gamma_\lambda/2\Gamma_\lambda)^\wedge$ where $K^* = K\backslash\{0\}$, $T_\lambda^* = T_\lambda\backslash\{0\}$, Γ_λ the value group of V_λ and where $G^\wedge$ denotes the Pontrjagin character group of the abelian group G.

iv) Choose P_0 with $\lambda_{P_0} = \lambda$ then

$$\Lambda^{-1}(\{\lambda\}) = \{\chi_{P_0} \cdot (\nu \circ v) \mid \nu \in (\Gamma_\lambda/2\Gamma_\lambda)^\wedge\}$$

where every order P is replaced by its character $\chi_P : K^* \to \{1, -1\}$, $a \mapsto 1$ or -1 according to whether $a \in P$ or $a \in -P$. This result follows from the Baer–Krull theorem, see [L 1], sect. 3, [K–S], Kap. II, §7 e.g. Consequently, $\wedge^{-1}(\{\lambda\})$ is finite iff $[\Gamma_\lambda : 2\Gamma_\lambda] < \infty$. If $[\Gamma_\lambda : 2\Gamma_\lambda] < \infty$ then $\# \wedge^{-1}(\{\lambda\}) = [\Gamma_\lambda : 2\Gamma_\lambda]$.

There is a further characterization of orderings in the same fiber of Λ. It makes use of the notion of a 2-primary chain of orderings (of higher) level or, equivalently, of a 2-primary chain of signature. In the sense of Harman [H], a 2-primary chain of orderings is a sequence $(P_i)_{i \geq 0}$ satisfying

1) P_0, P_1 are orders; P_i, $i \geq 1$ is an ordering of exact level 2^{i-1}, i.e. $[K^* : P_i^*] = 2^i$,

2) $P_i \cup -P_i = (P_0 \cap P_{i-1}) \cup -(P_0 \cap P_{i-1})$ for $i \geq 2$.

In the sense of N. Schwartz [Schw] a (non-trivial, normalized) 2-primary chain of signatures is given by an order P_0 and a character $\nu : \Gamma \to \mathbb{Z}_2$ with values in the group of dyadic integers $\mathbb{Z}_2$ such that $\nu(\Gamma) \not\subseteq 2\mathbb{Z}_2$ where Γ is the value group of $A(P_0)$. From the data (P_0, ν) one derives the chain in the sense of Harman in the following way

$$P_i = \{0\} \cup (\chi_{P_0} \cdot \nu \circ v)^{-1}(2^i\mathbb{Z}_2), i \geq 1$$

where χ_{P_0} denotes the character attached to P_0 and v the Krull-valuation belonging to $A(P_0)$. This assignment provides an equivalence between these two concepts. We say that the P_i's belong to (P_0, ν).

Proposition 2.14. *Let P and Q be distinct orders of K. Then the following statements are equivalent*

i) $\lambda_P = \lambda_Q$,

ii) *P and Q belong to a 2-primary chain of orderings (or signatures).*

Proof. ii) $\Rightarrow$ i) This is proved in [H], i) $\Rightarrow$ ii). Let v be the valuation attached to $\lambda_P = \lambda_Q = \lambda$ and Γ its value group. From $P \neq Q$ we infer the existence of a non-trivial character $\mu : \Gamma \to \{\pm 1\}$ satisfying $\chi_P = \chi_Q \cdot (\mu \circ v)$. We identify $\{\pm 1\}$ with $\mathbb{Z}/2\mathbb{Z}$ and understand μ as a character $\Gamma \xrightarrow{\mu} \mathbb{Z}/2\mathbb{Z}$. In order to prove our claim we will show that μ can be factored via a character $\hat{\mu} : \Gamma \to \mathbb{Z}_2$, i.e. will have a commutative diagram

$$\begin{array}{ccc} \Gamma & \xrightarrow{\hat{\mu}} & \mathbb{Z}_2 \\ & \mu \searrow \quad \swarrow \pi & \\ & \mathbb{Z}/2\mathbb{Z} & \end{array}$$

where π is the canonical epimorphism. To deduce the existence of $\hat{\mu}$ we first prove that the natural epimorphism $\mathbb{Z}/2^l\mathbb{Z} \xleftarrow{\pi_l} \mathbb{Z}/2^{l+1}\mathbb{Z}$ induces an epimorphism $\mathrm{Hom}(\Gamma, \mathbb{Z}/2^l\mathbb{Z}) \leftarrow \mathrm{Hom}(\Gamma, \mathbb{Z}/2^{l+1}\mathbb{Z})$ for every $l \in \mathbb{N}$. In fact, the exact sequence

$$0 \to \mathbb{Z}/2\mathbb{Z} \to \mathbb{Z}/2^{l+1}\mathbb{Z} \to \mathbb{Z}/2^l\mathbb{Z} \to 0$$

induces an exact sequence

$$\mathrm{Hom}(\Gamma, \mathbb{Z}/2^{l+1}\mathbb{Z}) \to \mathrm{Hom}(\Gamma, \mathbb{Z}/2^l\mathbb{Z}) \to \mathrm{Ext}^1(\Gamma, \mathbb{Z}/2\mathbb{Z}).$$

As Γ is torsion-free $\mathrm{Ext}^1(\Gamma, \mathbb{Z}/2\mathbb{Z}) = 0$, [C–E], Cor. 6.2, p. 136. Thus, our intermediate claim is proved. As an application we derive the existence of a sequence (μ_n), $\mu_n \in \mathrm{Hom}(\Gamma, \mathbb{Z}/2^n\mathbb{Z})$ satisfying $\pi_n \circ \mu_{n+1} = \mu_n$ for every n. Then $\widehat{\mu} := \varprojlim \mu_n : \Gamma \to \varprojlim \mathbb{Z}/2^n\mathbb{Z} = \mathbb{Z}_2$ is the required homomorphism. □

Using the statements above about the fibers of $\Lambda : X(K) \to M(K)$ and 2.14 we arrive at characterizations of the case that $\Lambda : X(K) \to M(K)$ is bijective.

Corollary 2.15. *The following statements are equivalent:*

i) *$\Lambda : X(K) \to M(K)$ is bijective,*

ii) *for every $a \in K$: $a^2 \in \sum K^4$,*

iii) *every residually real valuation ring of K has a 2-divisible value group: $\Gamma = 2\Gamma$,*

iv) *K does not admit 2-primary chain of orderings.*

If K satisfies one of these conditions then $M(K)$ and $X(K)$ are even homeomorphic, in particular $M(K)$ is a totally disconnected compact space.

Proof. What is not covered by previous arguments is the equivalence of ii) and iii). However, $a^2 \in \sum K^4$ implies $2v(a) \in 4\Gamma$, hence $v(a) \in 2\Gamma$, whenever v is the valuation of a residually real valuation ring. The converse is proved in [Be 1]. The equivalence of ii) and iv) also follows from [G 1] and [G 2]. □

Remark 2.16. Since a totally Archimedean field K has no residually real valuation ring besides the trivial one the condition iii) is satisfied in that case. Thus $X(K) \xrightarrow{\sim} M(K)$ if K is totally Archimedean, in particular, $M(K)$ is compact and totally disconnected. All this apply to real algebraic number fields as mentioned already at the end of section 1.

Remark 2.17. We had introduced $M(K)$ via the topological space $\mathrm{Hom}(H(K), \mathbb{R})$ There is a similiar interpretation of $X(K)$. Every $P \in X(K)$ induces the signature homomorphism

$$\mathrm{sgn}_P : W(K) \to \mathbb{Z}$$

where $W(K)$ denotes the Witt ring of K, [L 1]. Moreover $X(K) \to \mathrm{Hom}(W(K), \mathbb{Z})$, $P \mapsto \mathrm{sgn}_P$ is a bijection as stated by the Harrison–Lorenz–Leicht theorem. If $\mathrm{Hom}(W(K), \mathbb{Z})$ is endowed with the subspace topology inherited from $\mathbb{Z}^{W(K)}$ then the map above is even a homeomorphism. Thus, in graphical presentation:

$$\begin{array}{ccc} X(K) & \xrightarrow{\sim} & \mathrm{Hom}(W(K), \mathbb{Z}) \\ \downarrow \Lambda & & \downarrow \Lambda' \\ M(K) & \xrightarrow{\sim} & \mathrm{Hom}(H(K), \mathbb{R}) \end{array}$$

The map Λ' is described as follows. Let $r \in \mathbb{Q}$ and $a \in H(K)$. Then

i) $(\Lambda'(\sigma))(a) > r \Leftrightarrow r \neq a, \sigma(< a - r >) = 1,$

ii) $(\Lambda'(\sigma))(a) = \sup\{r \mid r \neq a, \sigma(< a - r >) = 1\}.$

To prove it set $\lambda = \Lambda'(\sigma)$, interpreted as a place $K \to \mathbb{R} \cup \infty$. On the unit group V_λ^* we get

$$\mathrm{sgn}\,\lambda(\epsilon) = \mathrm{sgn}_P(< \epsilon >) \text{ where } \lambda = \lambda_P, \sigma = \mathrm{sgn}_P .$$

This yields the claim i). The second one follows from i) once it is known that the set in question is bounded. Given $a \in H(K)$ choose $n \in \mathbb{N}$ such that $n - a \neq 0$, $n - a \in \sum K^2$. If $\sigma(< a - r >) = 1$ then the binary form $\rho =< n - a, a - r >$ has signature $\sigma(\rho) = 2$. From

$$\rho =< (n - a) + (a - r), (n - r)(n - a)(a - r) >$$

we derive $\sigma(< n - r >) = 1$, i.e. $r < n$.

3. The space of real places in real algebraic geometry

In this section we will be studying real algebraic geometry from a birational point of view. In other words, we will be concerned with function fields over $\mathbb{R}$. For general notions and results we refer to the book of Bochnak, Coste and Roy [BCR] if no other references are given.

Let $F \mid R$ be a function field and Y be a projective model of F, i.e. an integral projective $\mathbb{R}$-variety with function field $\mathbb{R}(Y) = F$. The Artin–Lang homomorphism theorem, in its geometric version, states that F is formally real if and only if Y contains a simple (= regular) real point, see [Be 2], p. 4 ff e.g. Let $Y(\mathbb{R})$ and $Y_{\mathrm{reg}}(\mathbb{R})$ denote the set of real, resp. of regular real (closed) points in Y. Then F is formally real iff $Y_{\mathrm{reg}}(\mathbb{R}) \neq \emptyset$. There is a natural Euclidean topology on $Y(\mathbb{R})$ in which it is compact (it stems from any embedding into some $\mathbb{P}^n(\mathbb{R})$).

Since every projective variety is complete every valuation ring V of F has a (unique) center $c(V)$ in V. This is, by definition, the unique point $x \in V$ whose local ring $\mathcal{O}_x$ is dominated by V. In particular, consider $\lambda \in M(F)$ with valuation ring V_λ. We set

$$c(\lambda) := c(V_\lambda)$$

and call $c(\lambda)$ the center of λ on Y.

Clearly, the residue field of $\mathcal{O}_{c(V)}$ is contained in that of V. In the case of $\lambda \in M(F)$ this means $c(\lambda) \in Y(\mathbb{R})$. Therefore, we are concerned with the center map

$$c = c_Y : M(F) \to Y(\mathbb{R}), \lambda \mapsto c(\lambda).$$

As usual, $x \in Y(\mathbb{R})$ is called central if $x = c(\lambda)$ for some $\lambda \in M(F)$. We set $Y_c(\mathbb{R}) = \{x \in Y(\mathbb{R}) \mid x \text{ central}\}$. There is the fundamental theorem of Dubois characterizing $Y_c(\mathbb{R})$, cf. [Du], p. 62, th. 3 or [Be 4], th. 1.10.

Theorem 3.1. *$Y_c(\mathbb{R})$ is the closure of $Y_{\mathrm{reg}}(\mathbb{R})$ (in the Euclidean topology of $Y(\mathbb{R})$).*

Since $M(F)$ and $Y(\mathbb{R})$ are topological spaces it is natural to ask whether the center map c is continuous. In fact we have

Proposition 3.2. *$c : M(F) \to Y(\mathbb{R})$ is continuous.*

Proof. We choose a representation $Y = \operatorname{Proj}\operatorname{Spec}\mathbb{R}[X_0, \dots, X_n]/I$, I a homogenous ideal. Then $F = \operatorname{quot}\left(\mathbb{R}\left[\frac{x_1}{x_0}, \dots, \frac{x_n}{x_0}\right]\right)$ where x_i denotes the residue class of X_i in $\mathbb{R}[X_0, \dots, X_n]/I$. We have a Zariski-open covering

$$Y = \bigcup_{i=0}^{n} \operatorname{Spec}\mathbb{R}\left[\frac{x_0}{x_i}, \dots, \frac{x_n}{x_i}\right] =: \bigcup_{i=0}^{n} V_i$$

which in turn provides an open covering $Y(\mathbb{R}) = \cup V_i(\mathbb{R})$ in the Euclidean topology. Set $U_i := \left\{\lambda \in M(F) \mid \lambda\left(\frac{x_k}{x_i}\right) \neq \infty \text{ for } k = 0, \dots, n\right\}$. Then U_i is open in $M(F)$

and the center map c maps U_i into $V_i(\mathbb{R})$. More precisely, we get

$$c : U_i \to V_i(\mathbb{R}), \lambda \mapsto \left(\ldots, \lambda\left(\frac{x_k}{x_i}\right), \ldots\right)_{k \neq i} \subseteq \mathbb{R}^n$$

where we naturally interpret $V_i(\mathbb{R})$ as a subspace of $\mathbb{R}^n$. On U_i the map c is obviously continuous. Hence, the proposition is proved. □

An alternative approach via real spectra

The study of the center map for projective varieties may be reduced to the investigation of affine varieties Z with compact real locus $Z(\mathbb{R})$. Indeed if the projective variety $Y = \operatorname{Proj}\operatorname{Spec}\mathbb{R}[X_0, \ldots, X_n]/I$ is given we pass to the affine subvariety $Z = D_+\left(\sum_0^n x_i^2\right)$, $x_i = X_i \bmod I$. Then $Z(\mathbb{R}) = Y(\mathbb{R})$, and $Z(\mathbb{R})$ turns out to be compact since Y is projective.

In the following considerations we therefore start with an arbitrary integral affine $\mathbb{R}$-variety Z with coordinate ring $\mathbb{R}[Z]$ and formally real function field $F = \mathbb{R}(Z)$. The natural mapping

$$\Phi_0 : Z(\mathbb{R}) \to \operatorname{Sper}\mathbb{R}[Z]$$

is a (continuous) dense embedding by the Artin–Lang theorem, cf. [Be 4, 1.13], e.g. In fact the image is a subspace of $\operatorname{Max}\operatorname{Sper}\mathbb{R}[Z]$, obviously a dense one.

In our present situation we can prove

Proposition 3.3. *The following statements* i), ii) *and* iii) *are equivalent.*

i) $Z(\mathbb{R})$ *is compact,*

ii) $Z(\mathbb{R}) \xrightarrow{\Phi} \operatorname{Max}\operatorname{Sper}\mathbb{R}[Z]$ *is surjective,*

iii) $Z(\mathbb{R}) \xrightarrow{\Phi} \operatorname{Max}\operatorname{Sper}\mathbb{R}[Z]$ *is a homeomorphism.*

If i) - iii) *hold then* iv) $\mathbb{R}[Z] \subseteq H(F)$.

Proof. i) ⇒ ii) Φ is continuous and $Z(\mathbb{R})$ compact. Hence, the image of Φ is a dense quasicompact subspace of the Hausdorff space $\operatorname{Max}\operatorname{Sper}\mathbb{R}[Z]$. Therefore, the surjectivity results. ii) ⇒ iii) Both spaces are compact and Φ is bijective. iii) ⇒ iv) Pick any $f \in \mathbb{R}[Z]$. Consider f as a function on the space $Z(\mathbb{R})$. This space is compact since it is homeomorphic to the compact space $\operatorname{Max}\operatorname{Sper}\mathbb{R}[Z]$. Hence, f is bounded on $Z(\mathbb{R})$, i.e. $n \pm f \geq 0$ on $Z(\mathbb{R})$ for some $n \in \mathbb{N}$. Invoking that Φ is a dense embedding we derive that f is bounded on $\operatorname{Sper}\mathbb{R}[Z]$. In particular, $n \pm f \in \sum F^2$, i.e. $f \in H(F)$. That iii) implies i) is trivial. □

To apply 3.3 we assume that Z is an integral affine $\mathbb{R}$-variety with a *compact real locus* $Z(\mathbb{R})$. Then $\mathbb{R}[Z] \subseteq H(F)$ which yields the restriction map Sper $H(F) \xrightarrow{\text{res}}$

$\operatorname{Sper}\mathbb{R}[Z]$. Using $\mathbb{R} \subseteq \mathbb{R}[Z]$ we even get the following commutative diagram:

$$\begin{array}{ccc} \operatorname{Max}\operatorname{Sper} H(F) & \xrightarrow{\text{res}} & \operatorname{Max}\operatorname{Sper}\mathbb{R}[Z] \\ \downarrow & /// & \Phi^{-1}\downarrow \\ M(F) & \xrightarrow{c} & Z(\mathbb{R}) \end{array}$$

where all maps are continuous, the vertical ones are homeomorphisms and the map c is the center map introduced above. To verify commutativity assume that the place λ be assigned to $x \in Z(\mathbb{R})$ via the upper route in the diagram above. From $\mathbb{R}[Z] \subseteq H(F)$ we first infer $\lambda(f) = f(x)$ for every $f \in \mathbb{R}[Z]$. Then obviously, $\mathcal{O}_x \subseteq V_\lambda$ as was to be shown.

To see examples we first restrict ourselves to function field F of curves, i.e. $\operatorname{tr}(F \mid \mathbb{R}) = 1$. There is a unique smooth projective model of F, say Γ. From the theory of algebraic curves we see that $M(F) \to \Gamma(\mathbb{R})$ is a bijection. Hence, in view of the last proposition and the fact that both spaces are compact we obtain the

Proposition 3.4. *In the situation above the center map is a homeomorphism between $M(F)$ and $\Gamma(\mathbb{R})$.*

In higher dimensions the situation gets much more complicated. This is mainly due to the fact that a function field of transcendence degree at least 2 admits a vast amount of smooth projective models, due to Hironaka's fundamental results. H. W. Schülting has clarified the relation between the space $M(F)$ and the various models. We are going to describe his results, see [Schü 2] for further references.

Let F be a formally real function field over $\mathbb{R}$. We study the set of all smooth projective models of F. Let Y be one of these. Set

$$\mathcal{L}_1(Y) = \{(Z,\varphi) \mid \varphi : Z \to Y \text{ is a birational } \mathbb{R}\text{-morphism}\}.$$

$\mathcal{L}_1(Y)$ can be ordered as follows: $(Z_1,\varphi_1) \geq (Z_2,\varphi_2)$ iff there exists a birational $\mathbb{R}$-morphism $\psi : Z_1 \to Z_2$ representing ${\varphi_2}^{-1} \circ \varphi_1$. Finally, let $\mathcal{L}(Y)$ be set of all pairs $(Z,\varphi) \in \mathcal{L}_1(Y)$ such that φ is a composition of monoidal transformations with regular centers.

Theorem 3.5. (H. W. Schülting)

i) *The center maps $c_Z : M(F) \to Z(\mathbb{R})$, $Z \in \mathcal{L}(Y)$ yield a homeomorphism*

$$M(F) \xrightarrow{\sim} \varprojlim_{\mathcal{L}(Y)} Z(\mathbb{R}).$$

ii) $H(F) = \bigcup_{\mathcal{L}(Y)} \mathcal{O}_{\mathbb{R}}(Z)$ *where* $\mathcal{O}_{\mathbb{R}}(Z) = \bigcap_{z\in Z(\mathbb{R})} \mathcal{O}_z(Z)$.

We will use Schülting's approach to study the fibers of the center map $c : M(F) \to Y(\mathbb{R})$ over central points. That one should do this, at least in the case of regular points, was first proposed by L. Bröcker. We will need the following lemma.

Lemma 3.6. *Let $p : X \to Y$ be a continuous surjective map between the compact spaces X and Y. If every fiber of p is connected then the following statements are true.*

i) *If Z is a closed connected subspace of Y then $p^{-1}(Z)$ is connected as well.*

ii) *p induces a bijection $p_* : \pi_0(X) \to \pi_0(Y)$.*

Proof. i) By replacing Y by Z and X by $p^{-1}(Z)$ we may assume from the beginning that Y is connected and that the connectedness of X is to be shown. Assume $X = C_1 \dot\cup C_2$ with C_i open and closed, $C_i \neq \emptyset$. Then $Y = p(C_1) \cup p(C_2)$. Since the C_i's are compact the images are closed in Y. The connectedness of Y yields $p(C_1) \cap p(C_2) \neq \emptyset$. Choose y in the intersection. Then, obviously, the fiber over y is not connected, contradicting our hypothesis. ii) This follows from i). □

The following result, in the case of regular points, is due to L. Bröcker. So far, no proof has appeared in the literature.

Proposition 3.7. *The fiber of the center map over a central point x is a compact space with finitely many connected components. If x is even regular then the fiber is connected.*

Proof. By Hironaka's result on the resolution of singularities there is a sequence of monoidal transformations

$$Y \leftarrow Y_i \leftarrow \ldots \leftarrow Y_r$$

with regular centers where all Y_i are projective models of F and Y_r is smooth in addition. Let $\mu : Y_r \to Y$ be the resulting morphism. It induces a continuous map $\mu : Y_r(\mathbb{R}) \to Y(\mathbb{R})$. The fiber $\mu^{-1}(x) \subseteq Y_r(\mathbb{R})$ is not empty. Indeed, $x = c_Y(\lambda)$ for some $\lambda \in M(F)$ and therefore $\mu(c_{Y_r}(\lambda)) = x$. The fiber $\mu^{-1}(x)$ in $Y_r(\mathbb{R})$ is a real algebraic set. Hence, $\pi_0(\mu^{-1}(x))$ is finite, cf. [BCR, 2.4],. Next consider the center map $M(F) \xrightarrow{c_r} Y_r(\mathbb{R})$. From $\mu \circ c_r = c_Y$ we first get $c_Y{}^{-1}(\{x\}) = c_r{}^{-1}(\mu^{-1}(\{x\}))$. In order to show

$$\mid \pi_0(c_Y{}^{-1}(\{x\})) \mid = \mid \pi_0(\mu^{-1}(\{x\})) \mid$$

we want to apply lemma 3.6, ii). This is possible since the fibers of $M(F) \to Y_r(\mathbb{R})$ are connected as we are now going to prove. Suppose a fiber F_y over some $y \in Y_r(\mathbb{R})$ is not connected. Since F_y is a closed subset of the compact space $M(F)$ then there is a function $\varphi : M \to \mathbb{R}$ such that

i) φ has no zeros on F_y,

ii) φ changes sign on F_y.

The representation theorem for $H(F)$ [Be 2, (2.20)] assures the existence of $a \in H(F)$ with $\|\Phi(a) - \varphi\| < \epsilon$ for any given $\epsilon > 0$. Now, F_y being compact, we can find such an element $a \in H$ with i)′ $\lambda(a) \neq 0$ for every $\lambda \in F_y$,

ii)′ there are $\lambda, \mu \in F_y$ with $\lambda(a) > 0$, $\mu(a) < 0$.

Invoking theorem 3.5 we have a sequence of monoidal transformations

$$Y_r \leftarrow Z_1 \leftarrow \ldots \leftarrow Z_s$$

with regular centers such that $a \in \mathcal{O}_{\mathbb{R}}(Z_s)$. Then $a(c(\lambda)) > 0$, $a(c(\mu)) < 0$ where $c(\lambda)$ and $c(\mu)$ are the centers of λ, μ on Z_s respectively. Both centers lie in the fiber F'_y of $Z_s(\mathbb{R}) \to Y_r(\mathbb{R})$ over y. Since every point in this fiber is, as a regular point, a center of a real place, which is necessarily in F_y, the property i)′ implies that the fiber F'_y is not connected. However, the fibers of $Z_{i+1}(\mathbb{R}) \to Z_i(\mathbb{R})$, $i = 0, \ldots, s-1$, $Z_0 = Y_r$ are connected. This follows from the fact that fibers over regular points relative to monoidal transformations with regular centers are projective spaces. An iterative application of lemma 3.6 now yields that F'_y has to be connected. Thus, a contradiction is achieved, and the fibers of $M(F) \to Y_r(\mathbb{R})$ are all connected. □

Remark. The proof has shown that the fibers $Z(\mathbb{R}) \to Y(\mathbb{R})$ over a central point have the same number of connected components as long as Z is smooth because this number equals the cardinality of the set of connected components of the fiber $M(F) \to Y(\mathbb{R})$ over the point in question.

Examples. Consider the two projective curves over $\mathbb{R}$:

- $\Gamma_1 : zy^2 = x^2(x+z)$
- $\Gamma_2 : zy^2 = x^3$.

In both cases $(0,0)$ is the only singular point. Let $\widetilde{\Gamma}_i$ denote the normalization of Γ_i. In the case of Γ_1 the fiber over $(0,0)$ consists of 2 points in $\widetilde{\Gamma}_1(\mathbb{R})$ whereas in the case of Γ_2 there is a unique point over $(0,0)$.

We finally apply the last theorem to derive the well-known fact that the number $\pi_0(Y(\mathbb{R}))$ is a birational invariant among the smooth projective varieties [BCR, 10.4]. In fact, if Y is of this type then the center map $c : M(F) \to Y(\mathbb{R})$ has connected fibers which results in a bijection $\pi_0(M(F)) \to \pi_0(Y(\mathbb{R}))$ by lemma 3.6, ii). But as already used every real algebraic set has only finitely many connected components. We state this in

Theorem 3.8. *Let Y be a smooth projective $\mathbb{R}$-variety with function field F. Then*

$$|\,\pi_0(M(F))\,| = |\,\pi_0(Y(\mathbb{R}))\,|\,.$$

We conclude this section by showing that the fiber of the center map $c : M(F) \to Y(\mathbb{R})$ allows an interpretation as the maximal real spectrum of some appropriate ring. To derive this let us consider, quite generally, any ring homomorphism

$$\varphi : A \to B$$

and its associated "restriction" map

$$\varphi_* : \operatorname{Sper} B \to \operatorname{Sper} A.$$

Given any $\alpha \in \operatorname{Max}\operatorname{Sper} A$ then the relative fiber

$$\varphi_*^{-1}(\alpha) \cap \operatorname{Max}\operatorname{Sper} B =: F_{\max}(\alpha)$$

is the space of closed points in the fiber φ_*^{-1}. Using [C–R, 4.3] or [Be 4, 2.17] we obtain a homeomorphism

$$F_{\max}(\alpha) \cong \operatorname{Max}\operatorname{Sper}(B \otimes_A k(\alpha))$$

where $k(\alpha)$ is the real closure of $(k(\wp), \bar{\alpha})$, $\alpha = (\wp, \bar{\alpha})$.

In our situation, Y is an affine $\mathbb{R}$-variety with compact real locus $Y(\mathbb{R})$. As shown at the beginning the center map can be identified with the restriction map

$$\text{res} : \operatorname{Max}\operatorname{Sper} H(F) \to \operatorname{Max}\operatorname{Sper} \mathbb{R}[Y]$$

induced by the inclusion $\mathbb{R}[Y] \hookrightarrow H(F)$. Assume

$$c(\lambda) = x \in Y(\mathbb{R}).$$

This means, setting $\alpha_\lambda = \lambda^{-1}(\mathbb{R}^2)$, $\alpha_x = \{f \in \mathbb{R}[Y] \mid f(x) \geq 0\}$,

$$\alpha_\lambda \cap \mathbb{R}[Y] = \alpha_x.$$

In view of $k(\alpha_x) = \mathbb{R}$, $\alpha_x \cap -\alpha_x = \mathcal{M}_x = \{f \in \mathbb{R}[X] \mid f(x) = 0\}$ we obtain

Proposition 3.9. *If x is a central point of $Y(\mathbb{R})$ then the fiber of the center map over x is homeomorphic to* $\operatorname{Max}\operatorname{Sper}(H(F)/\mathcal{M}_x H(F))$.

4. The connected components of $M(K)$

Given any topological space Y we denote by $\pi_0(Y)$ its topological space of connected components. The topology is defined as the quotient topology relative to the canonical map

$$\begin{cases} Y & \to & \pi_0(Y) \\ y & \mapsto & \text{component } C \text{ with } y \in C. \end{cases}$$

In case Y is a compact space $\pi_0(Y)$ turns out to be a compact totally disconnected space, in particular any two distinct connected components C_1 and C_2 of Y can be separated by an open and closed set $U : C_1 \subseteq U$, $C_2 \subseteq Y \setminus U$, cf. [W, ch. 8].

In section 1 we proved that the map $M(K) \to \mathrm{Hom}(H(K), \mathbb{R})$ is a homeomorphism. As was first observed by H. W. Schülting, cf. [Schü 1], sect. 4 there is an analogous result for $\pi_0(M(K))$ and the Witt ring $W(H(K))$. To facilitate the notation the reference to the field K will be dropped in the sequel.

Let A be any Prüfer ring with quotient field L. The inclusion $A \to L$ induces an injective map

$$W(A) \to W(L)$$

first proved by M. Knebusch [K], Th. 11.1.1, but see also [Ho]. If $\wp$ is any prime ideal of A we obtain a factorization of the map above:

$$W(A) \to W(A_\wp) \to W(L).$$

In addition, if $2 \in A^*$ then every non-degenerate quadratic form over the valuation ring $A_\wp$ can be diagonalized. We will make use of these facts.

If $\lambda \in M$ is given it induces a homomorphism $H \to \mathbb{R}$ and finally a homomorphism

$$\lambda_* : W(H) \to W(\mathbb{R}) = \mathbb{Z}.$$

Proposition 4.1. *The following statements are equivalent:*

i) λ *and* μ *are in the same connected component,*

ii) $\lambda_* = \mu_*$,

iii) $\operatorname{sgn} \lambda(\epsilon) = \operatorname{sgn} \mu(\epsilon)$ *for every* $\epsilon \in H^*$.

Proof. i) $\Rightarrow$ ii) We have to show $\lambda_*(\rho) = \mu_*(\rho)$ for every $\rho \in W(H)$. Let V be the valuation ring of λ, $\mathcal{M}$ its maximal ideal. Then λ_* is the compositum of the maps $W(H) \to W(V) \to W(\mathbb{R}) = \mathbb{Z}$. Over V we have $\rho \bigotimes_H V = < a_1, \ldots, a_n >$ for some $a_i \in V^*$. Then $\lambda_*(\rho) = \sum_i \operatorname{sgn}(\lambda(a_i))$. Set $U = \{\psi \in M \mid \psi(a_i) \neq 0, \infty, \operatorname{sgn} \psi(a_i) = \operatorname{sgn} \lambda(a_i)\}$. Obviously, U is an open neighborhood of λ. We claim that

$$\psi_*(\rho) = \sum_i \operatorname{sgn} \psi(a_i) = \lambda_*(\rho)$$

for each $\psi \in U$. To prove this let $\widetilde{V}$ be the valuation ring of ψ. The entries a_i are units in $\widetilde{V}$ and we get

$$< a_1, \ldots, a_n > \otimes_{\widetilde{V}} K = \rho \otimes_H K = (\rho \otimes_H \widetilde{V}) \otimes_{\widetilde{V}} K.$$

Using the injectivity of $W(\widetilde{V}) \to W(K)$ we conclude $\rho \otimes_H \widetilde{V} = < a_1, \ldots, a_n >$ which in turn yields the claim above. So far we have proved that, for every $\rho \in W(H)$, the map $\lambda \mapsto \lambda_*(\rho)$ is a locally constant function $M \to \mathbb{Z}$. It is therefore constant on each of the connected components of M. Varying ρ over $W(H)$ we derive $\lambda_* = \mu_*$ in our present situation.
ii) $\Rightarrow$ iii) Note $\lambda_*(< \epsilon >) = \operatorname{sgn} \lambda(\epsilon)$ since $\epsilon \in H^*$. ii) $\Rightarrow$ i) Assume λ, μ to be in

different components. Since M is a compact Hausdorff space there exists a subset A, open and closed, such that $\lambda \in A$ and $\mu \in M\backslash A$. From the representation theorem of the real holomorphy ring [Be 2, (2.20)] we deduce the existence of a unit $\epsilon \in H^*$ with $\lambda(\epsilon) > 0$, $\mu(\epsilon) > 0$ contradicting our assumption. □

In the case of M and the real holomorphy ring H we have found natural homeomorphisms

$$M \xrightarrow{\sim} \operatorname{Hom}(H, \mathbb{R}) \xrightarrow{\sim} \operatorname{Max\,Sper} H.$$

Correspondingly we will show that in our present situation we have homeomorphisms

$$\pi_0(M) \xrightarrow{\sim} \operatorname{Hom}(W(H), \mathbb{Z}) \xrightarrow{\sim} \operatorname{Max\,Sper} W(H).$$

To start we rephrase what has been proved above. For every $\rho \in W(H)$ the mapping $\lambda \mapsto \lambda_*(\rho)$ is a locally constant function from M into $\mathbb{Z}$. Therefore, considering $\mathbb{Z}$ as a discrete topological space and endowing

$$\operatorname{Hom}(W(H), \mathbb{Z}) \subseteq \mathbb{Z}^{W(H)}$$

with the subspace topology inherited from the product topology on $\mathbb{Z}^{W(H)}$ we end up with a continuous map

$$\sum : \pi_0(M) \to \operatorname{Hom}(W(H), \mathbb{Z}), [\lambda] \mapsto \lambda_*$$

where $[\lambda]$ denotes the connected component in which λ lies. From the last proposition we infer the injectivity of $\sum$. $\operatorname{Hom}(W(H), \mathbb{Z})$ is a compact Hausdorff space as can be seen as follows. The fact $\lambda_*(\rho) \in [-\dim \rho, \dim \rho]$ first implies that $\operatorname{Hom}(W(H), \mathbb{Z})$ is contained in $\prod\limits_{W(H)} ([-\dim \rho, \dim \rho] \cap \mathbb{Z})$ which is a compact space. Routine arguments then show that $\operatorname{Hom}(W(H), \mathbb{Z})$ is a closed subspace of this compact space. Hence, it is compact as well. To prove $\sum$ to be a homeomorphism it is therefore sufficient to deduce its surjectivity. To do this we start by proving

$$\{\ker \sigma \mid \sigma \in \operatorname{Hom}(W(H), \mathbb{Z})\} = \operatorname{Min\,Sper} W(H).$$

Proof of this claim. Since $W(H)$ is embedded in $W(K)$ and the latter ring is an integral extension of $\mathbb{Z}$ this holds for $W(H)$ as well. Given $\sigma \in \operatorname{Hom}(W(H), \mathbb{Z})$ we first get $\ker \sigma \cap \mathbb{Z} = (0)$, $W(H) = \mathbb{Z} \bigoplus \ker \sigma$. Since $W(H)$ is integral over $\mathbb{Z}$, $\ker \sigma$ has to be a minimal prime ideal of $W(H)$; moreover σ is determined by $\ker \sigma$. Conversely, if a minimal prime ideal $\wp$ of $W(H) \subseteq W(K)$ is given it can be extended to a minimal prime ideal $\wp$ of $W(K)$, cf. [Bo], ch. II, §2, prop. 6. Since K is formally real $P = \ker \tilde{\sigma}$ for some signature $\tilde{\sigma} = \operatorname{sgn}_P : W(K) \to \mathbb{Z}$, cf. [L 1]. Hence, $\wp = \ker \sigma$ where $\sigma = \tilde{\sigma}_{|W(H)}$. □

The arguments above have shown the following: given $\sigma \in \operatorname{Hom}(W(H), \mathbb{Z})$ there is an order P of K with $\sigma = \operatorname{sgn}_{P|W(H)}$. Let λ be the real place attached to P. We want to show $\sigma = \lambda_*$. Pick $\rho \in W(H)$. Then $\sigma(\rho) = \operatorname{sgn}_P < a_1, \ldots, a_n >$

where $\rho \bigotimes V_\lambda = < a_1, \ldots, a_n >$, $a_i \in V_\lambda^*$, V_λ the valuation ring of λ. From $a_i \in V_\lambda^*$ we infer $\operatorname{sgn}_P(a_i) = \operatorname{sgn} \lambda(a_i)$ which in turn yields $\sigma(\rho) = \lambda_*(\rho)$.

So far, we have proved $\pi_0(M) \xrightarrow{\sim} \operatorname{Hom}(W(H), \mathbb{Z})$. It remains to show that the latter space is homeomorphic to $\operatorname{Max}\operatorname{Sper} W(H)$. This can be deduced from 2.11 or directly as follows. We first investigate $\operatorname{Sper} W(H)$. As shown above $\operatorname{Min}\operatorname{Spec} W(H) = \{\ker \sigma \mid \sigma \in \operatorname{Hom}(W(H), \mathbb{Z})\}$. Since $W(H)/_{\ker \sigma} = \mathbb{Z}$ we see that each σ gives rise to the point

$$\alpha_\sigma := (\ker \sigma, \mathbb{Z}_+) \in \operatorname{Sper} W(H).$$

Moreover, any non-minimal prime ideal must have positive residue characteristic. This results in

$$\operatorname{Sper} W(H) = \{\alpha_\sigma \mid \sigma \in \operatorname{Hom}(W(H), \mathbb{Z})\}.$$

Two distinct minimal prime ideals are clearly incomparable. Hence,

$$\operatorname{Sper} W(H) = \operatorname{Max}\operatorname{Sper} W(H).$$

Furthermore, the map

$$\operatorname{Hom}(W(H), \mathbb{Z}) \to \operatorname{Max}\operatorname{Sper} W(H), \sigma \mapsto \alpha_\sigma$$

is bijective. Since both spaces are compact we only need to deduce that this map is continuous to find that it is a homeomorphism in fact. A subbasis in $\operatorname{Sper} W(H) = \operatorname{Max}\operatorname{Sper} W(H)$ is given by the sets $\{\alpha \mid \rho(\alpha) > 0\}$ where ρ ranges over $W(H)$. From $\rho(\alpha_\sigma) = \sigma(\rho)$ the continuity is readily shown.

We are going to summarize the results above.

Theorem 4.2. *The maps* $\pi_0(M) \to \operatorname{Hom}(W(H), \mathbb{Z})$, $[\lambda] \mapsto \lambda_*$ *and*

$$\operatorname{Hom}(W(H), \mathbb{Z}) \to \operatorname{Max}\operatorname{Sper} W(H), \ \sigma \mapsto \alpha_\sigma$$

are homeomorphisms.

Proposition 4.1 strongly hints at the role of the units of the real holomorphy ring in the study of connected components of $M(K)$. In fact we can quote the following result from [Be 3].

Theorem 4.3. *Let K be a formally real field and $H(K)$ its real holomorphy ring. Denote by $\mathbb{E}$ the set of units of $H(K)$ and set $\mathbb{E}^+ = \mathbb{E} \cap \sum K^2$. Then*

i) $|\pi_0(M(K))| < \infty \Leftrightarrow [\mathbb{E} : \mathbb{E}^+] < \infty$,

ii) *if $\pi_0(M(K))$ is finite then*

$$|\pi_0(M(K))| = \log_2[\mathbb{E} : \mathbb{E}^+].$$

We deduce the following

Corollary 4.4. $|\pi_0(M(K))| = 1 + \log_2[K^{*2} \cap \sum K^4 : (\sum K^{*2})^2]$.

Remark. As shown in [Be 1] $(\sum K^2)^2 \subseteq \sum K^4$, hence the index of the groups above is well defined.

Proof of 4.4. We show the existence of an isomorphism between $\mathbb{E}/\mathbb{E}^+ \cup -\mathbb{E}^+$ and $G := K^{*2} \cap \sum K^4/(\sum K^{*2})^2$ which gives the result in view of 4.3. Consider the homomorphism

$$\phi : \mathbb{E} \to \text{ class of } \epsilon^2 \text{ in } G.$$

We first prove that Φ is surjective. Let $x \in K^{*2} \cap \sum K^4$. From $x \in \sum K^4$ we get from [Be 3], 1.9 that $x = \epsilon q^2$ where $\epsilon \in \mathbb{E}$, $q \in \sum K^2$. The hypothesis $x = y^2$ leads to $\epsilon = (y/q)^2$. As a Prüfer ring $H(K)$ is integrally closed, thus $\epsilon = \nu^2$, $\nu \in \mathbb{E}$ and $x = \nu^2 \cdot q^2$ which is a representation of the desired type. Next, $\ker \phi = \mathbb{E}^+ \cup -\mathbb{E}^+$ is readily checked. □

We conclude this paper by presenting a criterion for the separation of connected components in terms of orders.

Theorem 4.5. *Let P and Q be distinct orders of K. Then the following statements are equivalent.*

i) *λ_P and λ_Q are in the same connected component,*

ii) *there is no $b \in P \cap -Q$, $b \neq 0$ satisfying*

$$b^2 \in \sum K^4.$$

Proof. i) $\Rightarrow$ ii) Assume on the contrary that $b \in P \cap Q$ satisfies $b \neq 0$, $b^2 \in \sum K^4$. From [Be 3, (1.8)] we first get $b^2 = \nu \cdot q^2$, $\nu \in \mathbb{E}$, $q \in \sum K^2$, or finally $\nu = \epsilon^2$, $\epsilon \in \mathbb{E}$. We may therefore assume $b = \epsilon \in \mathbb{E}$. The Harrison sets $H(\epsilon)$ and $H(-\epsilon)$ form a partition of $X(K)$. Applying the map Λ we obtain $M(K)$ as the union of the closed subsets $C_+ := \Lambda(H(\epsilon))$ and $C_- := \Lambda(H(-\epsilon))$. The first set contains λ_P, the second one λ_Q. By assumption, these two places lie in the same connected component which leads to $C_+ \cap C_- \neq \emptyset$. Let λ be in both sets. Then $\lambda = \lambda_{P_1}$, $\epsilon \in P_1$ and $\lambda = \lambda_{P_2}$, $\epsilon \in -P_2$. Since ϵ is a unit of $H(K)$, a fortiori of V_λ, we arrive at the contradiction $\operatorname{sgn} \lambda(\epsilon) = \operatorname{sgn}_{P_1}(\epsilon) = +1$, $\operatorname{sgn} \lambda(\epsilon) = \operatorname{sgn}_{P_2}(\epsilon) = -1$. ii) $\Rightarrow$ i) Assume that λ_P and λ_Q lie in distinct components. Then there exists an open and closed set U with $\lambda_P \in U$, $\lambda_Q \subseteq M(K) \backslash U$. The continuous characteristic function χ_U can be approximated by some $\epsilon \in H(K)$, i.e. $\chi_U(\lambda) \approx \lambda(\epsilon)$. Choose ϵ such that $| \chi_U(\lambda) - \lambda(\epsilon) | < \frac{1}{2}$ for every λ. Then, obviously, $\epsilon \in \mathbb{E}$. Using the fact that ϵ is a unit in $H(K)$, a fortiori in $V_{\lambda_P} = A(P)$ and in $A(Q)$, we derive $\epsilon \in P$, $\epsilon \in -Q$. Moreover, $\epsilon^2 \in \mathbb{E}^+ \subseteq \sum K^4$, cf. [Be 3, (1.6)]. □

References

[Be 1] Becker, E., Summen n-ter Potenzen in Körpern. J. Reine Angew. Math. **307/308** (1979), 8–30.

[Be 2] Becker, E., Valuations and real places in the theory of formally real fields. In: Lecture Notes in Math. **959** (1982), 1–40.

[Be 3] Becker, E., The real holomorphy ring and sums of $2n$-th powers. In: Lecture Notes in Math. **959** (1982), 139–181.

[Be 4] Becker, E., On the real spectrum of a ring and its applications to semi-algebraic geometry. Bull. Amer. Math. Soc. **15** (1986), 19–60.

[B–P] Becker, E. and Powers, V., Sums of powers in rings and the real holomorphy ring. Preprint 1993.

[BCR] Bochnak, J., Coste, M. and Roy, M.F., Géométrie Algébrique Réelle. Springer-Verlag, Berlin 1987.

[Bo] Bourbaki, N., Algèbre commutative. Paris 1964.

[C] Craven, T., The Boolean space of orderings of a field. Trans. Amer. Math. Soc. **209** (1975), 225–235.

[C–E] Cartan, H. and Eilenberg, S., Homological algebra. Princeton 1956.

[C–R] Coste, M. and Roy, M.F., La topologie spectre réel. Contemp. Math. **8** (1982), 27–59.

[Du] Dubois, D.W., Real commutative algebra, I. Places. Rev. Math. Hisp.-Amer. **39** (1979), 57–65.

[G 1] Gondard, D., Théorie du premier ordre des corps chaînable et des corps chaîne-clos. C.R. Acad. Sci. Paris t. 304, #16 (1987).

[G 2] Gondard, D., Chainable fields and real algebraic geometry. In: Lecture Notes in Math. **1420** (1990), 128–148.

[H] Harman, J., Chains of higher level orderings. Contemp. Math **8** (1982), 141–174.

[Ho] Hostmaelingen, C., Sur l'anneau de Witt d'un anneau de Prüfer. C.R. Acad. Sci. Paris, t. 280 (1975), 69–71.

[K] Knebusch, M., Grothendieck und Wittringe von nichtausgearteten symmetrischen Bilinearformen. Sitzungsber. Heidelb. Akad. Wiss. Math.-Natur. Kl. 1969/1970, 3. Abh.

[K–S] Knebusch, M. and Scheiderer, C., Einführung in die reelle Algebra. Braunschweig - Wiesbaden 1989.

[L 1] Lam, T.Y., Orderings, valuations and quadratic forms. CBMS-notes **52** (1983).

[L 2] Lam, T.Y., An introduction to real algebra. Rocky Mountain J. **14** (1984), 767–814.

[P] Prestel, A., Lectures on formally real fields. Lecture Notes in Math. **1093** (1984).

[Schü 1] Schülting, H.W., On real places of a field and their holomorphy ring. Comm. Algebra **10** (1982), 1239–1284.

[Schü 2] Schülting, H.W., Prime divisors on real varieties and valuation theory. J. Algebra **98** (1986), 499–514.

[Schw] Schwartz, N., Chain signatures and real closures. J. Reine Angew. Math. **347** (1984), 1–20.

[W] Willard, S., General toplogy. Reading 1970.

On real holomorphy rings

Ralph Berr

Introduction

Let F be a formally real field. A valuation ring $V \subset F$ is called real if the residue field of V is formally real. Now let $H(F)$ denote the (absolute) real holomorphy ring of F. By definition, $H(F)$ is the intersection of the real valuation rings of F. In his basic paper [3] Becker has shown that the ring $H(F)$ plays a crucial role in the study of sums of $2n$-th powers in fields. Moreover, if F is the function field of a surface over an archimedean real closed field then the geometrical description of $H(F)$ leads to a geometrical characterization of the sums of $2n$-th powers in fields ([18]). But as soon as one considers arbitrary torsion preorderings or varieties over arbitrary real closed fields one has to deal with certain (relative) real holomorphy rings, i.e. subrings $H \subset F$ which contain $H(F)$. Rings of this type play also an important role in the study of varieties over other formally real ground fields ([7]). Therefore it is quite natural to ask which properties of the ring $H(F)$ carry over to real holomorphy rings and to which arithmetical and geometrical aspects they are now related.

In the present paper we are concerned with this question. We will prove some properties of real holomorphy rings which are of interest for applications to the arithmetic of formally real fields and to real algebraic geometry. Thereby it will turn out that we may argue in a similar manner as it is known in the case of the absolute real holomorphy ring $H(F)$ (see [3]). Finally we will illustrate by some examples how the result of this paper may be applied to the fields mentioned above.

I wish to thank the referee for his helpful suggestions.

1. Sets of valuations and their holomorphy rings

In this section we will collect some basic facts concerning real holomorphy rings. For convenience we first state the following immediate consequence of [2, Theorem 2.16].

(1.1) Proposition. *Let $H \subset F$ be a real holomorphy ring. Then the following statements hold.*

(1) *H is a Prüfer ring.*
(2) *Any valuation ring $V \subset F$ which contains H is real.*

Given a valuation v of F we denote by A_v, m_v the valuation ring and the valuation ideal of v. If $H \subset F$ is a real holomorphy ring, then let $\mathcal{V}(H)$ denote the set of real valuations v of F with $H \subset A_v$. Since H is a Prüfer ring,

$$H = \bigcap_{v \in \mathcal{V}(H)} A_v.$$

Conversely, let $\mathcal{V}$ be any set of real valuations of F. Then

$$H(\mathcal{V}) := \bigcap_{v \in \mathcal{V}} A_v$$

is a real holomorphy ring with $\mathcal{V} \subset \mathcal{V}(H(\mathcal{V}))$.

In this section we will primarily investigate the relationship between a set $\mathcal{V}$ of real valuations of F and the associated holomorphy ring $H(\mathcal{V})$. We will do this from a topological point of view. To this end we use the valuation spectrum $\operatorname{Spv} F$ of F which is just a special case of the valuation spectrum of a ring, introduced by R. Huber ([13]).

By definition, the points of $\operatorname{Spv} F$ are the equivalence classes of valuations of F. In this paper we will not distinguish between a valuation and its equivalence class. The topology of $\operatorname{Spv} F$ is generated by the sets

$$U'(a) := \{v \in \operatorname{Spv} F \mid v(a) \geq 0\},$$

with $a \in F$. The space $\operatorname{Spv} F$ has many important properties. For example, $\operatorname{Spv} F$ is a spectral space. For this and further results concerning the valuation spectrum see [14]. Details about spectral spaces can be found in [12], [10].

If X is a spectral space, then $C \subset X$ is called constructible if C is an element of the Boolean algebra generated by the quasi-compact open sets. The topology generated by the constructible sets is called the constructible topology, whereas the usual topology of X is often called the spectral topology. Back to $\operatorname{Spv} F$, a subset $C \subset \operatorname{Spv} F$ is constructible if and only if C is an element of the Boolean algebra generated by the sets $\{U'(a)\}_{a \in F}$ ([14, Proposition 1.1.1]).

In this paper we are only interested in real valuations of F. Therefore let

$$\operatorname{Spv}_r F := \{v \in \operatorname{Spv} F \mid v \text{ is real }\}.$$

We consider $\operatorname{Spv}_r F$ as a subspace of $\operatorname{Spv} F$. Since $v \in \operatorname{Spv} F$ is non real if and only if there exists $q \in \sum F^2$ with $v(1+q) > 0$ it follows that $\operatorname{Spv}_r F \subset \operatorname{Spv} F$ is proconstructible, i.e. closed with respect to the constructible topology. Hence we have

(1.2) Proposition. $\operatorname{Spv}_r F$ *is a spectral space.*

Next we will have a closer look at the topology of $\operatorname{Spv}_r F$. Given $a \in F$ let

$$U(a) := \{v \in \operatorname{Spv}_r F \mid v(a) \geq 0\};$$

$$C(a) := \{v \in \operatorname{Spv_r} F \mid v(a) > 0\}.$$

Now let $a, b \in F$. Then

$$U(a) \cap U(b) = U(a^2 + b^2);$$

$$C(a) \cap C(b) = C(a^2 + b^2).$$

This proves the following result.

(1.3) Lemma. *The sets $U(a) \subset \operatorname{Spv_r} F$, $a \in F$, form a basis for the topology of $\operatorname{Spv_r} F$ and a subset $C \subset \operatorname{Spv_r} F$ is constructible if and only if C is a finite union of sets of the form $U(a_i) \cap C(b_i)$, with $a_i, b_i \in F$.*

Let $v, w \in \operatorname{Spv_r} F$. As usual w is called a specialization of v if w lies in the closure of v. Now apply (1.3) in order to get

(1.4) Corollary. *Let $v, w \in \operatorname{Spv_r} F$. Then w is a specialization of v if and only if $A_w \subset A_v$.*

We are now prepared to investigate the relationship between sets of real valuations and the associated holomorphy rings. Let $\mathcal{V} \subset \operatorname{Spv_r} F$. Then we have a map

$$\Phi_{\mathcal{V}} : \mathcal{V} \longrightarrow \operatorname{Spec} H(\mathcal{V}) : v \mapsto H(\mathcal{V}) \cap m_v.$$

Obviously, $\Phi_{\mathcal{V}}$ is injective. Finally, let $\operatorname{Spec}_{\mathcal{V}} H(\mathcal{V}) := \operatorname{Im}(\Phi_{\mathcal{V}})$. The following fact will be useful.

(1.5) Proposition. *Let $\mathcal{V} \subset \operatorname{Spv_r} F$. Then $\Phi_{\mathcal{V}} : \mathcal{V} \longrightarrow \operatorname{Spec}_{\mathcal{V}} H(\mathcal{V})$ is a homeomorphism with respect to the spectral and the constructible topology.*

Proof. Obviously $\Phi_{\mathcal{V}}$ is bijective. We first show that $\Phi_{\mathcal{V}}$ is continuous with respect to the spectral and constructible topology. Let $a \in H := H(\mathcal{V})$ and let $D(a) = \{\wp \in \operatorname{Spec}_{\mathcal{V}} H \mid a \notin \wp\}$. Then

$$\Phi_{\mathcal{V}}^{-1}(D(a)) = \{v \in \mathcal{V} \mid v(a) = 0\} = \mathcal{V} \cap U(a) \cap U(a^{-1})$$

is open. Next let $V(a) = \{\wp \in \operatorname{Spec}_{\mathcal{V}} H \mid a \in \wp\}$. Then $\Phi_{\mathcal{V}}^{-1}(V(a)) = \mathcal{V} \cap C(a)$ is constructible. Hence $\Phi_{\mathcal{V}}$ is continuous with respect to the spectral and constructible topology. It remains to show that $\Phi_{\mathcal{V}}$ is open with respect to these topologies. We first consider the spectral topology. For $a, b \in H$ let

$$U(a, b) = \{v \in \operatorname{Spv_r} F \mid v(a) \geq v(b)\}.$$

Let $c \in F^*$ with $U(c) \neq \emptyset$. Since F is the quotient field of H, there exist $a, b \in H \setminus \{0\}$ with $c = \frac{a}{b}$. Hence $U(c) = U(a, b)$. Thus it is sufficient to show that $\Phi_{\mathcal{V}}(U(a, b))$ is open for all $a, b \in H$ by (1.3). Now we define

$$S_{a,b} := \{h \in H \mid \exists g \in H : ha = gb\}$$

and

$$X := \bigcup_{h \in S_{a,b}} D(h) \cap \operatorname{Spec}_{\mathcal{V}} H.$$

We claim $\Phi_{\mathcal{V}}(U(a,b)) = X$. First let $\wp \in X$. Choose $v \in \mathcal{V}$ with $\Phi_{\mathcal{V}}(v) = \wp$. By assumption there exist $h, g \in H$ with

$$\begin{aligned} &(1) && ha = gb, \\ &(2) && h \notin \wp. \end{aligned}$$

From this we get $\frac{a}{b} \in A_v = H_\wp$. Hence $v(a) \geq v(b)$, i.e. $v \in U(a,b)$. Conversely, let $v \in U(a,b)$ and $\wp = \Phi_{\mathcal{V}}(v)$. Since $A_v = H_\wp$ we find $h, g \in H$ with

$$\begin{aligned} &(3) && h \notin \wp = H \cap m_v, \\ &(4) && ha = gb. \end{aligned}$$

Hence $\wp \in X$. This shows that $\Phi_{\mathcal{V}}$ is open with respect to the spectral topology. Next we consider the constructible topology. For $a, b \in H$ let

$$C(a,b) = \{v \in \operatorname{Spv_r} F \mid v(a) > v(b)\}.$$

As above we see that a subset $C \subset \operatorname{Spv_r} F$ is constructible if and only if C is a finite union of sets of the type $C(a,b) \cap U(c,d)$ with $a, b, c, d \in H$. Therefore it remains to show that $\Phi_{\mathcal{V}}(C(a,b))$ is constructible for any $a, b \in H \setminus \{0\}$. Let

$$Y = \bigcup_{h,g \in H : ha = gb} D(h) \cap V(g) \cap \operatorname{Spec}_{\mathcal{V}} H,$$

where $V(g) = \{\wp \mid g \in \wp\}$. Then Y is open with respect to the constructible topology and the same arguments as above prove $\Phi_{\mathcal{V}}(C(a,b)) = Y$. □

If $\mathcal{V} \subset \operatorname{Spv_r} F$ is proconstructible, then $\mathcal{V}$ is a spectral space. Hence (1.5) implies

(1.6) Corollary. *If $\mathcal{V} \subset \operatorname{Spv_r} F$ is proconstructible, then $\Phi_{\mathcal{V}} : \mathcal{V} \longrightarrow \operatorname{Spec}_{\mathcal{V}} H(\mathcal{V})$ is an isomorphism of spectral spaces.*

Let $H \subset F$ be a real holomorphy ring. Then the map $\Phi_{\mathcal{V}(H)} : \mathcal{V}(H) \longrightarrow \operatorname{Spec} H$ is surjective, by (1.1). Thus we have

(1.7) Corollary. *Let $H \subset F$ be a real holomorphy ring. Then $\mathcal{V}(H)$ is canonically homeomorphic to* Spec H. *In particular,* $\operatorname{Spv_r} F$ *is homeomorphic to* Spec $H(F)$.

Again let $H \subset F$ be a real holomorphy ring. Then the following diagram commutes:

$$\begin{array}{ccc} \operatorname{Spec} H & \longrightarrow & \operatorname{Spec} H(F) \\ \Phi_{\mathcal{V}(H)} \Big\uparrow & & \Big\uparrow \\ \mathcal{V}(H) & \longrightarrow & \operatorname{Spv_r} F \end{array}$$

Hence the preceding results imply

(1.8) Corollary. *Let $H \subset F$ be a real holomorphy ring. Then the inclusion $H(F) \hookrightarrow H$ induces an embedding* $\operatorname{Spec} H \hookrightarrow \operatorname{Spec} H(F)$ *of spectral spaces.*

Let $H \subset F$ be a real holomorphy ring. We say that $\mathcal{V} \subset \operatorname{Spv_r} F$ is related to H if $H = H(\mathcal{V})$. For several applications it is important to know which prime ideals belong in this situation to $\operatorname{Spec}_{\mathcal{V}} H$. In the next step we are concerned with this problem. Thereby we denote by $M - \operatorname{Spec} H$ the subspace of closed points of of $\operatorname{Spec} H$.

(1.9) Proposition. *Let $H \subset F$ be a real holomorphy ring and let $\mathcal{V} \subset \operatorname{Spv_r} F$ be related to H. Then the following statements are equivalent:*
(1) $\mathcal{V} \subset \operatorname{Spv_r} F$ *is quasi-compact;*
(2) $M - \operatorname{Spec} H \subset \operatorname{Spec}_{\mathcal{V}} H$.

Proof. (1) $\Rightarrow$ (2) : For $a_1, \ldots, a_n \in H$ let

$$C(a_1, \ldots, a_n) := \{v \in \mathcal{V} \mid v(a_1) > 0, \ldots, v(a_n) > 0\}.$$

Note that $C(a_1, \ldots, a_n) \subset \mathcal{V}$ is closed. Now let $m \subset H$ be a maximal ideal. We have $H = H(\mathcal{V})$, as $\mathcal{V}$ is related to H. Hence

$$(*) \qquad m \subset \bigcup_{\wp \in \operatorname{Spec}_{\mathcal{V}} H} \wp.$$

Let $a_1, \ldots, a_n \in m$, then $\sum a_i^2 \in m$. By $(*)$ there exists $v \in \mathcal{V}$ with $\sum a_i^2 \in m_v$. Thus $a_1, \ldots, a_n \in m_v$, as v is real. This shows $C(a_1, \ldots, a_n) = \cap C(a_i) \neq \emptyset$. Since $\mathcal{V}$ is quasicompact, there exists

$$v_0 \in \bigcap_{a \in m} C(a).$$

But then $\Phi_{\mathcal{V}}(v_0) = m$ by the maximality of m.
(2) $\Rightarrow$ (1): By (1.5), $\Phi_{\mathcal{V}} : \mathcal{V} \longrightarrow \operatorname{Spec}_{\mathcal{V}} H$ is a homoemorphism with respect to the spectral topology. But $\operatorname{Spec}_{\mathcal{V}} H$ is a quasi-compact subspace of $\operatorname{Spec} H$, as $M - \operatorname{Spec} H \subset \operatorname{Spec}_{\mathcal{V}} H$. □

Now let $\operatorname{Rad}(H)$ denote the Jacobson-radical of H. If we consider proconstructible subsets $\mathcal{V} \subset \operatorname{Spv_r} F$ rather than quasi-compact subsets, we can improve the last result as follows.

(1.10) Proposition. *Let $H \subset F$ be a real holomorphy ring and let $\mathcal{V} \subset \mathrm{Spv_r}\, F$ be related to H. If $\mathcal{V}$ is proconstructible, then for any $\wp \in \mathrm{Spec}\, H$ with $\mathrm{Rad}(H) \subset \wp$ there exist $\wp_1, \wp_2 \in \mathrm{Spec}_{\mathcal{V}}\, H$ with $\wp_1 \subset \wp \subset \wp_2$.*

Proof. Let $\wp \in \mathrm{Spec}\, H$ with $\mathrm{Rad}(H) \subset \wp$. From (1.9) we know $M - \mathrm{Spec}\, H \subset \mathrm{Spec}_{\mathcal{V}}\, H$. Hence there exists $\wp_2 \in \mathrm{Spec}_{\mathcal{V}}\, H$ with $\wp \subset \wp_2$. Furthermore, since $\mathrm{Rad}(H) \subset \wp$, by (1.9) we have

$$H \setminus \wp \subset \bigcup_{v \in \mathcal{V}} (H \setminus m_v).$$

For $a_1, \ldots, a_n \in H$ let $D(a_1, \ldots, a_n) = \{\wp \in \mathrm{Spec}\, H \mid a_1, \ldots, a_n \notin \wp\}$. Now assume $a_1, \ldots, a_n \in H \setminus \wp$. Then $\prod a_i \in H \setminus \wp$. Hence there exists $v \in \mathcal{V}$ with $\prod a_i \in H \setminus m_v$. Thus $\mathrm{Spec}_{\mathcal{V}}\, H \cap D(a_1, \ldots, a_n) \neq \emptyset$. By (1.6), $\mathrm{Spec}_{\mathcal{V}}\, H$ is a proconstructible subset of $\mathrm{Spec}\, H$. Hence there exists

$$\wp_1 \in \mathrm{Spec}_{\mathcal{V}}\, H \cap \bigcap_{a \in H \setminus \wp} D(a).$$

But then $\wp_1 \subset \wp$. □

We conclude this section by a generalization of a result due to Schülting [16, Corollary 3.3], [3, Lemma (1.8)]. In the proof we will use the same arguments as in [3].

(1.11) Proposition. *Let $\mathcal{V} \subset \mathrm{Spv_r}\, F$ be quasicompact and related to the real holomorphy ring $H \subset F$, let $n \in \mathbb{N}$ and let $S \subset F^*$ be a subgroup with $\sum F^{*2n} \subset S$. Given $a \in S$, then the following statements are equivalent:*
(1) $2n \mid v(a)$ for all $v \in \mathcal{V}$;
(2) $a \in (H^* \cap S) \cdot \sum F^{2n}$.

Proof. The implication (2) $\Rightarrow$ (1) is obvious. Therefore assume $2n | v(a)$ for all $v \in \mathcal{V}$. Given $v \in \mathcal{V}$ we find $b_v \in F, \varepsilon_v \in A_v^*$ with $a = \varepsilon_v b_v^{2n}$. Moreover, since H is a Prüfer ring with $H \subset A_v$, there exist $c_v, d_v \in H \setminus m_v$ with $\varepsilon_v = \frac{c_v}{d_v}$. But then $a = (c_v d_v^{2n-1}) b_v^{2n} d_v^{-2n}$. Hence, for any valuation v with $H \subset A_v$ we find a representation

$$(*) \qquad a = \varepsilon_v b_v^{2n},$$

with $\varepsilon_v \in A_v^* \cap H$. Now let $I \subset H$ be the ideal generated by $a b_v^{-2n}$ with $v \in \mathcal{V}$. Note that $\varepsilon_v \notin H \cap m_v$ for all v. Since $\mathcal{V}$ is quasicompact, (1.9) implies $I = H$. Hence there exist $a_1, \ldots, a_l \in F$ such that $H = I = a(a_1^{2n}, \ldots, a_l^{2n})$. Since H is a real holomorphy ring, we have the relation $(a_1^{2n}, \ldots, a_l^{2n}) = (a_1^{2n} + \cdots + a_l^{2n})$ as fractional ideals of H (cf. the proofs of (2.2) and (2.3) below). Thus there exists $\varepsilon \in H^*$ with

$$(**) \qquad a^{-1} = \varepsilon \cdot \sum_{i=1}^{l} a_i^{2n}.$$

But then $\varepsilon \in H^* \cap S$, as $a \in S$ and $\sum F^{*2n} \subset S$. This completes the proof. □

(1.12) Corollary. *Let $\mathcal{V} \subset \mathrm{Spv_r}\, F$ be quasicompact and related to $H(F)$. Given $n \in \mathbb{N}$ and $a \in \sum F^2$, then the following statements are equivalent:*
(1) $2n \mid v(a)$ for all $v \in \mathcal{V}$;
(2) $a \in \sum F^{2n}$.

Proof. By [3, Theorem 1.6] we have $H(F)^* \cap \sum F^2 \subset \sum F^{2n}$. Now apply (1.11). □

2. Preorderings and their holomorphy rings

A subset $T \subset F$ is called a preordering if

$$T + T \subset T, \quad 0, 1 \in T, \quad -1 \notin T$$

and $T^* = T \setminus \{0\}$ is a subgroup of F^*. If F^*/T^* is a torsion group, then T is called a torsion preordering. For a preordering $T \subset F$ let

$$A(T) := \{x \in F \mid \exists n \in \mathbb{N} : n \pm x \in T\};$$

$$I(T) := \{x \in F \mid \forall n \in \mathbb{N} : 1 \pm nx \in T\}.$$

The ring $A(T)$ plays an important role in the study of preorderings. For convenience we recall the following result of E. Becker (see [1, (2.7) and (3.7)]).

(2.1) Proposition. *For a torsion preordering $T \subset F$ the following statements hold.*
(1) $A(T)$ is a real holomorphy ring.
(2) $I(T) \subset \mathrm{Rad}(A(T))$.
(3) As a ring, $A(T)$ is generated by the elements $\frac{1}{1+t}$ with $t \in T$.

As an easy consequence we obtain the following useful fact.

(2.2) Lemma. *Let $T \subset F$ be a torsion preordering, $v \in \mathrm{Spv_r}\, F$ with $A(T) \subset A_v$ and let $t_1, \ldots, t_n \in T$. Then*

$$v(t_1 + \cdots + t_n) = \min\{v(t_i) \mid i \in \{1, \ldots, n\}\}.$$

Proof. We may assume w.l.o.g. that $v(t_1) = \min\{v(t_i) \mid i \in \{1, \ldots, n\}\}$. Let

$$s := \sum_{i=2}^{n} \frac{t_i}{t_1}.$$

Then $s \in T$, as T is a preordering. Hence $v(1+s) \leq 0$, by (2.1) (3). But $v(t_1) \leq v(t_1 + \cdots + t_n) = v(t_1) + v(1+s)$. Thus $v(1+s) = 0$. □

Let A be any subring of F. For $a_1, \dots, a_n \in F$ we denote by $(a_1, \dots, a_n)$ the fractional ideal of A generated by $a_1, \dots, a_n$. We will need the following result which can be proved as in the case $T = \sum F^2$ (see [16]).

(2.3) Lemma. *Let $T \subset F$ be a torsion preordering, $t_1, \dots, t_k \in T$ and $n \in \mathbb{N}$. Then in $A(T)$ the following relation holds:*

$$(t_1, \dots, t_k)^n = (t_1^n, \dots, t_k^n) = (t_1^n + \dots + t_k^n).$$

Proof. The first equality holds for arbitrary Prüfer rings (see [11, §24, Ex.]). Next let $i_0 \in \{1, \dots, k\}$. Then we have

$$\frac{t_{i_0}^n}{\sum t_i^n} = \frac{1}{1 + \sum_{i \neq i_0} \frac{t_i^n}{t_{i_0}^n}} \in A(T),$$

by (2.1) (3). Hence $t_{i_0}^n \in (t_1^n + \dots + t_k^n)$. □

(2.4) Corollary. *Let $T \subset F$ be a preordering and let $H \subset F$ be any ring. Then the following statements are equivalent:*
(1) $A(T) \subset H$;
(2) For all $t_1, \dots, t_k \in T$ and $n \in \mathbb{N}$, H satisfies the relation $(t_1^n, \dots, t_k^n) = (t_1^n + \dots + t_k^n)$.

Proof. The implication (1) ⇒ (2) follows as in the proof of (2.2). Next assume that (2) holds and let $t \in T$. Then $(1, t) = (1 + t)$. Hence there exists $h \in H$ with $1 = h(1 + t)$. But then $A(T) \subset H$, by (2.1) (3). □

In the remaining part of this section we are mainly concerned with the rings $A(T)$, where T is a quadratic preordering, i.e. $F^2 \subset T$. In the case $T = \sum F^2$, all the results of this section can be found in [2] and [3]. For a quadratic preordering $T \subset F$ let

$$X_T := \{P \in \operatorname{Sper} F \mid T \subset P\}.$$

Thus X_T is a compact space.

Now let $\mathbb{Q}$ denote the field of the rational numbers.

(2.5) Proposition. *Let $T \subset F$ be a quadratic preordering. Then*

$$A(T)^* \cap T = \left\{ q \frac{q' + t}{1 + t} \mid q, q' \in \sum \mathbb{Q}^2, t \in T \right\}.$$

Proof. For $P \in X_T$ let $A(P)$ denote the natural valuation ring of P. Then $A(T) = \cap A(P)$, where P runs through X_T, by [1, (3.7)]. Let w_P be the canonical valuation associated with $A(P)$ and let

$$x := q \frac{q' + t}{1 + t},$$

with $q, q' \in \sum \mathbb{Q}^2$ and $t \in T$. Then $w_P(x) = 0$ for all $P \in X_T$. Hence $x \in A(T)^* \cap T$. Conversely, assume $x \in A(T)^* \cap T$ and let $P \in X_T$. Then $x \in A(P)^* \cap P$. Hence, for all $P \in X_T$ there exists $n_P \in \mathbb{N}$ such that

$$\frac{1}{n_P} <_P x <_P n_P.$$

Since X_T is compact, we find $n \in \mathbb{N}$ with

$$\frac{1}{n} <_P x <_P n$$

for all $P \in X_T$. But then

$$\frac{n^2 - nx}{nx - 1} =: t \in T.$$

Now a simple computation shows

$$x = \frac{1}{n} \cdot \frac{n^2 + t}{1 + t}.$$

□

For the study of sums of $2n$-th powers in fields two results of E. Becker are of fundamental importance. Namely, for all $n \in \mathbb{N}$ we have

$$\begin{array}{ll} (1) & H(F) = A(F^2) = A(F^{2n}); \\ (2) & H(F)^* \cap \sum F^2 = H(F)^* \cap \sum F^{2n} \end{array}$$

(see [2, Theorem 3.3] and [3, Theorem 1.6]). We are now going to prove similar results for arbitrary preorderings $T \subset F$. However, at this point one has to clarify by which preorderings the sums of $2n$-th powers have to be replaced, i.e. which preorderings of higher level are associated with a quadratic preordering. This leads to the following notions.

For a subset $U \subset F$ let $\sum U$ denote the additive semigroup with 0 generated by U. Furthermore, if $n \in \mathbb{N}$ then let

$$U^n := \{u^n \mid u \in U\}.$$

Let $T \subset F$ be a quadratic preordering. We call a subgroup $S \subset T$ a generating set of T if

$$\begin{array}{ll} (1) & F^{*2} \subset S \\ (2) & T = \sum S. \end{array}$$

Now let $n \in \mathbb{N}$ and let $S \subset T$ be a generating set of T. Then $S_n := \sum S^n$ is a preordering. Namely, since $S_n \subset T$, we have only to show that $S_n^* = S \setminus \{0\}$ is a subgroup of F^*. But S_n^* is closed under multiplication, as $S \subset F^*$ is a subgroup. Moreover, $F^{*2} \subset S$ implies $F^{*2n} \subset S_n$. Given $s \in S_n$ this yields

$$s^{-1} = \frac{s^{2n-1}}{s^{2n}} \in S_n.$$

(2.6) Proposition. *Let $T \subset F$ be a quadratic preordering, $S \subset T$ a generating set of T and let $n \in \mathbb{N}$ be odd. Then*

$$A(T) = A(\sum S^n).$$

Proof. Since $\sum S^n \subset T$, we have $A(\sum S^n) \subset A(T)$. Let

$$T_{\text{odd}} := \{x \in F \mid x^k \in \sum S^n \text{ for some odd } k \in \mathbb{N}\}.$$

Then T_{odd} is a preordering with $A(T_{\text{odd}}) = A(\sum S^n)$, by [1, Proposition (4.6)]. But on the other hand we have $S \subset T_{\text{odd}}$, hence $T \subset T_{\text{odd}}$. But then $A(T) \subset A(T_{\text{odd}})$ and we are done. □

In order to prove the next result we need some further notions. Given a preordering $T \subset F$ let

$$X(T) := \{\varphi \in \text{Hom}(A(T), \mathbb{R}) \mid \varphi(T) \subset \mathbb{R}^+\}.$$

Then every $a \in A(T)$ determines a function

$$\hat{a} : X(T) \longrightarrow \mathbb{R} : \varphi \mapsto \varphi(a).$$

We give $X(T)$ the weak topology with respect to all functions $\hat{a}$, $a \in A(T)$. Then $X(T)$ is a compact space ([1, Lemma 1.2]) and we have a ring homomorphism

$$\Phi : A(T) \longrightarrow C(X(T), \mathbb{R}) : a \mapsto \hat{a}.$$

Now we can state

(2.7) Proposition. *Let $T \subset F$ be a quadratic preordering, $S \subset T$ a generating set of T and let $n \in \mathbb{N}$ be odd. Then*

$$A(T)^* \cap T = A(T)^* \cap \sum S^n.$$

Proof. Obviously, $A(T)^* \cap \sum S^n \subset A(T)^* \cap T$. For the converse direction we first show $X(T) = X(\sum S^n)$. Since $A(T) = A(\sum S^n)$ and $\sum S^n \subset T$, we have $X(T) \subset X(\sum S^n)$. Next let $\varphi \in X(\sum S^n)$ and $s \in S$. Then $\varphi(s^n) \geq 0$. Thus $\varphi(s) \geq 0$, as n is odd. But then $\varphi(T) = \varphi(\sum S) \subset \mathbb{R}^+$, i.e. $\varphi \in X(T)$. Thus $X(T) = X(\sum S^n)$. Now let $a \in A(T)^* \cap T$. Then $\hat{a}(\varphi) > 0$ for all $\varphi \in X(T) = X(\sum S^n)$. Since $X(\sum S^n)$ is compact we find $n \in \mathbb{N}$ such that $a - \frac{1}{n}$ is positive on $X(\sum S^n)$. By [5, Hauptsatz (ii)] there exists $k \in \mathbb{N}$ with $k(1 + n(a - \frac{1}{n})) = kna \in \sum S^n$. Hence $a \in \sum S^n$. □

As a consequence of the last result we get the existence of certain identities. The next lemma serves as preliminary for this. For the proof we use the same ideas as in [3, Proof of (1.8)].

(2.8) Lemma. *Let $T \subset F$ be a quadratic preordering, $S \subset T$ a generating set and let $s_1, \ldots, s_k \in S$. If n is odd, then*

$$(s_1 + \cdots + s_k)^n \in \sum S^n.$$

Proof. Let $s := s_1 + \cdots + s_k$. Let v be a valuation of F with $A(T) \subset A_v$. Then $v(s) = v(s_i)$ for some $i \in \{1, \ldots, k\}$, by (2.2). Hence we find $\varepsilon'_v \in A_v^*$ and $s_v \in S$ with $s^{-n} \cdot s_v^n = \varepsilon'_v$. Moreover, since $A(T)$ is a Prüfer ring with $A(T) \subset A_v$, there exist $a_v, b_v \in A(T) \setminus m_v$ with $\varepsilon'_v = \frac{a_v}{b_v}$. But then $s^{-n}(s_v b_v^2)^n = a_v b_v^{2n-1}$. Note that $s_v b_v^2 \in S$. Therefore, for any valuation v with $A(T) \subset A_v$ we find a representation

$$(*) \qquad s^{-n} s_v^n = \varepsilon_v,$$

with $\varepsilon_v \in A_v^* \cap A(T)$ and $s_v \in S$. Now let $I \subset A(T)$ be the ideal generated by $s^{-n} s_v^n$, where v ranges over the valuations of F with $A(T) \subset A_v$. Then $I = A(T)$, as $\varepsilon_v \notin A(T) \cap m_v$ for all v. Hence there exist $\tilde{s}_1, \ldots, \tilde{s}_l \in S$ such that $A(T) = I = s^{-n}(\tilde{s}_1^n, \ldots, \tilde{s}_l^n)$. Now apply (2.3) in order to see that there exists $\varepsilon \in A(T)^*$ with

$$(**) \qquad s^n = \varepsilon \cdot \sum_{i=1}^{l} \tilde{s}_i^n.$$

But then $\varepsilon \in A(T)^* \cap T$. By (2.7) we have $A(T)^* \cap T = A(T)^* \cap \sum S^n$. Now the claim follows from $(**)$. □

We now turn to the identities mentioned above. Let $\mathbb{Q}[X_1, \ldots, X_k]$ be the polynomial ring over $\mathbb{Q}$ in k indeterminates. Then we denote by $m_1, \ldots, m_{2^k} \in \mathbb{Q}[X_1, \ldots, X_k]$ the multilinear monomials.

(2.9) Theorem. *Let $X_1, \ldots, X_k$ be indeterminates over $\mathbb{Q}$ and let $n \in \mathbb{N}$ be odd. Then there exist natural numbers $s_1(k,n), \ldots s_{2^k}(k,n)$ and $f_{ij} \in \mathbb{Q}(X_1, \ldots, X_k)$, $i \in \{1, \ldots, 2^k\}$ and $j \in \{1, \ldots, s_i(k,n)\}$ such that*

$$(X_1 + \cdots + X_k)^n = \sum_{i=1}^{2^k} \sum_{j=1}^{s_i(k,n)} m_i(X_1, \ldots, X_k)^n f_{ij}^{2n}.$$

Proof. Let $F := \mathbb{Q}(X_1, \ldots, X_k)$ and let $T \subset F$ be the preordering generated by F^2 and $\{X_1, \ldots, X_k\}$. Then

$$S = \bigcup_{i=1}^{2^k} m_i(X_1, \ldots, X_k) \cdot F^2$$

is a generating set for T and the claim follows from (2.8). □

Now traditional arguments show that (2.9) carries over to arbitrary formally real fields. Given $x_1, \ldots, x_k \in F$ then let $T_F(x_1, \ldots, x_k) \subset F$ denote the semi ring generated by F^2 and $x_1, \ldots, x_k$. Keeping the notation of (2.9) fixed we get:

(2.10) Corollary. *Let F be a field, let $n \in \mathbb{N}$ be odd and let $x_1, \ldots, x_k \in F^*$. If $T_F(x_1, \ldots, x_k)$ is a preordering then there exist $y_{ij} \in F$, $i \in \{1, \ldots, 2^k\}$ and*

$j \in \{1, \ldots, s_i(k,n)\}$ *such that*

$$(x_1 + \cdots + x_k)^n = \sum_{i=1}^{2^k} \sum_{j=1}^{s_i(k,n)} m_i(x_1, \ldots, x_k)^n y_{ij}^{2n}.$$

Proof. By [3, Lemma 2.3] there exists a place

$$\lambda : \mathbb{Q}(X_1, \ldots, X_k) \longrightarrow F \cup \infty \text{ with } \lambda(X_i) = x_i, i = 1, \ldots, k.$$

Let $B \subset \mathbb{Q}(X_1, \ldots, X_k)$ be the valuation ring of λ and let $T \subset \mathbb{Q}(X_1, \ldots, X_k)$ be the quadratic preordering generated by $X_1, \ldots, X_k$. Since $T_F(x_1, \ldots, x_k) \subset F$ is a preordering, there exists a preordering $P \subset \mathbb{Q}(X_1, \ldots, X_k)$ which is compatible with B (i.e. $1 + m_B \subset P$) and contains T. Let $f_{ij} \in \mathbb{Q}(X_1, \ldots, X_k)$ be as in (2.9). By the choice of λ we have

$$\sum_{i=1}^{2^k} \sum_{j=1}^{s_i(k,n)} m_i(X_1, \ldots, X_k)^n f_{ij}^{2n} \in B.$$

Since P is comptible with B we even get $f_{ij} \in B$, i.e. $\lambda(f_{ij}) \neq \infty$ for all i, j. (Note that $x_1, \ldots, x_k \neq 0$ implies $m_i(X_1, \ldots, X_k) \in B^*$). Now the claim follows from (2.9). □

3. On the description of real holomorphy rings

For applications to real algebraic geometry there are two kinds of real holomorphy rings which are of special interest. First let $A \subset F$ be a subring and let

$$H(F/A) = \bigcap V,$$

where V ranges over the real valuation rings of F with $A \subset V$. Next let K be a subfield of F and let B be a valuation ring of K. Now let

$$H_B(F) := \bigcap V,$$

where V ranges over the real valuation rings of F which extend B. In order to treat these cases simultaneously, we introduce the following notion. Let $A \subset F$ be a ring and B a real valuation ring of the quotient field quot$(A) \subset F$. Then let

$$\mathcal{V}(A,B) := \{v \in \mathrm{Spv_r}\, F \mid A_v \cap A = B \cap A\};$$
$$X(A,B) := \{P \in \mathrm{Sper}\, F \mid P \sim v \text{ for some } v \in \mathcal{V}(A,B)\},$$

where $P \sim v$ means that P is compatible with v. We call

$$H_B(F/A) := \bigcap_{v \in \mathcal{V}(A,B)} A_v$$

the real holomorphy ring of (A, B). Note that, if B is the trivial valuation, then $H_B(F/A) = H(F/A)$ and if A is the quotient field of B, then $H_B(F/A) = H_B(F)$. In this section we will show how the ring $H_B(F/A)$ is determined by A and B. In order to simplify the notion, we keep A, B fixed and write $H, \mathcal{V}, X$ rather than $H_B(F/A), \mathcal{V}(A, B), X(A, B)$.

(3.1) Lemma. *$\mathcal{V}$ is proconstructible and $X \subset \operatorname{Sper} F$ is compact.*

Proof. That $\mathcal{V}$ is proconstructible follows immediately from the definition. Next let $P \in \operatorname{Sper} F \setminus X$ and let $v \in \mathcal{V}$. Then we find $a_v \in P$ with $v(a_v) = 0$ and $v(1 + a_v) > 0$. Since $\mathcal{V}$ is proconstructible there exists $a_1, \ldots, a_n \in P$ with

$$V \subset \bigcup_{i=1}^{n} \{v \in \mathcal{V} \mid v(a_i) = 0 \wedge v(1 + a_i) > 0\}.$$

Thus $P \in D(a_1, \ldots, a_n) = \{Q \in \operatorname{Sper} F \mid a_1 \in Q \wedge \cdots \wedge a_n \in Q\}$ and $X \cap D(a_1, \ldots, a_n) = \emptyset$. Hence, X is closed and, consequently compact. □

Recall that we have defined $\operatorname{Spec}_{\mathcal{V}} H := \{m_v \cap H \mid v \in \mathcal{V}\}$.

(3.2) Corollary. *Let $\wp \in \operatorname{Spec} H$. If $\operatorname{Rad}(H) \subset \wp$ then $\wp \in \operatorname{Spec}_{\mathcal{V}} H$.*

Proof. By (1.10) there exist $\wp_1, \wp_2 \in \operatorname{Spec}_{\mathcal{V}} H$ with $\wp_1 \subset \wp \subset \wp_2$. But then

$$A \cap B = A \cap H_{\wp_2} \subset A \cap H_{\wp} \subset A \cap H_{\wp_1} = A \cap B.$$

Thus $\wp \in \operatorname{Spec}_{\mathcal{V}} H$. □

Now let

$$T_H := \bigcap_{P \in X} P.$$

Finally, if $U \subset F$ is any subset, then we denote by $\langle U \rangle \subset F$ the ring generated by U. We next prove the characterization of H we are looking for.

(3.3) Theorem. *Let $A \subset F$ be a ring and B a real valuation ring of* $\operatorname{quot}(A)$. *Then the following rings coincide.*

(1) $H := H_B(F/A)$.

(2) $H_0 := \{x \in F \mid \exists t \in B^* \cap A : t \pm x \in T_H\}$.

(3) $H_1 := (B^* \cap A) \cdot A(T_H)$.

(4) $H_2 := \left\langle \frac{b}{1+t} \mid t \in T_H, b \in B^* \cap A \right\rangle$.

Proof. We first show $H = H_0$. If $\mathcal{V} = \emptyset$, then $X = \emptyset$ and $H = F = H_0$. So assume $\mathcal{V} \neq \emptyset$ and let $a \in H_0$. Then there exists $t \in B^* \cap A$ with $t \pm a \in T_H$. Let $v \in \mathcal{V}$ and let $P \in X$ be compatible with v. Then $a^2 <_P t^2$ as $t^2 - a^2 \in T_H$. Thus we have $0 \leq v(t) \leq v(a)$ as $t \in B^* \cap A \subset A_v$ and $v \in \mathcal{V}$ is compatible with P. Hence $H_0 \subset H$. Next let $a \in H$ and let $P \in X$. In the first step we show that there exists $t_P \in \sum(B^* \cap A)^2$ with $t_P \pm a \in P$. Let

$$A(P) = \{x \in F \mid \exists n \in \mathbb{N} : n \pm x \in P\}$$

be the natural valuation ring of P and let

$$V(P) = \{x \in F \mid \exists t \in \sum (B^* \cap A)^2 : t \pm x \in P\}.$$

Since B is a real valuation ring, $\sum(B^* \cap A)^2$ is a semi ring which implies that $V(P)$ is a ring. Hence $V(P)$ is a valuation ring as $A(P) \subset V(P)$. Next let $v \in \mathcal{V}$ with $v \sim P$. Then $A_v \cap A = B \cap A$. Hence $B \cap A$ is a P-convex subset of A. From this we conclude $V(P) \cap A = B \cap A$, i.e. $H \subset V(P)$. But then $a \in V(P)$ which shows that there exists $t_P \in \sum(B^* \cap A)^2$ with $t_P \pm a \in P$. Hence

$$X = \bigcup_{P \in X} \{Q \mid t_P \pm a \in Q\}.$$

By (3.1) we find $t_1, \ldots, t_n \in \sum(B^* \cap A)^2$ such that

$$X = \bigcup_{i=1}^{n} \{Q \mid t_i \pm a \in Q\}.$$

Now let $t := \sum t_i \in \sum(B^* \cap A)^2$. Then $t \pm a \in P$ for all $P \in X$. Hence $t \pm a \in T_H$ which shows $H = H_0$.
The remaining equalities are easy consequences of this result. Note that (2.1) (3) implies $H_1 \subset H_2 \subset H$. Therefore it is sufficient to show $H \subset H_1$. Let $a \in H$, hence $t \pm a \in T_H$ for some $t \in B^* \cap A$. Then $t \in T_H$ and we get $1 \pm at^{-1} \in T_H$, i.e. $at^{-1} \in A(T_H)$. Thus $a \in H_1$ which proves $H = H_1$. □

(3.4) Corollary. *Let A, B and H as in* (3.3), *$S \subset T_H$ a generating set of T_H and let $n \in \mathbb{N}$ be odd. Then*

$$H = \{x \in F \mid \exists t \in B^* \cap A : t \pm x \in \sum S^n\}.$$

Proof. Let H_n be the ring on the right hand side. Then $H_n \subset H$ by (3.3). Obviously $A(\sum S^n) \subset H_n$. Hence $A(T_H) \subset H_n$, by (2.6). By (3.3) it remains to show $(B^* \cap A) \subset H_n$. Let $b \in B^* \cap A$. Then $b^{2n} \pm b^{2n} \in \sum S^n$ which shows $b^{2n} \in H_n$. But H_n is integrally closed as $A(T_H) \subset H_n$. Hence $b \in H_n$. □

(3.5) Corollary. *Let A, B and H as above and assume $T_H = \sum F^2$. Then for any $n \in \mathbb{N}$ we have*

$$H = \{x \in F \mid \exists t \in B^* \cap A : t \pm x \in \sum F^{2n}\}.$$

Proof. Again let H_n denote the ring on the right hand side. It remains to consider the case that n is even. As in (3.4) we see $B^* \cap A \subset H_n$. Moreover, $A(\sum F^{2n}) \subset H_n$. But $A(\sum F^2) = A(\sum F^{2n})$ by [2, Theorem (3.3)]. Now (3.3) (3) implies $H = H_n$. □

The preceding results allow us to describe the "totally positive" units of the ring H. The formulation of the next result has been suggested by the referee.

(3.6) Corollary. *Let A, B and H be as above and assume $H \neq F$. Then we have*

$$H^* \cap T_H = \{\frac{1}{1+b}\frac{1+b'+t}{1+t} \mid b, b' \in \sum(B^* \cap A)^2, t \in T_H\}.$$

Proof. First let

$$x = \frac{1}{1+b}\frac{1+b'+t}{1+t}$$

with $b, b' \in \sum(B^* \cap A)^2$ and $t \in T_H$. Since $H \neq F$ we have $\mathcal{V} \neq \emptyset$. Let $v \in \mathcal{V}$. Then $b, b' \in A_v$. Hence $v(x) = 0$. This shows $x \in H^* \cap T_H$. Conversely, assume $x \in H^* \cap T_H$. Given $P \in X$ let

$$V(P) = \{x \in F \mid \exists b \in \sum(B^* \cap A)^2 : b \pm x \in P\}.$$

In the proof of (3.3) we have seen that $V(P)$ is a valuation ring of F with $H \subset V(P)$. Hence there exists $b_P \in \sum(B^* \cap A)^2$ such that

$$\frac{1}{b_P} <_P x <_P b_P.$$

Since X is compact, we find finitely many $b_1, \dots, b_n \in \sum(B^* \cap A)^2$ such that

$$X \subset \bigcup_{i=1}^{n}\{P \in X \mid \frac{1}{b_i} <_P x <_P b_i\}.$$

Now let $b := \sum b_i$. Then

$$\frac{1}{1+b} <_P x <_P 1+b$$

for all $P \in X$. Hence we have

$$t := \frac{1+b-x}{x-(1+b)^{-1}} \in T_H$$

and this implies

$$x = \frac{1}{1+b}\frac{(1+b)^2+t}{1+t}.$$

□

This result can be sharpened if $B^* \subset A^*$. In this case we have

(3.7) Corollary. *Let A, B and H be as in (3.6) and assume that $B^* \subset A^*$. Then we have*

$$H^* \cap T_H = \{\frac{1}{b}\frac{b'+t}{1+t} \mid b, b' \in \sum B^{*2}, t \in T_H\}.$$

Proof. Since B is a real valuation ring we have

$$1 + \sum(B^* \cap A)^2 \subset \sum(B^* \cap A)^2 \subset \sum B^{*2}.$$

Thus in view of (3.6) it remains to prove that $H^* \cap T_H$ contains the set on the right hand side. Let

$$x = \frac{1}{b}\frac{b'+t}{1+t}$$

with $b, b' \in \sum B^{*2}$ and $t \in T_H$. Let $v \in \mathcal{V}$. Since $B^* \subset A^*$ we have $B^* \subset A_v^*$. Hence $v(x) = 0$ and consequently $x \in H^* \cap T_H$. □

4. Some applications to real algebraic geometry

In this section we will illustrate by two examples how the results of the preceding sections can be applied in geometrical situations. Further applications can be found in [7]. However, we will begin with a result which is related only to the theory of the absolute holomorphy ring.

Let $\mathbb{R}$ denote the field of the real numbers and let $A_k(\mathbb{R})$ denote the k-dimensional affine space over $\mathbb{R}$. Given an algebraic set $V \subset A_k(\mathbb{R})$ we let $\mathcal{R}[V]$ denote the ring of regular functions on V.

(4.1) Proposition. *Let $V \subset A_k(\mathbb{R})$ be a bounded algebraic set and let $n \in \mathbb{N}$. Given $f \in \mathcal{R}[V]^*$, then the following statements are equivalent:*
(a) $f(x) > 0$ *for all* $x \in V$;
(b) $f \in \sum \mathcal{R}[V]^{2n}$.

Proof. We have only to show $(a) \Rightarrow (b)$. Let $\wp \subset \mathcal{R}[V]$ be a real prime ideal and let $R(\wp)$ denote the residue field of $\mathcal{R}[V]/\wp$. Let $\bar{f}$ denote the image of f with respect to the canonical projection $\pi : \mathcal{R}[V] \longrightarrow \mathcal{R}[V]/\wp \longrightarrow R(\wp)$. Finally let $W \subset V$ be the algebraic subset defined by $\wp$. Then $f_{|W}$ is positive definite on W. Hence the Positivstellensatz over real closed fields implies $\bar{f} \in \sum R(\wp)^2$. Moreover, $\mathcal{R}[V]/\wp \subset H(R(\wp))$, as W is bounded. Hence $\bar{f} \in H(R(\wp))^* \cap \sum R(\wp)^2$. Now [3, Theorem 1.6] shows $\bar{f} \in \sum R(\wp)^{*2n}$. Thus we have proved

$$(*) \qquad \text{Given any real prime ideal } \wp \subset \mathcal{R}[V] \text{ then } \bar{f} \in R(\wp)^{*2n}.$$

Now let $P \subset \mathcal{R}[V]$ be an ordering of level n, i.e. there exists a real prime ideal $\wp \subset \mathcal{R}[V]$ and an ordering $\bar{P} \subset R(\wp)$ of level n with $P = \pi^{-1}(\bar{P})$. (For details concerning orderings of higher level in rings see [4]). By $(*)$ we have $\bar{f} \in \sum R(\wp)^{*2n}$. Hence $\bar{f} \in \bar{P} \setminus \{0\}$ which shows $f \in P^+ := P \setminus (P \cap -P)$. Thus we get

$$f \in \bigcap P^+,$$

where P ranges over all orderings of level n of $\mathcal{R}[V]$. By the intersection theorem for orderings of higher level [6, Theorem 6] we find $s, d \in \sum \mathcal{R}[V]^{2n}$ with $fs = 1 + d$. But $1 + d \in \mathcal{R}[V]^*$. Thus $f \in \sum \mathcal{R}[V]^{2n}$. □

Next we will use the results of section 2 in order to obtain a similar characterization of regular functions which are positive definite on certain semi-algebraic sets. Again let $V \subset A_k(\mathbb{R})$ be an algebraic set. We fix $f_1, \ldots, f_m \in \mathcal{R}[V]$ and set

$$A := \{x \in V \mid f_1(x) \geq 0 \wedge \cdots \wedge f_m \geq 0\}.$$

Given any ring B and a subset $M \subset B$ we let $\langle M \rangle \subset B$ denote the semiring with 1 generated by M. Let $n \in \mathbb{N}$. Now we consider

$$S := \langle \mathcal{R}[V]^2, f_1, \ldots, f_m \rangle \subset \mathcal{R}[V]$$

$$S_n := \langle \mathcal{R}[V]^{2n}, f_1^n, \ldots, f_m^n \rangle \subset \mathcal{R}[V].$$

Next let $\wp \subset \mathcal{R}[V]$ be a prime ideal. Then let

$$T(\wp) := \langle R(\wp)^2, \bar{f}_1, \ldots, \bar{f}_m \rangle \subset R(\wp)$$

$$T(\wp)_n := \langle R(\wp)^{2n}, \bar{f}_1^n, \ldots, \bar{f}_m^n \rangle \subset R(\wp),$$

where $R(\wp)$ denotes as above the quotient field of $\mathcal{R}[V]/\wp$. Next assume that our semi-algebraic set A is Zariski-dense in V. Let $\wp \subset \mathcal{R}[V]$ be a minimal prime ideal. Then $S, T(\wp)$ are preorderings of $\mathcal{R}[V]$, $R(\wp)$. Let $\tilde{T}(\wp) \subset R(\wp)^*$ be the multiplicative subgroup generated by $R(\wp)^{*2}$ and $\bar{f}_1, \ldots, \bar{f}_m$. Then $\tilde{T}(\wp)$ is a generating set for $T(\wp)$ and for odd $n \in \mathbb{N}$ we have $T(\wp)_n = \sum \tilde{T}(\wp)^n$.

(4.2) Proposition. *Let $V \subset A_k(\mathbb{R})$ be an algebraic set and let*

$$A := \{x \in V \mid f_1(x) \geq 0 \wedge \cdots \wedge f_m \geq 0\}$$

be Zariski-dense in V, where $f_1, \ldots, f_m \in \mathcal{R}[V]$. Assume that A is bounded. If $n \in \mathbb{N}$ is odd, then for $f \in \mathcal{R}[V]$ the following statements are equivalent:
(1) $f(x) > 0$ for all $x \in A$;
(2) $fs = 1 + d$ for some $s, d \in S_n$.

Proof. Again we have only to show (1) $\Rightarrow$ (2). Recall that a prime ideal $\wp \subset \mathcal{R}[V]$ is called S_n-convex if for all $s_1, s_2 \in S_n$ we have

$$s_1 + s_2 \in \wp \Rightarrow s_1, s_2 \in \wp.$$

Consequently, if $\wp$ is S_n-convex, then $T(\wp)_n \subset R(\wp)$ is a preordering. We claim that $T(\wp)$ is a preordering as well. Namely,

$$T_{\text{odd}} := \{x \in R(\wp) \mid x^k \in T(\wp)_n \text{ for some odd } k \in \mathbb{N}\}$$

is a preordering, by [1, Proposition (4.6)]. By the definition of $T(\wp)_n$ we see $T(\wp) \subset T_{\text{odd}}$. Thus $T(\wp)$ is a preordering of $R(\wp)$. In the next step we will prove:

(∗) Given any S_n-convex prime ideal $\wp \subset \mathcal{R}[V]$, then $\bar{f} \in T(\wp)_n^*$.

Let $\wp \subset \mathcal{R}[V]$ be S_n-convex and let $W \subset V$ be the algebraic subset defined by $\wp$. We have just seen that $T(\wp)$ is a preordering of $R(\wp)$. Since f is positive definite on A, the model completeness of the theory of real closed fields implies

$$\bar{f} \in T(\wp)^* \text{ and } W \cap A \neq \emptyset.$$

We next show $\bar{f} \in A(T(\wp))^*$. Since $W \cap A$ is bounded, we find some $n \in \mathbb{N}$ such that

$$\frac{1}{n} < f(x) < n$$

for all $x \in A \cap W$. Again the model completeness of the theory of real closed fields implies

$$n \pm \bar{f} \in T(\wp) \text{ and } \bar{f} \pm \frac{1}{n} \in T(\wp).$$

But this just means $\bar{f} \in A(T(\wp))^*$. Moreover, we have already seen that $\bar{f} \in T(\wp)^*$. Now (2.7) implies $\bar{f} \in A(T(\wp))^* \cap T(\wp)^* \subset T(\wp)_n^*$. Let $P \subset \mathcal{R}[V]$ be an ordering of level n with $S_n \subset P$ and let $\wp := P \cap -P$. Then $\wp$ is an S_n-convex prime ideal. Hence $(*)$ implies $f \in P^+ := P \setminus \wp$. Therefore we have

$$f \in \bigcap P^+,$$

where P ranges over the orderings of level n of $\mathcal{R}[V]$ which contain S_n. By the intersection theorem [6, Theorem 6] there exist $s, d \in S_n$ with $fs = 1 + d$. □

(4.3) Example. Let $a < b \in \mathbb{R}$ and let $f \in \mathbb{R}[X]$ be a univariate polynomial. Suppose that f is positive definite on $[a, b]$. Let

$$h = (X - a)(b - X).$$

Then

$$[a, b] = \{x \in \mathbb{R} \mid h(x) \geq 0\}$$

Hence the last result implies that for any odd $n \in \mathbb{N}$ there exist $s, t \in \mathbb{N}$ and $f_1, \ldots, f_s, g_1, \ldots, g_t \in \mathbb{R}(X)$ with

$$f = \sum_{i=1}^{s} f_i^{2n} + h^n \cdot \sum_{i=1}^{t} g_i^{2n}.$$

Finally let C be a smooth irreducible projective curve over a real closed field R. Given a regular function $f \in \mathcal{R}[C]$ we let $\mathrm{div}_{\mathrm{real}}(f)$ denote the real divisor of f. We conclude this section by a description of regular functions $f \in \mathcal{R}[C]$ which are positive semidefinite on $C(R)$ and whose real divisor is divisible by $2n$. In the case that R is archimedean it is known that this is equivalent to the condition that f is a sum of $2n$-th powers in the function field of C. This fails to be true as soon as R is non-archimedean (see [15]).

(4.4) Proposition. *Let C be a smooth irreducible projective curve over a real closed field R with function field F. Given $n \in \mathbb{N}$ and a positive semidefinite*

function $f \in \mathcal{R}[C]$, *then the following statements are equivalent:*
(1) $2n \mid \mathrm{div}_{\mathrm{real}}(f)$.
(2) *There exist* $r \in R$ *and* $q \in \sum F^2$ *such that*

$$f \in \frac{r^2+q}{1+q} \cdot \sum F^{2n}.$$

Proof. Let $\mathcal{V}$ be the set of real prime divisors of $F \mid R$ enlarged by the trivial valuation and let $H := H(\mathcal{V})$. Then $\mathcal{V}$ is a proconstructible subset of $\mathrm{Spv}\, F$. Since f is positive semidefinite on $C(R)$ we get $f \in \sum F^2$. Hence (1) is equivalent to

$$(2') \quad f \in (H^* \cap \sum F^2) \cdot \sum F^{2n},$$

by (1.11). Moreover, with the notions of section 3 we have $T_H = \sum F^2$ and $H = H_R(F, R)$. Now the claim follows from (3.7). □

References

[1] E. Becker, Partial orders on a field and valuation rings. Comm. Algebra **7**, 1933–1976 (1979).

[2] E. Becker, Valuations and real places in the theory of formally real fields. In: Géométrie Algébrique Réelle et Formes Quadratiques, Lecture Notes in Math. **959**, 1–40.

[3] E. Becker, The real holomorphy ring and sums of $2n^{th}$ powers. In: Géométrie Algébrique Réelle et Formes Quadratiques, Lecture Notes in Math. **959**, 139–181.

[4] E. Becker, D. Gondard, On rings admitting orderings and 2-primary orderings of higher level. Manuscripta Math. **65**, 63–82 (1989).

[5] E. Becker N. Schwartz, Zum Darstellungssatz von Kadison–Dubois. Arch. Math. **40**, 421–428 (1983).

[6] R. Berr, The intersection theorem for orderings of higher level in rings. Manuscripta Math. **75**, 273–277 (1992).

[7] R. Berr, Null- and Positivstellensätze for generalized real closed fields. To appear.

[8] J. Bochnak, M. Coste, M.F. Roy, Géométrie algébrique réelle. Ergeb. Math. Grenzgeb. (3), Band **12**.

[9] L. Bröcker, H.W. Schülting, Valuations of function fields from the geometrical point of view. J. Reine Angew. Math. **365** 12–32 (1986).

[10] M. Carral, M. Coste, Normal spectral spaces and their dimensions. J. Pure Appl. Algebra **30**, 227–235 (1983).

[11] R. Gilmer, Multiplicative ideal theory. Kingston 1968.

[12] M. Hochster, Prime ideal structure in commutative rings. Trans. Amer. Math. Soc. **142**, 43–60 (1969).

[13] R. Huber, Bewertungsspektrum und rigide Geometrie. Habilitationsschrift, Universität Regensburg (1990).

[14] R. Huber, M. Knebusch, On valuation spectra. Preprint 1992.

[15] A. Prestel, Model theory of fields: An application to positive semidefinite polynomials. Mém. Soc. Math. France (N.S.) **16**, 53–65 (1984).

[16] H.W. Schülting, On real places of a field and their holomorphy ring. Comm. Algebra **10**, 1239–1284 (1982).

[17] H.W. Schülting, Real holomorphy rings in real algebraic geometry. In: Géométrie Algébrique Réelle et Formes Quadratiques, Lecture Notes in Math. **959**, 433–442.

[18] H.W. Schülting, Prime divisors on real varieties and valuation theory. J. Algebra **98**, 499–514 (1986).

Some practical tools for algebraic geometry

*John F. Canny**

Abstract. In this paper, we present a collection of techniques which improve the efficiency of sparse elimination and calculations over real closed fields. The impetus for the work is the construction of an algebraic-geometric toolkit, written in C, which can solve elimination and equation-solving problems in algebraically closed fields, or for semi-algebraic sets. We have implemented several of the lowest layers of the toolkit, and we are seeing large speedups over existing methods such as Gröbner bases, and for small examples, over homotopy methods. Two of the techniques are taken from recently published work, and the other two are new, unpublished work in progress.

1. Introduction

In the late 1980's, Gel'fand and his colleagues began the study of discriminants and resultants of sparse polynomial systems. This work is of practical significance because most of the polynomial systems encountered in robotics and vision exhibit this kind of sparseness. Sparseness leads to a lowering of effective degree, and the sparse theory provides a simple direct method for proving bounds on the number of solutions. Sparseness can also be exploited to speedup equation solving and elimination of variables. The first efficient algorithms for the sparse resultant were described by the author and a collaborator in [CE93], [EC93].

Since the work of Schwartz and Sharir and Grigor'ev et al., also in the 1980s, there has been a steadily increasing interest in algorithms for semi-algebraic sets. A semi-algebraic representation is convenient for the shape of physical objects, and this has fueled interest in semi-algebraic approaches to problems such as robot motion planning, geometric modeling and vision. Much recent work [HRS90], [Ren92] [GV92] has focussed on improving the asymptotic complexity of these algorithms, but unfortunately, most of it ignores the complexity in practice. The work of the author [Can88a], [Can91b], [Can91a], on the other hand, has been specifically directed at practical algorithms. We here summarize this work and explain where the practical gains come from.

*Supported by a David and Lucile Packard Foundation Fellowship and by NSF Presidential Young Investigator Grant IRI-8958577

In this paper, we present a collection of techniques which improve the efficiency of sparse elimination and calculations over real closed fields. The impetus for the work is the construction of an algebraic-geometric toolkit, written in C, which can solve elimination and equation-solving problems in algebraically closed fields, or for semi-algebraic sets. We have implemented several of the lowest layers of the toolkit, and we are seeing large speedups over existing methods such as Gröbner bases, and for small examples, over homotopy methods.

The paper is organized as follows: Section 2 summarizes our sparse elimination algorithms. Section 3 describes a practically efficient method for solving existential formulae over the reals. Section 4 reviews an improvement of the [BOKR86] method for sign determination. Section 5 describes a new method for efficiently computing in infinitesimal real extensions. Sections 3 and 5 describe new work, while the other sections are reviews of earlier work.

2. Sparse systems

In a moment we will be able to say precisely what we mean by a sparse polynomial system. But even before that, we should answer the question: Why study sparse systems? For us, the reason is that sparse systems are ubiquitous in robotics, vision and modeling. This is not a theorem that we can prove, but our experience shows that almost all the systems that arise in vision, robotics and modeling are sparse. And we can exploit this sparseness in two ways: (i) To prove tight or tighter bounds for the number of solutions (ii) To compute those solutions in a time that depends on the sparse bounds from (i), not on the classical degree bound (Bezout's bound) which is usually much larger.

Typical examples of sparse systems are those that describe the inverse kinematics for a 6R robot [MC92b], forward kinematics for the Stewart platform, camera motion from point matches, and intersections of tensor product surface patches. As the dimension of the problem increases, the difference between the sparse and non-sparse bounds increases dramatically. Very few algebraic problems with more than 4 variables can be solved with classical resultant methods, which ignore sparseness, but the inverse kinematics problem has 6 variables and only 16 solutions, and these can be found in on the order of 15 milliseconds [MC92b].

Gröbner basis algorithms also exploit sparseness, and this explains their speed on certain problems. But when a sparse resultant was known for a problem, such as the Dixon resultant for tensor product surface implicitization [MC92a], the solution using resultants could be computed much faster. For example, for implicitizing a bicubic surface patch, the Dixon resultant can be used to compute the solution in about 30 minutes, and provides the correct integer coefficients. We have tried this problem on all the major Gröbner algorithms, and we know of none which can finish the problem without running out of memory after some days, *even when working on a homomorphic image of the problem in a finite field.*

So ideally, one would like a sparse resultant for any polynomial system. The first efficient (polynomial in the size of the answer) algorithms were given in [CE93] and [EC93]. We will refer to those later in this section. For now we move on to the characterization of sparse systems.

Suppose we are given $n+1$ polynomials $f_1, \ldots, f_{n+1}$ in $x_1, \ldots, x_n$ with complex coefficients. We use x^e to denote the monomial $x_1^{e_1} \cdots x_n^{e_n}$, where $e = (e_1, \ldots, e_n) \in \mathbb{Z}^n$ is a multi-exponent. Let $\mathcal{A}_i = \{a_{i1}, \ldots, a_{im_i}\} \subseteq \mathbb{Z}^n$ denote the set of exponents occurring in f_i, then

$$f_i = \sum_{j=1}^{m_i} c_{ij} x^{a_{ij}} \ , \qquad \text{for } i = 1, \ldots, n+1 \ , \tag{1}$$

and we suppose $c_{ij} \neq 0$ so that $\mathcal{A}_i$ is uniquely defined given f_i.

Sparse systems may have trivial solutions with some $x_i = 0$ for choices of coefficients, and this makes it difficult to perform variable elimination or equation solving. We discount these solutions and look only for solutions $x = \xi$ with $\xi \in (\mathbb{C}^*)^n$, where $\mathbb{C}^* = \mathbb{C} - \{0\}$. Under this assumption, we can deal with the more general case of f_i's which are polynomials in the x_i and their reciprocals, the *Laurent* polynomials $\mathbb{C}[x_1, x_1^{-1}, \ldots, x_n, x_n^{-1}]$.

Definition 2.1. The finite set $\mathcal{A}_i \subset \mathbb{Z}^n$ of all monomial exponents appearing in f_i is the *support* of f_i. The *Newton Polytope* of f_i is $Q_i = \mathrm{Conv}(\mathcal{A}_i) \subset \mathbf{R}^n$, the convex hull of $\mathcal{A}_i$.

A polynomial system is *unmixed* if all supports $\mathcal{A}_i$ are the same for $i = 1, \ldots, n+1$, otherwise it is *mixed*.

Definition 2.2. The *Minkowski Sum* $A + B$ of convex polytopes A and B in $\mathbf{R}^n$ is the set

$$A + B = \{a + b | a \in A, b \in B\} \ .$$

$A + B$ is a convex polytope. Let $\mathrm{Vol}(A)$ denote the usual n-dimensional volume of A.

Definition 2.3. Given convex polytopes $A_1, \ldots, A_n \subseteq \mathbf{R}^n$, there is a unique real-valued function $MV(A_1, \ldots, A_n)$ called the *Mixed Volume* which is multilinear with respect to Minkowski sum, such that $MV(A_1, \ldots, A_1) = n!\mathrm{Vol}(A_1)$. Equivalently, if $\lambda_1, \ldots, \lambda_n$ are scalars, then $MV(A_1, \ldots, A_n)$ is precisely the coefficient of $\lambda_1\lambda_2 \cdots \lambda_n$ in $\mathrm{Vol}(\lambda_1 A_1 + \cdots + \lambda_n A_n)$ expanded as a polynomial in $\lambda_1, \ldots, \lambda_n$.

The Newton polytopes capture the combinatorial properties of the system in a remarkable way. We have the following bound on the number of roots of a system of n polynomials in n variables, see [Ber75], [Kus76], [Kho78].

Theorem 2.4. [Ber75] *Let $f_1, \ldots, f_n \in \mathbb{C}[x_1, x^{-1}, \ldots, x_n, x_n^{-1}]$. The number of common zeros in $(\mathbb{C}^*)^n$ is either infinite, or does not exceed $MV(Q_1, \ldots, Q_n)$. For almost all specialization of the coefficients c_{ij} the number of solutions is exactly $MV(Q_1, \ldots, Q_n)$.*

We refer to a polynomial system as "sparse" if the mixed volume of its Newton polytopes is small, say compared to the Bezout bound for the system.

For systems of $n+1$ polynomials in n unknowns, there are generically no solutions, and there is an algebraic condition on the coefficients for a solution to exist. That is, a solution exists whenever a certain polynomial in the coefficients of the system is zero. This polynomial is called the resultant of the system.

We adopt the following definition for the *sparse* resultant from [PS91]; it was first defined as the $(\mathcal{A}_1, \ldots, \mathcal{A}_{n+1})$*-resultant* in [GKZ90]. Regard a polynomial f_i as a point $(c_{i1}, \ldots, c_{im_i}) \in \mathbb{C}^{m_i}$ in the space of all possible polynomials with the given set of exponents $\mathcal{A}_i$. Then the input system is a point $c = (c_{11}, \ldots, c_{1m_1}, \ldots, c_{(n+1)1}, \ldots, c_{(n+1)m_{n+1}})$ in $\mathbb{C}^{m_1} \times \cdots \times \mathbb{C}^{m_{n+1}}$. Let $Z_0 = Z_0(\mathcal{A}_1, \ldots, \mathcal{A}_{n+1})$ be the set of all points c such that the system has a solution in $(\mathbb{C}^*)^n$, and let $Z = Z(\mathcal{A}_1, \ldots, \mathcal{A}_{n+1})$ denote the closure of Z_0 in the product of projective spaces. It is known that Z is an irreducible algebraic set, see [PS91].

Definition 2.5. The *sparse resultant* $R(\mathcal{A}_1, \ldots, \mathcal{A}_{n+1})$ of the system (1) is an irreducible polynomial in $\mathbb{Z}[c]$. If $\text{codim}(Z) = 1$ then $R(\mathcal{A}_1, \ldots, \mathcal{A}_{n+1})$ is the defining polynomial of the hypersurface Z. If $\text{codim}(Z) > 1$ then $R(\mathcal{A}_1, \ldots, \mathcal{A}_{n+1}) = 1$.

Proposition 2.6. [PS91] *The sparse resultant is separately homogeneous in the coefficients* $(c_{i1}, \ldots, c_{im_i})$ *of each* f_i *and its degree in these coefficients equals the mixed volume of the other* n *Newton polytopes* $MV(Q_1, \ldots, Q_{i-1}, Q_{i+1}, \ldots, Q_{n+1})$.

This implies that the total degree $\deg R$ of the resultant equals the sum of all $n+1$ n-fold Mixed Volumes.

The simplest and most efficient way to define the resultant is as the determinant of matrix whose entries are the coefficients of the polynomials, or zero. This is a generalization of the classical Sylvester formula for the resultant of two polynomials. In general, the sparse resultant cannot be expressed as the determinant of a single matrix, but in practice it is just as good to define a matrix whose determinant is a multiple of the resultant.

Once a resultant matrix is defined, it can be used for numerical equation solving or for symbolic variable elimination. For the first use, we construct from the resultant matrix another matrix whose eigenvectors define the solutions of the system. For the second use, we repeatedly evaluate the resultant for specializations of the coefficients, and use Chinese remaindering and sparse interpolation to reconstruct the answer. These uses are well described in [MC93].

We will not describe the construction of our resultant matrices because of space limitations, but refer the reader to [CE93] and [EC93] which describe our two algorithms. For summary purposes, we will refer to the two algorithms as Sparse1 and Sparse2 respectively.

Sparse1 constructs a matrix M whose determinant is a multiple of the resultant, and [CE93] describes several ways to recover the resultant without factoring. The rows and columns of M are indexed by the integer points in the Minkowski sum $Q = Q_1 + \cdots + Q_{n+1} + \delta$, where δ is a single point (chosen at random to move the Minkowski sum into general position). Each row of M contains coefficients of a particular f_i. To a first order then, the size of M (number of rows or columns) is the same as the volume of the Minkowski sum Q. The degree of the resultant on the other hand, equals the mixed volume of the Q_i's, which is somewhat smaller. The size of M is however, polynomial in the mixed volume, hence in the size of the resultant, for reasonable systems.

Sparse2 constructs a matrix M' whose determinant is a multiple of the resultant, and whose size is at most the size of M. Previously, some formula were given for the sparse resultant as an exact determinant (not a multiple) in [SZ93]. But these formulae apply only to a very restricted class of systems. For the systems in this class, Sparse2 also constructs optimal formulae. In general, optimal formulae do not exist for sparse systems, but we believe that the size of the matrix M' will be much closer to the mixed volume than to the Minkowski sum volume, we think within a small constant factor of the former. We will be endeavoring to prove this claim in the near future. In any case, Sparse2 gives very good performance on the examples we have tried.

3. Non-emptiness of semi-algebraic sets

Semi-algebraic (SA) sets are a versatile class that includes the forms of most familar objects in $\mathbb{R}^3$, such as cones, cylinders, spheres, and combinations of these. SA sets in 3D are slightly more general than CSG (constructive solid geometry) models, for example, SA sets include sets of mixed dimension, and sets which are not topologically closed. Both of these are forbidden in CSG.

SA sets are defined in any dimension, and can also be used to represent the set of legal (e.g. obstacle-avoiding) configurations of a mechanical system, such as a robot, or the set of camera transformations consistent with some geometric constraints, or the set of positions at a given distance from a surface to guide a milling machine. The defining formulae for SA sets are the real analogue of SAT formulae, and they provide the ability to search over a space of real values for a solution satisfying some property.

A semi-algebraic set can be defined as the set of points in $\mathbb{R}^n$ satisfying a certain predicate of the form $B(A_1, \ldots, A_k)$ where $B : \{0,1\}^k \to \{0,1\}$ is a boolean function and each A_i is an atomic formula of one of the following types:

$$(F_i = 0) \quad (F_i \neq 0) \quad (F_i > 0) \quad (F_i < 0) \quad (F_i \geq 0) \quad (F_i \leq 0) \qquad (2)$$

with each F_i a polynomial in $x_1, \ldots, x_n$ with rational (for our computational purposes) coefficients. In the method that follows it will be helpful to assume (wlog) a certain form for the defining predicate:

Definition 3.1. A formula $B(A_i, \ldots, A_k)$ is said to be in *monotone standard form* if the boolean function B is monotone, and all atomic formulae A_i are either $(F_i = 0)$ or $(F_i > 0)$.

Predicate complexity. We measure the complexity of a predicate with four quantities, the number of polynomials k, the number of variables n, the maximum degree of the polynomials d, and the maximum coefficient length c of the coefficients of the polynomials.

3.1. Stratifications

Definition. A *stratification* $\underline{\underline{S}}$ of a set $S \subset \mathbb{R}^n$ is a partition of S into a finite number of disjoint subsets S_i called *strata* such that each S_i is a manifold.

A regular stratification satisfies some additional conditions which are well described in [GWDL76]. There are several ways to construct regular stratifications. We will only need two:

- **Taking products.** If $\underline{\underline{C}}$ and $\underline{\underline{D}}$ are regular stratifications of the spaces C and D respectively, then the product $\underline{\underline{C}} \times \underline{\underline{D}}$ is a regular stratification of $C \times D$.
- **Preimage of a transversal map.** If $F : M \to N$ is transversal to $\underline{\underline{D}}$ for a regular stratification $\underline{\underline{D}}$ of a subset $D \subseteq N$, then $F^{-1}(\underline{\underline{D}})$ is a regular stratification of $F^{-1}(D)$.

where $F \pitchfork \underline{\underline{D}}$ means F is transversal to all the strata of $\underline{\underline{D}}$, and $F^{-1}(\underline{\underline{D}})$ is the set $\{F^{-1}(\sigma) \mid \sigma \in \underline{\underline{D}}\}$. If we now define

Definition. Let $F_i \in \mathbb{Q}[x]$, $i = 1, \ldots, k$ be a collection of polynomials that define a map $F : \mathbb{R}^n \to \mathbb{R}^k$. A *sign sequence* σ is an element of $\{\mathbb{R}^-, \{0\}, \mathbb{R}^+\}^k$. The sets $F^{-1}(\sigma)$ are called *sign-invariant sets* of F.

Then we can view a semi-algebraic set S as a finite union of sign-invariant sets of some polynomial map F. The sign partition $(\underline{\underline{\mathbb{R}}})^k$ of $\mathbb{R}^k$ is a regular stratification of $\mathbb{R}^k$. So if a map $F : \mathbb{R}^n \to \mathbb{R}^k$ is transversal to $(\underline{\underline{\mathbb{R}}})^k$, then the preimage $F^{-1}((\underline{\underline{\mathbb{R}}})^k)$, which is the collection of sign-invariant sets of F, is a regular stratification.

3.2. Infinitesimals

We will make extensive use of extensions of real fields by infinitesimals. This process is simple to implement computationally, and has been well formalized in [BCR87] using the real spectrum. An elementary description of the use of infinitesimal elements is given in [Can88b] in an algorithm for the existential theory of the reals.

Definition. For a given real field R, we say that an element ϵ is *infinitesimal* with respect to R if the extension $R(\epsilon)$ is ordered such that ϵ is positive, but smaller than any positive element of R.

We will have cause to make use of towers of such field extensions. We will use the suggestive notation $\delta \gg \epsilon$ for two infinitesimals to mean that ϵ is infinitesimal with respect to the real closure of the field $R(\delta)$.

3.3. Transformation algorithm

The following algorithm takes the formula B defining a semi-algebraic set S and transforms it to a new formula B_0. The formula B_0 defines a set S_0 which is non-empty if and only if S is. The set S_0 is compact and regularly stratified by the polynomials defining it, so it is simpler to decide its non-emptiness. The size of B_0 is larger than B by at most a constant factor (2 if B is in standard form, 4 otherwise).

- Convert the input formula to monotone standard form.
- Add to the formula a conjunction with the polynomial inequality $\sum_{j=1}^{n} x_j^2 \leq 1/\rho^2$ (converted to standard form), where ρ is an infinitesimal. Let the resulting formula be $B(A_1, \ldots, A_k)$, it defines a bounded set in the extension field $\mathbb{R}(\rho)$.
- Choose an $a \in (\mathbb{R} - \{0\})^k$ at random, or let $a_1 \gg a_2 \gg \cdots \gg a_k > 0$ be a series of infinitesimals.
- Construct a formula B_0 from the input formula $B(A_1, \ldots, A_k)$ as follows. For each atomic predicate A_i, replace A_i with

$$\begin{array}{llll} \text{if } A_i \text{ is} & F_i = 0 & \text{then} & (F_i + \epsilon a_i \geq 0) \wedge (F_i - \epsilon a_i \leq 0) \\ \text{if } A_i \text{ is} & F_i > 0 & \text{then} & (F_i - \delta a_i \geq 0) \end{array} \tag{3}$$

  where $\delta \gg \epsilon$ are infinitesimals. Then the set S_0 defined by this formula is closed and bounded, therefore compact. By the results of [Can91a] the connected components of the sign-invariant sets of B_0 are in one-to-one correspondence with those of B. Return B_0.

The correctness of the steps of the transformation was proved in [Can91a].

3.4. Non-emptiness for compact, regular sets

Once we know that the set S_0 is compact and regularly stratified by the signs of polynomials F_i in B_0, it is easy to decide if it is non-empty. We choose a linear map $\pi : \mathbb{R}^n \rightarrow \mathbb{R}$ and find all the critical points of this map restricted to the strata of S_0. If S_0 is non-empty, π will attain a maximum value on it at some point p, which will be a critical point. By determining the signs of the polynomials F_i at some *candidate* critical point p, and evaluating the predicate B_0, we can check if p in fact lies in S_0.

By testing all critical points in this way, we are guaranteed to find a "witness" p to $S_0 \neq \emptyset$ if it exists. Conversely, if S_0 is empty, no critical point will satisfy the formula.

3.5. Enumerating critical points

Since we know the polynomials F_i are in general position, the intersection of any $k+1$ of them in k dimensions will be null. Any $j \leq k$ of them will intersect in a manifold of dimension $k-j$. Let p be an extremal point of π in the set S_0. Then p is also an extremal of π in a manifold M which is the set of zeros of some polynomials $F_{i_1}, \ldots, F_{i_j}$. These polynomials are precisely the F_i which are zero at p.

So to enumerate all potential witness points, we enumerate the critical points of π on the set of common zeros of $F_{i_1}, \ldots, F_{i_j}$, for every set of $j \leq k$ polynomials. First we define a polynomial

$$g = \sum_{l=1}^{j} F_{i_l}^2$$

and solve the system

$$g = \mu \qquad \frac{\partial g}{\partial x'_2} = \cdots = \frac{\partial g}{\partial x'_k} = 0 \tag{4}$$

"in the limit" as $\mu \to 0$. The coordinates $x'_2, \ldots, x'_k$ are a basis for the $k-1$ dimensional linear space which is the kernel of π. The process of solving this system in the limit is described in [Can88a] and [Can91a], and involves computing the u-resultant of the system, arranging it in powers of μ, and retaining the lowest degree coefficient. The result is a polynomial $P(s)$ and rational functions $(Q_1, \ldots, Q_k)$ such that the solutions to the system (4) are all the tuples $(Q_1(\alpha_i), \ldots, Q_k(\alpha_i))$ where α_i are the roots of $P(s) = 0$.

To compute the signs of the other polynomials at these critical points, we substitute $x_i \mapsto Q_i(s)$ for $i = 1, \ldots, k$, and the set of signs we are looking for is precisely the sign sequences of

$$(F_l(Q(s))) \text{ for } l \in \{1, \ldots, n\} - \{i_1, \ldots, i_j\} \quad \text{at roots of} \quad P(s) = 0$$

and to find these we simply apply the sign determination algorithm of the next section. In summary, we have:

- Enumerate all subsets of $j \leq k$ polynomials $\{F_{i_1}, \ldots, F_{i_j}\}$. Do this in order of increasing j, so that "easy" witness points will be found early.
- For each subset, construct a representation of the critical points of π as a polynomial $P(s)$ and rational functions $Q_1(s), \ldots, Q_k(s)$.
- Substitute $Q_i(s)$ for x_i in the remaining $n-j$ polynomials, and determine their signs at roots of $P(s) = 0$ using the algorithm of the next section.

- Substitute these signs into the formula B_0 to check if the corresponding critical point lies in S_0. If yes, S_0 is non-empty, if not, continue until no more critical points, and then return "empty".

3.6. Complexity analysis

The total number of $j \leq k$-tuples of polynomials is $O(n^k)$. Constructing the representation for the solutions of the system (4) can be done in time $d^{O(k)}$, [Can91a]. Applying Bezout's theorem to this system, we see that the polynomial $P(s)$ has degree $O((2d)^k)$, and the numerator and denominator of the Q_i's have the same bound. Substituting Q_i's into the other F_i's gives $n - j$ polynomials of degree $O(d(2d)^k)$ in s.

The sign determination step, using the bounds of the next section, takes $O(nk^3d(2d)^{2k})$ arithmetic steps. Substituting each sign sequence and evaluating the predicate B_0 takes $O(n)$ time. Multiplying by the number of polynomial tuples, we get an overall bound of

$$n^{k+1}d^{(O(k))}$$

we must be careful of course, because we are counting arithmetic steps, when the arithmetic is happening over a field extension $\mathbb{Q}[\epsilon, \delta, \mu, \rho]$. As we explain in section 5 however, in most situations, we can perform this arithmetic for about the same cost as integer arithmetic.

There are two reasons why this method offers significantly better practical efficiency than other methods. The first is the way it deals with infinitesimals, as just mentioned. The second is in the form of the bound, which has a "combinatorial" part n^{k+1} and an algebraic part $d^{O(k)}$. The algebraic part is the same as the methods of [HRS90], [Ren92] and [GV92] although the constant hidden in the exponent is slighly better in our method because of the use of improved sign determination algorithm of the next section.

But the major difference is in the combinatorial part. The other schemes achieve combinatorial complexities only of $n^{O(k)}$, where the hidden constant is the same as for the algebraic part, and is at least 4. In typical geometric problems, n is large compared to d and k, so n^{k+1} is much smaller than n^{4k}. e.g. If $n = 10$, $k = 3$, we have $n^{k+1} = 10^4$, but $n^{4k} = 10^{12}$. The lower bound for the combinatorial complexity is n^k, so we have a near-optimal upper bound. We can improve this upper bound to $O(n^k \log n)$ using the techniques from [Can91a], but the algorithm becomes much more complicated, and the constant factors "downstairs" are much larger, so it would likely run slower on typical inputs.

4. Sign determination

Most of the recent work on real algebraic algorithms makes use of a sign-determination lemma due to Ben-Or, Kozen and Reif [BOKR86]. This lemma, henceforth called "BKR", takes a univariate polynomial $p(x)$, and polynomials $q_1(x), \ldots, q_n(x)$, and returns k sign sequences $\sigma \in \{-, 0, +\}^n$, where k is the number of real roots of $p(x)$. Each sign sequence σ corresponds to a particular root α of $p(x)$, in such a way that $\sigma_i = \mathrm{sign}(q_i(\alpha))$.

Given the importance of BKR, it is natural to look for improvements and simplifications. In this section, we summarize an improvement of BKR described in [Can91b], with better asymptotic (and practical) complexity. It also eliminates many of the matrix rank tests of BKR, replacing them with a simple recursive algorithm.

4.1. Sturm queries

If $f(x)$ and $g(x)$ are polynomials, let the Sturm sequence of f and g be denoted $r_0, r_1, \ldots, r_k$, where $r_0(x) = f(x)$, $r_1(x) = g(x)$, and the r_i satisfy:

$$r_{i+1} = q_i r_i - r_{i-1} \tag{5}$$

where q_i is the quotient of the polynomial division of r_{i-1} by r_i. In practice, pseudo-remainders [BT71] are often used so that the r_i have integer coefficients. For a real value v, let $\mathrm{SA}(f, g, v)$ denote the number of changes in sign in the sequence $r_0(v), r_1(v), \ldots, r_k(v)$.

Let $\mathrm{SC}(f, g)$ denote the quantity $\mathrm{SA}(f, g, +\infty) - \mathrm{SA}(f, g, -\infty)$, where signs at infinity are limits for large real values. The classical Sturm theorem states that $\mathrm{SC}(f, f')$ equals the number of real roots of $f(x)$. The most general form of the theorem states that $\mathrm{SC}(f, f'g)$ gives the sum of the number of real roots of $f(x)$ where $g(x) > 0$, minus the number where $g(x) < 0$, common roots making no difference to the count. A useful application of this result is to consider $\mathrm{SC}(f, f'g^2)$ which counts real roots of f where $g \neq 0$. This observation was used by [RS90] to simplify the BKR sign determination lemma.

Lemma 4.1. [RS90] *Let r^+ denote the set of roots of $f(x)$ where $g(x)$ is positive, and similarly for r^0 and r^-. Then the numbers of roots in each set satisfy the following identity*

$$\begin{bmatrix} 1 & 1 & 1 \\ 1 & -1 & 0 \\ 1 & 1 & 0 \end{bmatrix} \begin{bmatrix} |r^+| \\ |r^-| \\ |r^0| \end{bmatrix} = \begin{bmatrix} \mathrm{SC}(f, f') \\ \mathrm{SC}(f, f'g) \\ \mathrm{SC}(f, f'g^2) \end{bmatrix}. \tag{6}$$

4.2. Sign determination

Now we can apply this lemma to find the signs of a system of polynomials $q_1(x), \ldots, q_n(x)$ at the roots of a single polynomial $p(x)$. We start by applying the above lemma to $p(x)$ and $q_1(x)$. We can rewrite the above identity in matrix form $H_1 R_1 = S_1$, where H_1 is the 3x3 matrix, R_1 is the vector of root counts for q_1 and S_1 is the vector of Sturm query counts for q_1:

$$\begin{bmatrix} 1 & 1 & 1 \\ 1 & -1 & 0 \\ 1 & 1 & 0 \end{bmatrix} \begin{bmatrix} |r_1^+| \\ |r_1^-| \\ |r_1^0| \end{bmatrix} = \begin{bmatrix} \mathrm{SC}(p, p') \\ \mathrm{SC}(p, p'q_1) \\ \mathrm{SC}(p, p'q_1^2) \end{bmatrix}. \tag{7}$$

There is a more general identity which we can write as $H_n R_n = S_n$ where we define H_n inductively as

$$H_{n+1} = \begin{bmatrix} H_n & H_n & H_n \\ H_n & -H_n & 0 \\ H_n & H_n & 0 \end{bmatrix} \tag{8}$$

and where R_{n+1} is the 3^{n+1} vector obtained by listing the elements of R_n with r_{n+1}^+ appended to each conjunction, then the elements of R_n with r_{n+1}^- appended, and finally a last copy of R_n with r_{n+1}^0. Similarly, S_{n+1} is obtained by listing first a copy of S_n, then another copy of S_n with q_{n+1} multiplying the derivative in each Sturm sequence, and then a third copy of S_n with q_{n+1}^2. We can write these definitions schematically as

$$R_{n+1} = \begin{bmatrix} R_n \cap r_{n+1}^+ \\ R_n \cap r_{n+1}^- \\ R_n \cap r_{n+1}^0 \end{bmatrix} \qquad S_{n+1} = \begin{bmatrix} S_n \\ S_n \cdot q_{n+1} \\ S_n \cdot q_{n+1}^2 \end{bmatrix} \tag{9}$$

and it is easy to verify that $H_{n+1} R_{n+1} = S_{n+1}$. It is also easy to see that H_{n+1} is non-singular. Simply add the middle rows to the last rows to give a block diagonal matrix whose determinant is non-zero if the determinant of H_n is.

H_n is a very large matrix, of size $3^n \times 3^n$. Fortunately, BKR showed that for sign determination, only a small non-singular submatrix of H_n is needed. They showed how to compute this submatrix inductively from submatrices of $H_{n/2}$.

We give here a direct method for constructing this submatrix of H_n. The advantage of our construction is that it only makes Sturm queries of small degree, thereby reducing the complexity of answering the queries.

4.3. A small Sturm query matrix

Our new matrix is denoted $K_{n,m}$ which indicates that it has 3^n columns and any m of them are linearly independent. Thus it is defined only if $m \leq 3^n$. If $m = 1$, $K_{n,m}$ is just the first row of H_n, i.e. a row of all ones. If $m > 3^{n-1}$, then $K_{n,m}$ is just H_n. Otherwise, it is defined recursively as follows:

$$K_{n,m} = \begin{bmatrix} K_{(n-1),m} & K_{(n-1),m} & K_{(n-1),m} \\ K_{(n-1),\lfloor\frac{m}{2}\rfloor} & -K_{(n-1),\lfloor\frac{m}{2}\rfloor} & 0 \\ K_{(n-1),\lfloor\frac{m}{3}\rfloor} & K_{(n-1),\lfloor\frac{m}{3}\rfloor} & 0 \end{bmatrix} \tag{10}$$

and as shown in [Can91b], $K_{n,m}$ has the property that any m of its columns are linearly independent.

Observation. $K_{n,m}$ is a submatrix of H_n, consisting of rows of H_n.

This observation is easily proved by induction using the recursive definitions of H_n and $K_{n,m}$ in equations (8) and (10).

Next we make a simple observation which provides good bounds on the degrees of the polynomials in the Sturm queries. Since $K_{n,m}$ is a submatrix consisting of rows of H_n, the product $K_{n,m}R_n$ is a subvector of S_n. Let $\Gamma_{n,m}$ be this subvector, which consists of rows of S_n which correspond to the rows of $K_{n,m}$. Then

$$\Gamma_{n,m} = K_{n,m}R_n \tag{11}$$

and we observe the following:

Lemma 4.2. *The maximum number of q_i's occuring in any Sturm query in $\Gamma_{n,m}$ is $\lfloor \log_2(m) \rfloor$.*

4.4. Sign-determination algorithm

The input is the polynomial $p(x)$, and polynomials $q_1(x), \ldots, q_n(x)$ with rational coefficients, and of degree at most d. The output is the set of sign sequences $\{(q_1(\alpha_j), \ldots, q_n(\alpha_j))\}$ at the $m \leq d$ real roots α_j of $p(x)$. Note that the set of sign sequences is an unordered set, and the algorithm does not output the sign sequences in the order of the α_j as real numbers.

1. We assume at the ith step that the algorithm knows for each sign sequence of $q_1, \ldots, q_{i-1}$, how many roots of $p(x)$ produce this sign sequence. Most of these are zero, and the algorithm only stores the sign sequences with at least one root, and the number of these is $m_{i-1} \leq d$.

2. There are $3m_{i-1}$ possible sign sequences for $q_1, \ldots, q_i$, and each of these defines a column of the matrix $K_{i,3m_{i-1}}$. These columns are linearly independent, so there are $3m_{i-1}$ rows which together with the specified columns, define a square submatrix J_i of $K_{i,3m_{i-1}}$.

3. We solve the $3m_{i-1} \times 3m_{i-1}$ system corresponding to this matrix, to find the actual sign sequences of $q_1, \ldots, q_i$, and repeat the above steps for $i = 1, \ldots, n$.

The first task then, is to give a procedure that accepts a list of m columns of the matrix $K_{n,m}$, and returns a list of m rows, such that the resulting $m \times m$ matrix is non-singular. This procedure needs to run in polynomial time in n and m and not the size of $K_{n,m}$.

The second task, which is very easy, is to determine the entries of this submatrix of $K_{n,m}$. Again this must be in polynomial time in m, so we cannot afford to construct all of $K_{n,m}$. This task reduces to determining the value of a single element of $K_{n,m}$ given its row and column indices.

Both steps can be solved with simple recursive procedures whose running time is $O(nm)$ and $O(n)$ respectively. Refer to [Can91b] for details.

4.5. Complexity analysis

Let m denote the number of real roots of $p(x)$, then m is not more than d, the degree of $p(x)$. The algorithm takes n steps to compute the signs of all q_i's and each step involves inverting a $3m \times 3m$ matrix and the calculation of $3m$ Sturm queries. Each Sturm query involves $p(x)$ and a product of at most $2\log_2(m)$ of the q_i's. The number of arithmetic operations for each Sturm query is $O(d \log m \log^2(d \log m))$, which simpifies to $O(d \log m \log^2 d)$ because d is at least as large as m. So the total time per step for the Sturm calculation is

$$O(md \log m \log^2 d) \tag{12}$$

or $O(d^2 \log^3 d)$ in the worst case when the number of roots is roughly d. The cost of inverting the matrix is $O(m^{2.376})$ and so the overall running time is:

$$O(n(md \log m \log^2 d + m^{2.376})). \tag{13}$$

The cost of inverting the matrix dominates if m is close to d. However, the actual number of real roots is often much less than d, and is $O(\log d)$ for a random polynomial. So the Sturm query time will often be the larger in practice, especially if naive algorithms are used.

The improvement over the original BKR algorithm is that we have replaced a factor of n, being the maximum number of polynomials in a Sturm query, with $\log m$. Since the actual number of real roots of a polynomial is small compared to its degree, the complexity of BKR without this improvement would typically be n^2d times some log factors. With the improvement, we typically get nd times some logs.

5. Computing with infinitesimals

In the algorithms of [HRS90], [Ren92], [GV92] and [Can91a], various singularities are dealt with by perturbing the input polynomials with infinitesimals. This moves the problem away from the singularity, and when done carefully preserves the important properties (like connectivity or non-emptiness) of the input. Computations with an infinitesimal ϵ are done in the rational field $\mathbb{Q}(\epsilon)$. That is, each number a or b in this extension field is a rational function (a quotient of polynomials) in ϵ. To perform arithmetic, we use the usual rules for arithmetic on rational functions. To determine the sign of such an element, we exclusive-or the signs of its numerator and denominator, which are polynomials in ϵ. To determine the sign of a polynomial in ϵ, we use the sign of the lowest degree non-zero coefficient.

But it is very expensive to compute with explicit rational functions. For example, in the extension $\mathbb{R}(\mu, \epsilon, \delta, \rho)$ that we have been using, an element of degree 10 would have several hundred coefficients. But the sign of the element, which is all we need for the sign-determination algorithm, is determined by just one of these coefficients. This element is the lowest degree element under the lexicographic ordering $\mu \prec \epsilon \prec \delta \prec \rho$

If we knew that this element was say $\mu^4\epsilon\delta^2\rho$, we could find it by computing modulo the ideal $(\mu^5, \epsilon^2, \delta^3, \rho^2)$, which effectively discards higher-degree terms. Since we dont know the degree, we would have to do some search, gradually increasing degree until we obtain a non-zero term. Rather than doing this repeatedly, we can obtain the lowest degree term by differenting a straight-line program.

5.1. Straight-line programs

Suppose we have computed an element $a \in \mathbb{R}$ from some other real values $b_1, \ldots, b_m$ via a series of arithmetic operations. For example, such a could be a coefficient of $P(s)$ described in section 3, or a coefficient of one of the Sturm query polynomials. We can represent a as a straight-line program (DAG whose vertices represent the arithmetic operations) rooted at the values $b_1, \ldots, b_m$. Now suppose that b_1 is specialized to the infinitesimal value ϵ, and that the other b_i take on integer values. We would like to know the sign of a. For simplicity we assume that $a = a(\epsilon)$ is a polynomial in ϵ. This is all we need in our applications.

We could substitute $\epsilon = 0$ and evaluate the straight-line program over the rationals. If we are lucky, $a(0)$ will have a non-zero value, and this gives the sign of $a(\epsilon)$. If not, we can construct a straight-line program for the derivative $\frac{da(\epsilon)}{d\epsilon}$. This has roughly double the size of the original straight-line program. Now evaluating this program at $\epsilon = 0$ gives us a_1, the coefficient of ϵ in $a(\epsilon)$. If this is non-zero, it gives us the sign of $a(\epsilon)$, otherwise we compute the second derivative, and continue. The extra program for the kth derivative is about $k+1$ times the size of the original program, and it uses nodes from the first $k-1$ derivatives. The *total* program size to compute the kth derivative is $\binom{k+2}{2}$ times the original.

This process generalizes easily to multivariate elements, using randomization. For example, to find the sign of $a(\mu, \epsilon, \delta)$ with $\mu \ll \epsilon \ll \delta$, we first substitute random integer values for ϵ and δ. With high probability, this doesnt change the degree of the lowest degree term in μ. Then we apply the procedure above to obtain a straight-line program for the first non-zero derivative at $\mu = 0$. Let

$aa(\mu, \epsilon, \delta)$ denote this derivative. Then $aa(0, \epsilon, \delta)$ is the lowest-degree coefficient of a in μ, times the constant $k_1!$, where k_1 is the order of the derivative.

We iterate the process, and set δ to a random integer, μ to zero, and run the univariate procedure on the straight-line program for aa as a polynomial in ϵ. This gives us a straight-line program for the first non-zero derivative at $\epsilon = 0$, which we denote $aaa(\mu, \epsilon, \delta)$.

Finally, we run the univariate routine on the straight-line program aaa with μ and ϵ both set to zero. Evaluating the resulting program at $\mu = \epsilon = \delta = 0$ gives the sign of the lexicographically first term, which is what we need.

Some simple analysis shows that the straight-line program for computing the sign when the lowest degree term is $\mu^{k_1}\epsilon^{k_2}\delta^{k_3}$ is $k_1^2k_2^2k_3^2$ times the original. More generally we have

Proposition 5.1. *Let $P(\epsilon_1, \ldots, \epsilon_m)$ be a polynomial, represented as a straight-line program of size L. If $\epsilon_1 \prec \cdots \prec \epsilon_m$ are infinitesimals, and if the lexicographically first term in P is $c\epsilon^{k_1} \cdots \epsilon^{k_m}$, then a straight-line program for this term can be constructed having size $\leq Lk_1^2 \cdots k_m^2$, in the same number of steps.*

We claim this method is useful in practice because the k_is are typically small constants, independent of the degree of a in μ, ϵ, δ. Each infinitesimal is used to perturb away from a possibly singular input, and the degree in that infinitesimal is a measure of the multiplicity of the singularity. Where the input is not singular at all, the degree in that infinitesimal will be zero. Most of the time, we expect small multiplicities, and the cost of working over the infinitesimal extension should be only a small constant factor more than integer arithmetic, this factor being the increase in the straight-line program size.

References

[BCR87] J. Bochnak, M. Coste, and M-F. Roy, *Géométrie algébrique réelle*. Ergeb. Math. Grenzgeb. (3) 12. Springer-Verlag, Berlin, 1987.

[Ber75] D. N. Bernstein, The number of roots of a system of equations. *Funktsional. Anal. i Prilozhen.*, 9(3):1–4, Jul-Sep 1975.

[BOKR86] M. Ben-Or, D. Kozen, and J. Reif, The complexity of elementary algebra and geometry. *J. Comput. System Sci.*, 32:251–264, 1986.

[BT71] W.S. Brown and J.F. Traub, On Euclid's algorithm and the theory of subresultants. *Journal of the ACM*, 18:505–514, 1971.

[Can88a] J.F. Canny, *The Complexity of Robot Motion Planning*. M.I.T. Press, Cambridge, 1988.

[Can88b] J.F. Canny. Some algebraic and geometric computations in PSPACE. In: *ACM Symposium on Theory of Computing*, 460–467, 1988.

[Can90] J.F. Canny, Computing roadmaps of semi-algebraic sets. Technical report, University of California, Berkeley, 1990.

[Can91a] J.F. Canny, Computing roadmaps of general semi-algebraic sets. In: *AAECC-91*, New Orleans, 1991.

[Can91b] J.F. Canny, An improved sign determination algorithm. In: *AAECC-91*, New Orleans, 1991.

[CE93] J. Canny and I. Emiris. An efficient algorithm for the sparse mixed resultant. In: *Conference on Algebraic Algorithms and Error-Correcting Codes*, Puerto Rico, 1993.

[EC93] I. Emiris and J. Canny, A practical algorithm for the sparse mixed resultant. In: *International Symposium on Symbolic and Algebraic Computation*, 1993 (submitted).

[GKZ90] I. M. Gel'fand, M. M. Kapranov, and A. V. Zelevinsky, Discriminants of polynomials in several variables and triangulations of Newton polytopes. *Algebra i Analiz*, 2:1–62, 1990. (English Translation in *Leningrad Math. J.* 2 (1991).)

[GV92] D.Y. Grigor'ev and N.N. Vorobjov, Counting connected components of a semialgebraic set in subexponential time. *Comput. Complexity*, 2:133–186, 1992.

[GWDL76] C.G. Gibson, K. Wirthmüller, A.A. Du Plessis, and E.J.N. Looijenga, *Topological Stability of Smooth Mappings*. Lecture Notes in Math. 552, Springer-Verlag, New York, 1976.

[HRS90] J. Heintz, M.F. Roy, and P. Solerno, Complexité du principe de Tarski–Seidenberg. *Bull. Soc. Math. France*, 118:101–126, 1990.

[Kho78] A.G. Khovanskii, Newton polyhedra and the genus of complete intersections. *Funktsional. Anal. Prilozhen.*, 12(1):51–61, Jan-Mar 1978.

[Kus76] A.G. Kushnirenko, Newton polytopes and the Bezout theorem. *Funktsional. Anal. Prilozhen.*, 10(3), Jul-Sep 1976.

[MC92a] D. Manocha and J.F. Canny, The implicit representation of rational parametric surfaces. *J. Symbolic Comput.*, 13:485–510, 1992.

[MC92b] D. Manocha and J.F. Canny, Real time inverse kinematics of general 6R manipulators. In: *IEEE Conference on Robotics and Automation*, 383–389, 1992.

[MC93] D. Manocha and J. Canny, Multipolynomial resultant algorithms. *J. Symbolic Comput.*, 1993. (To appear).

[PS91] P. Pedersen and B. Sturmfels, Product formulas for sparse resultants. Manuscript, 1991.

[Ren92] J. Renegar, On the computational complexity of the first-order theory of the reals, parts I, II, III. *J. Symbolic Comput.*, 13(3):255–352, 1992.

[RS90] M.F. Roy and A. Szpirglas, Complexity of computation on real algebraic numbers. *J. Symbolic Comput.*, 10(1):39–52, 1990.

[SZ93] B. Sturmfels and A. Zelevinsky, Multigraded resultants of sylvester type. *J. Algebra*, 163:115–127, 1994.

[Tar48] A. Tarski, *A Decision Method for Elementary Algebra and Geometry*. University of California Press, Berkeley, 1948.

Thom's first isotopy lemma: a semialgebraic version, with uniform bound

Michel Coste and Masahiro Shiota

Abstract. We show that, if the data in Thom's first isotopy lemma are semialgebraic, then one can construct a *semialgebraic* trivialization (without integrating a vector field). Moreover one has a uniform recursive bound on the degree of this semialgebraic trivialization, in terms of the complexity of the data.

1991 Mathematics Subject Classification: 14P10, 58A35, 32S60, 14P20.

1. Introduction and statement of the results

Our aim in this paper is to construct semialgebraic isotopies. Let us introduce some notations for the statement of our results, which will also be used throughout the paper.

We work over an arbitrary real closed field R. Let $p : X \to R^l$ be an application. If Y is a subset of X, we shall use $Y(0)$ to denote $Y \cap p^{-1}(0)$. A *trivialization* of p (over R^l) is a homeomorphism

$$h = (h_0, p) : X \longrightarrow X(0) \times R^l$$

such that $h_0(x) = x$ for all $x \in X(0)$. The trivialization h is said to be *compatible* with $Y \subset X$ when

$$h(Y) = Y(0) \times R^l.$$

In the following, $X \subset R^n$ will always be a locally closed semialgebraic subset, together with a finite semialgebraic stratification satisfying the regularity conditions a and b of Whitney. We denote by X^i the union of the i-th dimensional strata. It is a Nash manifold, in general not connected. It will be more convenient to deal with it as if it were *one* stratum.

Theorem 1. *Let p be a Nash mapping from a neighborhood of X in R^n to R^l. We suppose that the restriction of p to X is proper, and that the restriction of p to each stratum X^i is a Nash submersion. Then there is a semialgebraic trivialization of p over R^l, compatible with the strata X^i,*

$$h = (h_0, p_{|X}) : X \longrightarrow X(0) \times R^l$$

such that h restricted to X^i is a Nash diffeomorphism onto $X^i(0) \times R^l$, for any i.

So in this theorem we are in the situation of Thom's first isotopy lemma, with semialgebraic data (cf. for instance [G-al]). But, even in this case, Thom's first isotopy lemma does not give a semialgebraic trivialization of p. The reason is that the main tool used is the integration of a (controlled) vector field, and this does not produce a semialgebraic result. On the other hand, there is a very useful result about semialgebraic trivialization, due to Hardt [H]. In the situation of Theorem 1, it gives semialgebraic trivializations of p over a finite semialgebraic partition of R^l, but nothing is said about what happens when passing from one piece of this partition to an other.

It may be useful to have semialgebraic isotopies, since being semialgebraically homeomorphic is stronger than being homeomorphic (for compact polyhedra, it is the same as being PL isomorphic, cf. [SY]). Also, we prove the result over any real closed field, and as a consequence we get a uniform bound for the degree of the trivialization in terms of the "complexity" of the data. The *degree* of a semialgebraic set

$$S = \bigcup_{j=1}^{m} \{x \in R^n \ ; \ f_{j,0}(x) = 0, f_{j,1}(x) > 0, \ldots, f_{j,l_j} > 0\}$$

is defined to be the sum $\sum_{j,k} \deg(f_{j,k})$ of the degrees of the polynomials appearing in its description (so one should rather speak of the degree of the description). The degree of a semialgebraic mapping is the degree of its graph.

Theorem 2. *There is a recursive function $\chi(n,d)$ such that, if the degrees of p and of the strata X^i are bounded by d, then in Theorem 1 we can find a trivialization h with degree at most $\chi(n,d)$.*

The main tool we use to replace integration of vector fields is the Nash trivialization of Nash proper submersions from [CS]. We recall this result. First we introduce a notation. A function is of class $\mathcal{S}^r$ when it is semialgebraic and of class C^r, for some integer r. We shall always suppose that $r > 1$. With this we can define $\mathcal{S}^r$ manifolds. For more details we refer to [S], where $\mathcal{S}^r$ is called "C^r Nash".

Theorem (Nash trivialization). *Let M be a Nash (resp. $\mathcal{S}^r$) manifold, and let $p : M \longrightarrow R$ be a proper surjective Nash (resp. $\mathcal{S}^r$) submersion. Then there is a trivialization of p*

$$h = (h_0, p) : M \longrightarrow M(0) \times R$$

such that h is a Nash (resp. $\mathcal{S}^r$) diffeomorphism.

In Section 2, we extend this result to the case of a manifold with boundary, with p a submersion onto R^l, as well as its restriction to the boundary. In Section 3 we explain how to construct semialgebraic control data when we have a semialgebraic stratification satisfying Whitney's regularity conditions. Section 4 is devoted to the construction of a semialgebraic isotopy, going up in the dimensions of the strata. This is very like the construction of a controlled vector field in the usual proof of

Thom's first isotopy lemma, but with extra complications which come from the fact that it is more difficult to glue trivializations than to glue vector fields. The glueing procedure works only in the $\mathcal{S}^r$ category, so we have to use approximation in Section 6 to get trivializations which are Nash on strata. Before that, we derive in Section 5 the semialgebraic local triviality along strata from the first isotopy lemma. The last section is devoted to the proof of Theorem 2 on uniform recursive bounds.

2. The case of a manifold with boundary

In this section, $\overline{M}$ is an $\mathcal{S}^r$ manifold with boundary N and interior M. We want to extend the Nash trivialization theorem of [CS] to the following result.

Theorem 3. *Let $p : \overline{M} \to R^l$ be a proper $\mathcal{S}^r$ submersion, such that the restriction of p to the boundary N is also a submersion. Then there is a trivialization of p which is an $\mathcal{S}^r$ diffeomorphism $h = (h_0, p) : \overline{M} \to \overline{M}(0) \times R^l$.*

Remark. If $\overline{M}$ and p are Nash, then using the approximation theorems in [S] (cf. Theorem VI.2.2 and Lemma VI.2.3, which hold on any real closed field although the argument in [S] works only for $\mathbb{R}$), we can actually transform the trivialization h into a Nash trivialization.

We shall need in the following $\mathcal{S}^r$ tubular neighborhoods of an $\mathcal{S}^r$ manifold A, with retractions compatible with an $\mathcal{S}^r$ mapping which is submersive on A.

Lemma 4. *Let A be an $\mathcal{S}^r$ submanifold of R^n, and let f be an $\mathcal{S}^r$ mapping from a semialgebraic neighborhood of A to an $\mathcal{S}^r$ manifold B. Suppose that the restriction of f to A is submersive. Then there exist a smaller semialgebraic neighborhood T of A and an $\mathcal{S}^r$ retraction $\pi : T \to A$ such that $f \circ \pi = f_{|T}$.*

Proof. Let a be the dimension of A, and b the dimension of B. We have an $\mathcal{S}^{r-1}$ mapping $\phi : A \to \mathbb{G}_{n,n-a+b}(R)$ which associates to a point $x \in A$ the normal space to $A \cap f^{-1}(f(x))$ at x in R^n. We can approximate ϕ by an $\mathcal{S}^r$ mapping ψ. Consider now the vector bundle

$$E = \{(x, \xi) \in A \times R^n \ ; \ \xi \in \psi(x)\}$$

with projection $\eta : E \to A$. If ψ is close enough to ϕ, the mapping

$$\begin{aligned} E &\longrightarrow R^n \times B \\ (x, \xi) &\longmapsto (x + \xi, f(x)) \end{aligned}$$

induces an $\mathcal{S}^r$ diffeomorphism, say Θ, from an open semialgebraic neighborhood of the zero section in E onto an open semialgebraic neighborhood of the graph of $f_{|A}$ in $R^n \times B$. Then there is a semialgebraic neighborhood T of A in R^n on which the formula

$$\pi(x) = \eta \circ \Theta^{-1}(x, f(x))$$

defines an $\mathcal{S}^r$ retraction $\pi : T \to A$, with $f \circ \pi = f$. □

Now we come back to the situation of Theorem 3. We can suppose, replacing p by its graph, that $\overline{M} \subset R^n \times R^l$ and that p is the restriction of the projection $R^n \times R^l \to R^l$ (which we also call p). We begin, using Lemma 4, by choosing a tubular neighborhood T of N in $R^n \times R^l$ with an $\mathcal{S}^r$ retraction $\pi : T \to N$ with $p \circ \pi = p_{|T}$.

We can find an $\mathcal{S}^r$ function τ on $\overline{M}$, which is 0 on N, positive on M, and submersive in a neighborhood of N. There is a neighborhood W of N in $\overline{M} \cap T$ and a strictly positive $\mathcal{S}^r$ function ϕ on N such that $x \mapsto (\pi(x), \tau(x))$ is an $\mathcal{S}^r$ diffeomorphism onto the set of $(y,t) \in N \times R$ such that $0 \le t < \phi(y)$. Define $\rho : W \to [0,1[$ by $\rho(x) = \tau(x)/\phi(\pi(x))$. Then $\rho^{-1}(0) = N$ and $(\pi, \rho)_{|W}$ is an $\mathcal{S}^r$ diffeomorphism from W onto $N \times [0,1[$.

We shall prove the following strengthening of the theorem.

Proposition 5. *With (T, π, W, ρ) as above, we can choose the $\mathcal{S}^r$ trivialization h of Theorem 3 so that $\rho(h_0(x)) = \rho(x)$ and $\pi(h_0(x)) = h_0(\pi(x))$ for any $x \in W$ with $0 \le \rho(x) < \frac{1}{2}$.*

The proof of the proposition is by induction on l. We first prove the induction step, and then we shall prove the result for $l = 1$.

Before the statement of the next lemma, we introduce some notation. We suppose that $l > 1$ and we denote by

$$p = (p', p_l) : R^n \times R^l \longrightarrow R^l = R^{l-1} \times R$$
$$q : R^n \times R^{l-1} \longrightarrow R^{l-1}$$

the projections. If S is any subset of $R^n \times R^l$, we set

$$S' = \{x' \in R^n \times R^{l-1} \; ; \; (x', 0) \in S\}.$$

If $\delta : R^{l-1} \to R$ is a strictly positive function, we set

$$S' \times]-\delta, \delta[= \{(x', t) \in S' \times R \; ; \; |t| < \delta(q(x'))\},$$
$$S_\delta = \{x \in S \; ; \; |p_l(x)| < \delta(p'(x))\}.$$

Lemma 6. *We keep the hypotheses and the notation of Proposition 5. There is a strictly positive Nash function $\delta : R^{l-1} \to R$ and an $\mathcal{S}^r$ diffeomorphism $g = (g', p_l) : \overline{M}_\delta \to \overline{M}' \times]-\delta, \delta[$ such that $p \circ g = p_{|\overline{M}_\delta}$ and that $\rho(x) = \rho(g'(x), 0)$ for all $x \in (W)_\delta$ such that $0 \le \rho(x) < \frac{2}{3}$.*

Proof. Choose a strictly positive Nash function $\delta : R^{l-1} \to R$ such that T contains $N' \times]-\delta, \delta[$ and that $\pi_{|N' \times]-\delta,\delta[}$ is an $\mathcal{S}^r$ diffeomorphism onto N_δ. Call

$$\nu = (\nu', p_l) : N_\delta \longrightarrow N' \times]-\delta, \delta[$$

the inverse diffeomorphism. Then we get an $\mathcal{S}^r$ diffeomorphism

$$\gamma = (\gamma', p_l) : W_\delta \longrightarrow W' \times]-\delta, \delta[,$$

defined by $\rho(\gamma'(x),0)=\rho(x)$ and $\pi(\gamma'(x),0)=(\nu'(\pi(x)),0)$.

Now choose, using Lemma 4, a tubular neighborhood U of M in $R^n\times R^l$ with an $\mathcal{S}^r$ retraction μ such that $p\circ\mu=p$ on U. Let us consider

$$\Lambda = M\setminus\{x\in W\ ;\ \rho(x)<\frac{1}{3}\}.$$

If δ is sufficiently small, we can suppose that $\Lambda'\times]-\delta,\delta[$ is contained in U, that μ induces an $\mathcal{S}^r$ imbedding

$$\psi:\Lambda'\times]-\delta,\delta[\longrightarrow M_\delta,$$

and also that the image of ψ contains $M_\delta\setminus\{x\in W_\delta\ ;\ \rho(x)<\frac{1}{2}\}$. Call $\phi=(\phi',p_l)$ the inverse of ψ, defined on the image of ψ. Now we proceed to gluing γ and ϕ. For that, choose an increasing $\mathcal{S}^r$ function $\lambda:R\to R$ which is 0 on a neighborhood of $]-\infty,\frac{2}{3}]$ and 1 on a neighborhood of $[1,+\infty[$. Define

$$g:\overline{M}_\delta\longrightarrow\overline{M}'\times]-\delta,\delta[$$

by

$$g(x)=\begin{cases}\gamma(x) & \text{if } x\in W_\delta \text{ and } \rho(x)\le\frac{2}{3},\\ \phi(x) & \text{if } x\notin T_\delta,\\ (\phi'\circ\gamma^{-1}(\gamma'(x),\lambda(\rho(x))\cdot p_l(x)),p_l(x)) & \text{if } x\in W_\delta \text{ and } \rho(x)\ge\frac{2}{3}.\end{cases}$$

Remark that $g(x,0)=(x,0)$ for $x\in\overline{M}'$, and that $p_l\circ g=p_l$. So g is actually a diffeomorphism from $\overline{M}_\delta$ onto $\overline{M}'\times]-\delta,\delta[$, if δ has been chosen small enough. The compatibility with p comes from the similar compatibility for γ and ϕ, and the compatibility with ρ comes from the similar compatibility for γ. □

We use in the following proofs a little of the real spectrum, actually only the real spectrum $\widetilde{R}$ of the affine line. A point α of $\widetilde{R}$ may be identified with an ultrafilter $\mathcal{U}_\alpha$ of semialgebraic subsets of R. To a semialgebraic subset $S\subset R$ corresponds the subset $\widetilde{S}\subset\widetilde{R}$ of those α such that $S\in\mathcal{U}_\alpha$. To $\alpha\in\widetilde{R}$ is associated a real closed extension $\kappa(\alpha)$ of R, which is the inductive limit of the rings of semialgebraic continuous functions on S, for $\alpha\in\widetilde{S}$. If A is a semialgebraic subset of $R^q\times R$, its fiber $A_\alpha\subset\kappa(\alpha)^q$ consists of the q-uples represented by semialgebraic continuous functions $(f_1,\dots,f_q)$ such that for all x in S, $(f_1(x),\dots,f_q(x))\in A$, for some S such that $\alpha\in\widetilde{S}$. The fiber of a semialgebraic function at α may be defined through the fiber of its graph. The philosophy is that something (expressible by a formula of the theory of real closed fields) holds in the fiber at α if and only if it holds over some S with $\alpha\in\widetilde{S}$. For more details we refer to [BCR] 7.4 and 9.6, and to [CS].

Proof of the induction step from $l-1$ to l, for $l>1$. Consider $p:\overline{M}\to R^l=R^{l-1}\times R$. We can see this situation as parametrized by R, and then for $\alpha\in\widetilde{R}$ we have a proper submersion $p_\alpha:\overline{M}_\alpha\to\kappa(\alpha)^{l-1}$, whose restriction to the boundary N_α is also a submersion, and we have also $(T_\alpha,\pi_\alpha,W_\alpha,\rho_\alpha)$. The induction hypothesis gives us an $\mathcal{S}^r$ trivialization $\overline{M}_\alpha\to\overline{M}_\alpha(0)\times\kappa(\alpha)^{l-1}$, compatible with ρ_α near N_α as stated in Proposition 5. This means that there is a semialgebraic subset $S \subset R$, with $\alpha \in \widetilde{S}$, and an $\mathcal{S}^r$ diffeomorphism

$$\overline{M} \cap p^{-1}(R^{l-1} \times S) \to (\overline{M} \cap p^{-1}(\{0\} \times S)) \times R^{l-1},$$

compatible with ρ near N. If S is an open interval, the case $l = 1$ of Proposition 5 gives us an $\mathcal{S}^r$ trivialization of p over $\{0\} \times S$, compatible with ρ near N, and so we have an $\mathcal{S}^r$ trivialization of p over $R^{l-1} \times S$. Taking into account the compactness of $\widetilde{R}$, we find a finite number of points $a_1 < \ldots < a_s$ in R, with an $\mathcal{S}^r$ trivialization of p, compatible with ρ near N over any band $R^{l-1} \times]a_i, a_{i+1}[$ for $i = 0, \ldots, s$ ($a_0 = -\infty$ and $a_{s+1} = +\infty$). We also have an $\mathcal{S}^r$ trivialization of p, compatible with ρ near N, over any hyperplane $\{a_i\} \times R^{l-1}$, $i = 1, \ldots, s$. Applying Lemma 6, we get strictly positive Nash functions $\delta_i : R^{l-1} \to R$ such that we can extend the latter trivializations to $\mathcal{S}^r$ trivializations over $\{(u,t) \in R^{l-1} \times R \; ; \; |t - a_i| < \delta_i(u)\}$, keeping the compatibility with ρ near N. Then we have to glue all these trivializations. Let us see for instance how to glue a trivialization

$$g_1 = (g_{10}, p) : \overline{M} \cap p^{-1}(R^{l-1} \times]0,1[) \longrightarrow P \times R^{l-1} \times]0,1[$$

with a trivialization

$$g_2 = (g_{20}, p) : \overline{M}_\delta \longrightarrow P \times R^{l-1} \times]-\delta, \delta[.$$

Here $\delta : R^l \to R$ is, as for Lemma 6, a strictly positive Nash function, $P = \overline{M} \cap p^{-1}(0, \frac{1}{3}\delta(0))$ and $(g_{10})_{|P}$ and $(g_{20})_{|P}$ are both the identity. Choose $\beta : R^{l-1} \times R \to R^{l-1} \times R$ an $\mathcal{S}^r$ mapping such that
- $\beta(u,t) = (0, \frac{1}{3}\delta(0))$ for $t \leq \frac{1}{3}\delta(u)$,
- $\beta(u,t) = (u,t)$ for $t \geq \frac{2}{3}\delta(u)$ and,
- in the other cases, $\beta(u,t) \in R^{l-1} \times]0, \delta[$.

Now define the trivialization

$$g : \overline{M} \cap p^{-1}(R^{l-1} \times]-\delta, 1[) \longrightarrow P \times R^{l-1} \times]-\delta, 1[$$

by

$$g(x) = \begin{cases} g_2(x) & \text{if } p_l(x) \leq \frac{1}{3}\delta(p'(x)), \\ g_1(x) & \text{if } p_l(x) \geq \frac{2}{3}\delta(p'(x)), \\ (g_{10} \circ g_2^{-1}(g_{20}(x), \beta(p(x))), p(x)) & \text{in the other cases.} \end{cases}$$

□

Proof of Proposition 5 *in case* $l = 1$. Consider $p : \overline{M} \to R$ as a family of compact $\mathcal{S}^r$ manifolds with boundary, parametrized by R. For any point α of the real spectrum $\widetilde{R}$, $\overline{M}_\alpha$ is a compact $\mathcal{S}^r$ submanifold of $\kappa(\alpha)^n$ with boundary N_α. Theorem 3.1 in [CS] tells us that M_α is $\mathcal{S}^r$ diffeomorphic to the extension to $\kappa(\alpha)$ of an $\mathcal{S}^r$ manifold A defined over R. Then the results contained in [CS], Section 4 (valid on any real closed field), about existence and uniqueness of the compactification of affine $\mathcal{S}^r$ manifolds without boundary, give us a compact affine $\mathcal{S}^r$ manifold $\overline{A}$ with boundary B, defined over R, and an $\mathcal{S}^r$ function τ on a neighborhood of B in $\overline{A}$, also defined over R, such that there is an $\mathcal{S}^r$ diffeomorphism θ from $\overline{M}_\alpha$ to the extension $\overline{A}_{\kappa(\alpha)}$ of $\overline{A}$ to $\kappa(\alpha)$, such that $\tau_{\kappa(\alpha)} \circ \theta = \rho_\alpha$ on $\rho_\alpha^{-1}([0, \frac{2}{3}[)$. This means that we have a semialgebraic subset $S \subset R$, with $\alpha \in \widetilde{S}$, and an $\mathcal{S}^r$ trivialization of p over S, compatible with ρ near N. Taking into account the compactness of $\widetilde{R}$, we get a finite number of points $a_1 < \ldots < a_s$ in R, with an $\mathcal{S}^r$ trivialization of p over any interval $]a_i, a_{i+1}[$ for $i = 0, \ldots, s$ ($a_0 = -\infty$ and $a_{s+1} = +\infty$). Following Lemma 6 (where we set $l = 1$), we have, for $i = 0, \ldots, s$, an $\mathcal{S}^r$ trivialization of p over an open interval $]a_i - \delta\, , \, a_i + \delta[$, compatible with ρ near N. Now we have to glue these trivializations in order to obtain one over the whole of R. We proceed as above (it is actually simpler). □

Remark. The trivialization h that we have constructed is compatible with ρ. Recall that ρ was constructed from some function τ, before Proposition 5; it is not hard to modify h, using the diffeomorphism from W onto $N \times]0, 1[$, in order that it is compatible with the original τ in a neighborhood of N.

It is possible to extend a trivialization of p already given in a neighborhood of the boundary.

Proposition 7. *Let $\overline{M}$, N and $p : \overline{M} \to R^l$ be as in Theorem 3. Suppose that there is a semialgebraic neighborhood W of N in $\overline{M}$ and an $\mathcal{S}^r$ trivialization*

$$\sigma = (\sigma_0, p) : W \longrightarrow W(0) \times R^l.$$

Then it is possible to find an $\mathcal{S}^r$ trivialization

$$h = (h_0, p) : \overline{M} \longrightarrow \overline{M}(0) \times R^l$$

which coincides with σ on a smaller neighborhood $W' \subset W$, such that $\sigma(W') = W'(0) \times R^l$.

Proof. Up to replacing W by a smaller neighborhood, we can find an $\mathcal{S}^r$ function $\rho : W \to [0, 1[$, with $\rho^{-1}(0) = N$, and an $\mathcal{S}^r$ retraction $\pi : W \to N$, such that (π, ρ) is an $\mathcal{S}^r$ diffeomorphism from W onto $N \times [0, 1[$, that ρ is compatible with σ, and that $\pi \circ \sigma_0 = \sigma_0 \circ \pi$. Indeed, one may first choose π and ρ on the fiber $p^{-1}(0)$ as was done before Proposition 5, and then push them using the trivialization σ. Now apply Proposition 5 to the $\mathcal{S}^r$ manifold $\overline{M}_1 = M \setminus \rho^{-1}([0, \frac{1}{2}[)$ with boundary $N_1 = \rho^{-1}(\frac{1}{2})$, with diffeomorphism

$$(\pi_1, \rho_1) : \overline{M}_1 \cap W \longrightarrow N_1 \times [0, 1[$$

defined by $\pi \circ \pi_1 = \pi$ and $\rho_1 = 2(\rho - \frac{1}{2})$. We get a trivialization $h_1 = (h_{1,0}, p) : \overline{M}_1 \to \overline{M}_1(0) \times R^l$ of p, which is compatible with ρ on $\rho^{-1}([\frac{1}{2}, \frac{3}{4}])$. Let $\lambda : R \to [0, 1]$ be an increasing $\mathcal{S}^r$ function which is 0 in a neighborhood of $]-\infty, \frac{1}{2}]$ and 1 in a neighborhood of $[\frac{3}{4}, +\infty[$. The trivialization h of p defined by

$$h(x) = \begin{cases} \sigma(x) & \text{if } x \in \rho^{-1}([0, \frac{1}{2}]) \\ h_1(x) & \text{if } x \notin \rho^{-1}([0, \frac{3}{4}]) \\ (h_{1,0} \circ \sigma^{-1}(\sigma_0(x), \lambda(\rho(x))p(x)), p(x)) & \text{otherwise} \end{cases}$$

does the job. □

We end this section by proving a very special case of the second isotopy lemma in a semialgebraic version.

Theorem 8. *Let P be a Nash manifold, and let $\overline{M}$ be a Nash manifold with boundary N. Let*

$$(\pi, p) : \overline{M} \longrightarrow P \times R^l$$

be a proper S^r submersion, which is also submersive in restriction to the boundary N. Denote as usual $\overline{M}(0) = \overline{M} \cap p^{-1}(0)$. Then there is an S^r diffeomorphism

$$h = (h_0, p) : \overline{M} \longrightarrow \overline{M}(0) \times R^l$$

such that h_0 is the identity on $\overline{M}(0)$ and that $\pi \circ h_0 = \pi$. Suppose moreover that τ is a non negative S^r function on a neighborhood of N in $\overline{M}$, with 0 as regular value and such that $\tau^{-1}(0) = N$. Then we can ask that $\tau \circ h_0 = \tau$ on a neighborhood of N in $\overline{M}$.

Proof. We can find a finite covering of P by semialgebraic opens which are S^r diffeomorphic to the affine space R^k, and we prove the result by induction on the number of opens in the cover. So we are reduced to the case where $P = U \cup V$, U and V semialgebraic opens in P, such that the conclusion of the theorem holds for $\pi^{-1}(U) \cap \overline{M} \to U \times R^l$ and that V is S^r diffeomorphic to R^k.

So we have $g = (g_0, p) : \pi^{-1}(U) \to \pi^{-1}(U)(0) \times R^l$, an S^r diffeomorphism such that g_0 is the identity on $\pi^{-1}(U)(0)$, that $\pi \circ g_0 = \pi$, and that $\tau \circ g_0 = \tau$ in a neighborhood of $N \cap \Pi^{-1}(U)$. On the other hand, according to Proposition 5, we have an S^r diffeomorphism

$$\Theta = (\Delta, \pi, p) : \pi^{-1}(V) \longrightarrow Q \times V \times R^l$$

where $Q = (\pi, p)^{-1}(v, 0)$ for some $v \in V$, such that Δ is the identity on Q and that $\tau \circ \Delta = \tau$ in a neighborhood of $N \cap \pi^{-1}(V)$ in $\pi^{-1}(V)$ (for this last part, see the remark after the proof of Proposition 5). We call Θ_0 the S^r diffeomorphism

$$\Theta_0 = (\Delta, \pi)_{|\pi^{-1}(V)(0)} : \pi^{-1}(V)(0) \longrightarrow Q \times V.$$

From these data we build an S^r diffeomorphism

$$\begin{aligned} Q \times (U \cap V) \times R^l &\longrightarrow Q \times (U \cap V) \times R^l \\ (\xi, y, t) &\longmapsto (\tau_{y,t}(\xi), y, t) \end{aligned}$$

defined by $\tau_{y,t}(\xi) = \Delta(g_0(\Theta^{-1}(\xi, y, t)))$. Note that $\tau_{y,0}$ is the identity of Q, for any y. Now let $(\phi, 1 - \phi)$ be an S^r partition of unity on P, subordinate to the covering (U, V). Define h_0 on $\pi^{-1}(U \cap V)$ by

$$h_0(x) = \Theta_0^{-1}\left(\tau_{y,\phi(y)t}(\Delta(x)), y\right) \quad \text{where} \quad y = \pi(x) \quad \text{and} \quad t = p(x).$$

If $\phi = 0$ we have $h_0(x) = \Theta_0^{-1}(\Delta(x), \pi(x))$ and if $\phi = 1$ we have $h_0(x) = \Theta_0^{-1}(\Delta(g_0(x)), \pi(x)) = g_0(x)$. So we can extend h_0 to X by

$$h_0(x) = \begin{cases} \Theta_0^{-1}(\Delta(x), \pi(x)) & \text{when } \pi(x) \in V \setminus U, \\ g_0(x) & \text{when } \pi(x) \in U \setminus V. \end{cases}$$

Then it is easy to check that $h = (h_0, p)$ is the desired diffeomorphism. □

3. Semialgebraic control data

We consider a semialgebraically stratified set $X = \bigcup_i X^i \subset R^n$ as above, satisfying Whitney's regularity conditions. Let ρ_i be the square of the distance to X^i. The regularity conditions imply that there is a semialgebraic open neighborhood T_i of X^i in R^n such that

i) ρ_i is a Nash function on T_i,

ii) $\rho_{i|T_i \cap X^j}$ is a submersion for any $j > i$,

iii) for any $j > i$, if $(x, \tau) \in X^i \times \mathbb{G}_{n,j-1}(R)$ is in the adherence of the graph of the mapping

$$\begin{aligned} T_i \cap X^j &\longrightarrow \mathbb{G}_{n,j-1}(R) \\ y &\longmapsto \ker(d_y \rho_{i|T_i \cap X^j}), \end{aligned}$$

then the tangent space $T_x X^i$ is contained in τ.

The point which will be useful for us is the following: suppose that we have an $\mathcal{S}^r$ retraction π of T_i onto X^i; then, on some smaller neighborhood $T' \subset T_i$, the mapping $(\pi, \rho_i)_{|T' \cap X^j}$ is a submersion to $X^i \times R$, for any $j > i$ (this is true more generally for C^1 retractions without the hypothesis of semialgebraicity, as shown by Mather [M]). Indeed, if it were not the case, there would be a semialgebraic path $\gamma : [0, 1[\rightarrow X$ with $\gamma(0) = x \in X^i$ and $\gamma(]0, 1[) \subset X^j$, such that $(\pi, \rho_i)_{|X^j}$ is a submersion at $\gamma(t)$ for no $t \in]0, 1[$, i.e., $d_{\gamma(t)}\pi$ restricted to $T_{\gamma(t)} X^j \cap \ker d_{\gamma(t)}\rho$ is never surjective onto $T_{\pi(\gamma(t))} X^i$. Let τ be the limit of $T_{\gamma(t)} X^j \cap \ker d_{\gamma(t)}\rho$ as $t \rightarrow 0$. By the semicontinuity of rank, $d_x \pi$ restricted to τ cannot be surjective onto $T_x X^i$. This is absurd, since by the regularity condition $\tau \supset T_x X^i$ and $d_x \pi$ is the identity on $T_x X^i$.

Remark. We could have started with a stratification satisfying the condition c of [B]. This condition asks that for each X^i, there is a semialgebraic neighborhood T_i in R^n and an $\mathcal{S}^r$ *control function* $\rho_i : T_i \rightarrow [0, +\infty[$ such that $\rho_i^{-1}(0) = X^i$, satisfying the conditions ii) and iii) above. Theorem 1 can be proved in this situation by the same proof as below. For Theorem 2, one has to take into account the degrees of the control functions.

Suppose now that p is an $\mathcal{S}^r$ mapping from a neighborhood of X to R^l, such that $p_{|X}$ is proper and that $p_{|X^i}$ is submersive for each stratum X^i. We want to have semialgebraic control data compatible with p, which means the following.

Proposition 9. *Up to reducing the semialgebraic neighborhood T_i of X^i (the tube), we can build an $\mathcal{S}^r$ retraction $\pi_i : T_i \to X^i$ for each i, satisfying the following conditions:*
1) *For any $j < i$,*

$$\pi_j(\pi_i(x)) = \pi_j(x)$$
$$\rho_j(\pi_i(x)) = \rho_j(x)$$

whenever both sides are defined.
2) *For any i, $p \circ \pi_i(x) = p(x)$ whenever both sides are defined.*

The construction is done as usual (for instance in [G-al]), once we have the following lemma, which is the continuation of Lemma 4. One proceeds by induction on i. For a fixed i one constructs π_i on the intersection of T_i with a neighborhood of $\bigcup_{j<k<i} X^k$, by downward induction on j; here the f of the following lemma is (π_j, ρ_j). Then one constructs π_i on T_i; here the f is p. □

Lemma 10. *Let A be an $\mathcal{S}^r$ submanifold of R^n, and let $f : U \to B$ be an $\mathcal{S}^r$ mapping defined on a semialgebraic neighborhood $U \subset R^n$ of A. Suppose that $f_{|A}$ is submersive. Suppose that we have an $\mathcal{S}^r$ retraction $\pi_1 : V_1 \to V_1 \cap A$, where V_1 is open in U, such that $f \circ \pi_1 = f$. Let V be an open semialgebraic subset of A whose closure in A is contained in $V_1 \cap A$. Then there exists a neighborhood $T \subset R^n$ of A and an $\mathcal{S}^r$ retraction $\pi : T \to A$ such that $f \circ \pi = f$, that $\pi^{-1}(V) \subset V_1$ and that $\pi = \pi_1$ on $\pi^{-1}(V)$.*

Proof. Recall that in Lemma 4 we built an $\mathcal{S}^r$ vector bundle $\eta : E \to A$ such that the mapping

$$E \longrightarrow R^n \times B$$
$$(x, \xi) \longmapsto (x + \xi, f(x))$$

induces an $\mathcal{S}^r$ diffeomorphism, say Θ, from an open semialgebraic neighborhood of the zero section in E onto an open semialgebraic neighborhood Ω of the graph of $f_{|A}$ in $R^n \times B$.

Let λ be an $\mathcal{S}^r$ function on A, $0 \leq \lambda \leq 1$, which is 1 on a neighborhood of $\operatorname{clos} V$ and 0 on a neighborhood of $X \setminus V_1$. There is a semialgebraic neighborhood T of A in R^n on which the formula

$$\pi_2(y) = \eta \circ \Theta^{-1}(y, f(y))$$

defines an $\mathcal{S}^r$ retraction $\pi_2 : T \to A$, with $f \circ \pi_2 = f$. We can choose T so small that, if $\pi_2(y)$ is in the support of λ, then $y \in V_1$ and $\big(y + \lambda(\pi_2(y))(\pi_1(y) - y), f(y)\big) \in \Omega$. Hence the formula

$$\pi(y) = \eta \circ \Theta^{-1}\big(y + \lambda(\pi_2(y))(\pi_1(y) - y), f(y)\big)$$

defines an $\mathcal{S}^r$ retraction $\pi : T \to A$ with $f \circ \pi = f$. If T is small enough we have $\lambda(\pi_2(y)) = 1$ when $\pi(y) \in V$, hence $\pi^{-1}(V) \subset V_1$ and $\pi = \pi_1$ on $\pi^{-1}(V)$. □

4. Construction of a semialgebraic isotopy

We are now in the situation of Proposition 9, with semialgebraic control data.

Proposition 11. *In this situation we can build a semialgebraic trivialisation* $h = (h_0, p) : X \to X(0) \times R^l$ *of* p *such that* $h_{|X^i}$ *is an* $\mathcal{S}^r$ *diffeomorphism onto* $X^i(0) \times R^l$ *for each stratum* X^i. *This trivialisation will moreover be compatible with the control data, i.e., it will satisfy* $\pi_i(h_0(x)) = h_0(\pi_i(x))$ *and* $\rho_i(h_0(x)) = \rho_i(x)$, *for all* i *and whenever both sides of identities are defined.*

We denote by $X^{\leq i}$ (resp. $X^{<i}$) the union of strata of X of dimension at most i (resp. smaller than i). The construction will be by induction: we suppose that we have already $h_{<i} = (h_{<i,0}, p)$ on $X^{<i}$, and we build $h_{\leq i}$ on $X^{\leq i}$.

First step. We extend $h_{<i}$ to a neighborhood of $X^{<i}$ in $X^{\leq i}$; this we do by downward induction. Consider $k < i$. Suppose that we have already Ω_{k+1}, the intersection of a semialgebraic neigborhood of $X^{<i} \setminus X^{\leq k}$ with X^i, and an $\mathcal{S}^r$ embedding

$$g_{k+1} = (g_{k+1,0}, p) : \Omega_{k+1} \to \Omega_{k+1}(0) \times R^l$$

such that $g_{k+1,0}$ is the identity on $\Omega_{k+1}(0)$ and that the following compatibilities are satisfied, for all j, with $k < j < i$, whenever both sides of the identities are defined:

$$\pi_j(g_{k+1,0}(x)) = h_{<i,0}(\pi_j(x)),$$
$$\rho_j(g_{k+1,0}(x)) = \rho_j(x).$$

We suppose also that we have a positive $\mathcal{S}^r$ function τ_{k+1} on Ω_{k+1}, which extends continuously by 0 to $(X^{<i} \setminus X^{\leq k})(0)$, such that $\tau_{k+1} \circ g_{k+1,0} = \tau_{k+1}$, and such that for any j, with $k < j < i$, if $(x, L) \in X^j \times \mathbb{G}_{n,i-1}$ is in the adherence of the set of $(y, \ker d_y(\tau_{k+1}))$ for $y \in \Omega_{k+1}$, then $L \supset T_x(X^j)$. We want to define Ω_k, g_k and τ_k with the same properties (note that if $k = i - 1$, we have $\Omega_i = \emptyset$). We can choose an integer q and a positive Nash function ε on X^k such that:

1) The set

$$U = \{(x,t) \in (\Omega_{k+1}(0) \cap T_k(0)) \times R^l\ ;\quad \rho_k(x) < \varepsilon(h_{<i}^{-1}(\pi_k(x), t)) \text{ and } \tau_{k+1}(x) < (\rho_k)^q(x)\}$$

is contained in $g_{k+1}(\Omega_{k+1} \cap T_k)$; we denote $V = g_{k+1}^{-1}(U)$.

2) Set, for $x \in V$, $\mu(x) = \tau_{k+1}(x)/(\rho_k)^q(x)$. The mapping (π_k, ρ_k, μ) is a proper submersion from V to $X^k \times]0, \varepsilon[\times]0, 1[$, where $X^k \times]0, \varepsilon[$ denotes the set of $(y, r) \in X^k \times R$ such that $r < \varepsilon(y)$.

3) If $(x, L) \in X^k \times \mathbb{G}_{n,i-1}$ is in the closure of the set of $(y, \ker d_y(\tau_{k+1}))$ for $y \in V$, then $L \supset T_x(X^k)$.

For this we use the following lemma, whose proof will be postponed to the end of the section.

Lemma 12. *Let* A, B *and* C *be disjoint semialgebraic subsets of* R^p, *such that* $A \cup B \cup C$ *is locally closed, that* $B \cup C$ *is closed in* $A \cup B \cup C$, *and that* C *is closed* *in $B\cup C$. Let ρ (resp. τ) be a non negative continuous semialgebraic function on $A\cup B\cup C$ (resp. $A\cup B$) whose zero set is C (resp. B). Let $\mathcal{A}$ be a semialgebraic subset of $A\times R^n$. Set $\mathcal{B}=(\operatorname{clos}\mathcal{A})\cap(B\times R^n)$ and $\mathcal{C}=(\operatorname{clos}\mathcal{B})\cap(C\times R^n)$. Then there exists an integer q such that*

$$\mathcal{C}=\operatorname{clos}\left(\{(x,t)\in\mathcal{A}\subset A\times R^n\ ;\ \tau(x)<\rho^q(x)\}\right)\cap(C\times R^n).$$

We note also that, for $x\in V$, the compatibility conditions

$$\begin{aligned}\pi_k(g_{k+1,0}(x))&=h_{<i,0}(\pi_k(x))\\ \rho_k(g_{k+1,0}(x))&=\rho_k(x)\end{aligned}$$

hold. Indeed, if $x\in T_j$ for some $j, k<j<i$, then

$$\begin{aligned}\pi_k(g_{k+1,0}(x))&=\pi_k(\pi_j(g_{k+1,0}(x)))=\pi_k(h_{<i,0}(\pi_j(x)))\\ &=h_{<i,0}(\pi_k(\pi_j(x))=h_{<i,0}(\pi_k(x)),\end{aligned}$$

and a similar calculation gives the compatibility with respect to ρ_k.

Now consider the $\mathcal{S}^r$ manifold

$$\overline{M}=\{x\in X^i\cap T_k\ ;\ \rho_k(x)<\varepsilon(\pi_k(x))\text{ and }(x\notin V\text{ or }\mu(x)\geq\frac{1}{2})\}$$

with boundary

$$N=\{x\in V\ ;\ \mu(x)=\frac{1}{2}\}.$$

Set $\rho'=\rho_k/(\varepsilon\circ\pi_k)$. The $\mathcal{S}^r$ mapping (π_k,ρ') is a proper submersion onto $X^k\times]0,1[$, which is also submersive in restriction to the boundary N. We know that $h_{<i}$ induces an $\mathcal{S}^r$ diffeomorphism from X^k onto $X^k(0)\times R^l$. So by Theorem 8 there is an $\mathcal{S}^r$ diffeomorphism

$$\Theta=(\theta,\rho',p):\overline{M}\longrightarrow Q\times]0,1[\times R^l,$$

where $Q=\overline{M}(0)\cap\rho'^{-1}(1/2)$, θ is the identity on Q, and $\pi_k\circ\theta=h_{<i,0}\circ\pi_k$. We can moreover suppose that if $x\in\overline{M}\cap V$ and $\mu(x)\leq 2/3$, then $\mu(\theta(x))=\mu(x)$. Let $f_0:\overline{M}\to\overline{M}(0)$ be defined by $\theta(f_0(x))=\theta(x)$ and $\rho_k(f_0(x))=\rho_k(x)$. Then $f=(f_0,p)$ is an $\mathcal{S}^r$ embedding of $\overline{M}$ into $\overline{M}(0)\times R^l$, which satisfies also $\pi_k(f_0(x))=h_{<i,0}(\pi_k(x))$, and $\mu(f_0(x))=\mu(x)$ if $x\in\overline{M}\cap V$ and $\mu(x)\leq 2/3$.

Now we can define Ω_k and g_k. we can suppose that $\overline{M}\cap\Omega_{k+1}=\overline{M}\cap V$, possibly reducing Ω_{k+1}. We set $\Omega_k=\Omega_{k+1}\cup\overline{M}$. The mapping $g_{k,0}:\Omega_k\to\Omega_k(0)$ is defined by

$$g_{k,0}(x)=\begin{cases}g_{k+1,0}(x) & \text{if } x \text{ is in } \Omega_{k+1}\setminus\overline{M},\\ f_0(x) & \text{if } x \text{ is in } \overline{M}\setminus V \text{ or if } x \text{ is in } V \text{ and } \mu(x)\geq 2/3,\\ g_{k+1,0}(y) & \text{if } x \text{ is in } V \text{ and } 1/2\leq\mu(x)<2/3,\end{cases}$$

where in the last case $y\in V\cap\overline{M}$ is such that $f_0(y)=f_0(x)$ and $p(y)=\phi(\mu(x))p(x)$, ϕ an $\mathcal{S}^r$ function which is 1 on a neighborhood of $]-\infty,1/2]$ and 0 on a neighborhood of $[2/3,+\infty[$.

It remains to define τ_k. For that we follow [B], Theorem 2. We choose an $\mathcal{S}^r$ submersive function ψ from $]0,+\infty[\times]0,+\infty[$ to $]0,+\infty[$ which satisfies $\psi(x,y)=x$ if $2x/3 \le y$ and $\psi(x,y)=y$ if $y \le x/2$. Then set

$$\tau_k(x)=\begin{cases}\tau_{k+1}(x) & \text{if } x \text{ is in } \Omega_{k+1}\setminus\overline{M},\\ (\rho_k)^q(x) & \text{if } x \text{ is in } \overline{M}\setminus V \text{ or if } x \text{ is in } V \text{ and } \mu(x)\ge 2/3,\\ \psi((\rho_k)^q(x),\tau_{k+1}(x)) & \text{if } x \text{ is in } V \text{ and } 1/2\le\mu(x)<2/3.\end{cases}$$

One has to check that if $(x,L)\in X^k\times\mathbb{G}_{n,i-1}$ is in the closure of the set of $(y,\ker d_y(\tau_k))$ for $y\in\Omega_k$, then $L\supset T_x(X^k)$. This is done as in [B].

Let us make precise the point where we have arrived.

Lemma 13. *There is an open semialgebraic subset $\Omega\subset X^i$, the intersection of X^i with a neighborhood of $X^{<i}$, and an $\mathcal{S}^r$ embedding $g=(g_0,p):\Omega\longrightarrow\Omega(0)\times R^l$ such that g_0 is the identity on $\Omega(0)$, and that the compatibility conditions $\pi_j(g_0(x))=h_{<i,0}(\pi_j(x))$ and $\rho_j(g_0(x))=\rho_j(x)$ hold for all $j<i$ whenever both sides are defined.* □

Second step. Now we proceed to the construction of $h_{\le i}$. Choose a strictly positive Nash function τ_0 on $X^i(0)$, which can be extended continuously by 0 on $X^{<i}(0)$ (cf. [BCR], 2.7.4). There is a strictly positive Nash function $\phi: R^l\to]0,+\infty[$ such that
1) $\tau_0^{-1}(]0,\phi[):=\{(x,t)\in X^i(0)\times R^l\ ;\ \tau_0(x)<\phi(t)\}\subset g(\Omega)$,
2) $\phi(R^l)$ is contained in an interval $]0,\varepsilon]$ which contains no critical value of τ_0,
3) for all $t\in R^l$, $\phi(t)\le\phi(0)$.
Set $\Omega'=g^{-1}(\tau_0^{-1}(]0,\phi[))$. The second condition on ϕ implies that there is a Nash diffeomorphism

$$\lambda=(\lambda_0,\tau_0):\tau_0^{-1}(]0,\varepsilon])\longrightarrow\tau_0^{-1}(\varepsilon)\times]0,\varepsilon].$$

Choose an $\mathcal{S}^r$ function θ on $\{(c,t)\in R\times R^l\ ;\ 0\le c\le\phi(t)\}$,which is an increasing function of c for fixed t, such that $\theta(c,t)=c$ for $0\le c\le\frac{1}{2}\phi(t)$ and that $\theta(\phi(t),t)=\phi(0)$. Define $\tau:\Omega'\longrightarrow R$ by $\tau(x):=\theta(\tau_0(g_0(x)),p(x))$. Then we have an $\mathcal{S}^r$ diffeomorphism

$$\begin{aligned}\sigma_i:\Omega'&\longrightarrow\tau_0^{-1}(]0,\phi(0)[)\times R^l\subset X^i(0)\times R^l\\ x&\longmapsto(y,p(x))\end{aligned}$$

where y is defined by $\lambda_0(y)=\lambda_0(g_0(x))$ and $\tau_0(y)=\tau(x)$. Remark that σ_i and g coincide on $g^{-1}(\tau_0^{-1}(]0,\frac{1}{2}\phi[))$. Note that (τ,p) is submersive, and so
- $\overline{M}:=X^i\setminus\tau^{-1}(]0,\frac{1}{4}\phi(0)[)$ is an $\mathcal{S}^r$ manifold with boundary $N:=\tau^{-1}(\frac{1}{4}\phi(0))$,
- p is proper submersive on $\overline{M}$, and its restriction to N is also submersive,
- on the neighborhood $W:=\tau^{-1}([\frac{1}{4}\phi(0),\phi(0)[)$ of N in $\overline{M}$, we have an $\mathcal{S}^r$ diffeomorphism $\sigma_{i|W}=(\sigma_{i,0},f)$ onto $W(0)\times R^l$.

In this situation, we can apply Proposition 7 to obtain a trivialization $\sigma_i':\overline{M}\to\overline{M}(0)\times R^l$ of p, which coincides with σ_i in some neighborhood $W'\subset W$ of N in $\overline{M}$. The restriction of the desired trivialization $h_{\le i}:X^i\longrightarrow X^i(0)\times R^l$ is given by σ_i' on $\overline{M}$ and σ_i on $X^i\setminus\overline{M}$. □

Proof of Lemma 12. The inclusion of the left-hand side in the right-hand side is obvious. Let us prove the reverse inclusion. Consider $(z,v) \in C \times R^n \setminus \mathcal{C}$. There is a positive ε_1 such that the intersection of the closed ball $\overline{B}((z,v),\varepsilon_1)$ with $\mathcal{B}$ is empty. For $r > 0$, set

$$K_r = \{(x,t) \in \mathcal{A} \cap \overline{B}((z,v),\varepsilon_1) \ ; \ \rho(x) = r\}.$$

Let

$$\delta(r) = \inf\{\tau(x) \ ; \ (x,t) \in K_r\}$$

if K_r is not empty, and $\delta(r) = 1$ otherwise. It is obvious that δ is a semialgebraic function of r, and it cannot be 0. Otherwise there would be a semialgebraic path $(x(s),t(s))$ in K_r with $\lim_{s\to 0_+} \tau(x(s)) = 0$. Then one should have $\lim_{s\to 0_+}(x(s),t(s)) \in \mathcal{B}$, which is impossible since $\overline{B}((z,v),\varepsilon_1) \cap \mathcal{B}$ is empty. Hence there is an integer q and $\varepsilon_2 > 0$ such that, if $0 < r < \varepsilon_2$, then $\delta(r) > r^q$. So the following formula is valid:

$$\exists \varepsilon > 0 \ \forall x \in R^p \ \forall t \in R^n \quad (\|(z,v) - (x,t)\| < \varepsilon \text{ and } \tau(x) < \rho^q(x)) \Longrightarrow (x,t) \notin \mathcal{A}.$$

Actually, one can take ε smaller than ε_1, and so small that any $(x,t) \in \mathcal{A} \cap B((z,v),\varepsilon)$ satisfies $\rho(x) < \varepsilon_2$. The integer q a priori depends upon (z,v). But since the proof is elementary (i.e. valid for any real closed field), a standard compactness argument shows that there is an integer q which works for any $(z,v) \notin \mathcal{C}$. For this q we have

$$\forall (z,v) \in (C \times R^n) \setminus \mathcal{C} \ \exists \varepsilon > 0 \quad B((z,v),\varepsilon) \cap \{(x,t) \in \mathcal{A} \ ; \ \tau(x) < \rho^q(x)\} = \emptyset,$$

which is what we wanted. □

5. Semialgebraic triviality along strata

We draw from Theorem 1 the semialgebraic local triviality along strata. Up to now we have not completed the proof of Theorem 1, and the trivialization that we have built in Proposition 11 is only $\mathcal{S}^r$ on each stratum. But the next theorem will not be used in the proof of Theorem 1. We consider the following situation. Let X be a locally closed semialgebraic set in R^n, with a semialgebraic stratification $X = \bigcup_i X^i$ satisfying Whitney's regularity conditions (or regularity condition (c) of [B]). Let V be a semialgebraic open subset of X^i, Nash diffeomorphic to R^i, and let $x_0 \in V$. Let π be an $\mathcal{S}^r$ (or Nash) retraction from an open neighborhood T of V onto V, with $T \cap X^i = V$.

Theorem 14. *Up to reducing the neighborhood T of V, there is a semialgebraic homeomorphism*

$$\sigma = (\sigma_0, \pi) : T \longrightarrow (T \cap \pi^{-1}(x_0)) \times V,$$

such that $\sigma_0(x) = x$ on $T \cap \pi^{-1}(x_0)$ and that, for all $j \geq i$, the restriction of σ to $X_j \cap T$ is an $\mathcal{S}^r$ (or Nash) diffeomorphism onto $(X_j \cap T \cap \pi^{-1}(x_0)) \times V$.

Proof. Possibly reducing T, we know that (π, ρ_i) is an $\mathcal{S}^r$ (or Nash) submersion from each $X^j \cap T$ to $V \times R$, for $j > i$. Still reducing T, and multiplying ρ_i by a positive $\mathcal{S}^r$ (or Nash) function of the form $\varphi \circ \pi$, we can actually suppose that (π, ρ_i) is a proper map from $T \setminus V$ onto $V \times]0,1[$, which is submersive in restriction to each stratum $X^j \cap T$. So we can apply Proposition 11 (or Theorem 1) to get a semialgebraic homeomorphism

$$h = (h_0, \pi, \rho_i) : T \setminus V \longrightarrow \left(\pi^{-1}(x_0) \cap \rho_i^{-1}(\frac{1}{2})\right) \times V \times]0,1[,$$

such that h restricted to each stratum $X^j \cap T$ is a diffeomorphism onto $(X^j \cap \pi^{-1}(x_0) \cap \rho_i^{-1}(\frac{1}{2})) \times V \times]0,1[$. From this h we define the desired σ by setting $h(\sigma_0(x)) = (h_0(x), x_0, \rho_i(x))$ if $x \in T \setminus V$, and $\sigma_0(x) = x_0$ if $x \in V$. □

We remark that we get from the proof that $\pi^{-1}(x_0)$ has actually a conic structure: it is semialgebraically homeomorphic to the open cone over $\pi^{-1}(x_0) \cap \rho_i^{-1}(\frac{1}{2})$. This structure is compatible with ρ in an obvious way.

We shall need, for the completion of the proof of Theorem 1, a result which expresses some uniformity of the local semialgebraic trivialization along a stratum X^i. Before stating the result, we have to specify some neighborhood U of the diagonal in $X^i \times X^i$. Let T be a Nash tubular neighborhood of X^i in R^n, with a Nash retraction $\pi : T \to X^i$ and a distance function $\rho : T \to R$. We suppose that there is a strictly positive Nash function $\sigma : X^i \to R$ such that the restriction of (π, ρ) to $T \setminus X^i$ is a proper surjective Nash submersion onto $X^i \times]0, \sigma[$ which is the set of (x, s) in $X^i \times R$ such that $0 < s < \sigma(x)$. We suppose also that T is so small that the restriction of (π, ρ) to each $X^j \cap T$, with $j > i$, is a submersion. We can find a neighborhood U of the diagonal in $X^i \times X^i$ such that
1) for any $(x, x') \in U$, one has $\sigma(x') \geq \sigma(x)/2$,
2) there is a finite covering of X^i by semialgebraic open subsets V_ν, and for each such V_ν we have Nash diffeomorphisms η and η_1 making the following diagram commutative:

$$\begin{array}{ccc}
U \cap (\mathrm{pr}_1)^{-1}(V_\nu) & \xrightarrow{\ \eta\ } & R^i \times R^i \\
\Big\downarrow {\scriptstyle \mathrm{pr}_1} & & \Big\downarrow {\scriptstyle \mathrm{pr}_1} \\
V_\nu & \xrightarrow{\ \eta_1\ } & R^i
\end{array}$$

where pr_1 is the projection on the first factor, and such that $\eta(x, x) = (\eta_1(x), 0)$ for all x in V_ν. We can set $\eta(x) = (\eta_1(x), \eta_2(x, x'))$.

With U as above, define

$$T_1 = \{(y, x') \in (X \cap T) \times X^i \ ; \ (\pi(y), x') \in U \text{ and } \rho(y) < \sigma(\pi(y))/2\}.$$

We set, for $(y,x') \in T_1$, $\pi_1(y,x') = (\pi(y),x')$ and $\mathrm{pr}_1(y,x') = y$. Define also

$$T_2 = \{(x,y') \in X^i \times (X \cap T)\ ;\ (x,\pi(y')) \in U \text{ and } \rho(y') < \sigma(x)/2\}$$

and, for $(x,y') \in T_2$, $\pi_2(x,y') = (x,\pi(y'))$ and $\mathrm{pr}_2(x,y') = y'$.

We remark that the stratification on X induces stratifications on T_1 and T_2, and that these stratifications satisfy Whitney's regularity conditions.

Proposition 15. *With the notations above, there is a semialgebraic homeomorphism $\psi : T_2 \to T_1$, inducing an S^r diffeomorphism from each stratum of T_2 to the corresponding stratum of T_1, such that $\pi_1 \circ \psi = \pi_2$ and that $\psi(x,y) = (y,x)$ when $\pi(y) = x$.*

Proof. The construction goes along the same line as in the proof of Theorem 8. The homeomorphism ψ is easy to construct over each $W_\nu = U \cap (\mathrm{pr}_1)^{-1}(V_\nu)$. We have a proper Nash mapping

$$\begin{aligned}(\eta \circ \pi_2, \rho \circ \mathrm{pr}_2) : (\pi_2)^{-1}(W_\nu) \setminus W_\nu &\longrightarrow R^i \times R^i \times]0,\sigma/2[\\ (x,y') &\longmapsto (\eta(x,\pi(y')), \rho(y'))\end{aligned}$$

which is a submersion in restriction to each stratum. Applying Proposition 11 to this situation, we get a semialgebraic trivialization of this mapping, which is an S^r diffeomorphism on each stratum. Then it is easy to construct a semialgebraic homeomorphism

$$\begin{aligned}\psi_\nu : (\pi_2)^{-1}(W_\nu) &\longrightarrow (\pi_1)^{-1}(W_\nu) \\ (x,y') &\longmapsto \psi_\nu(x,y') = (\theta(x,y'), \pi(y'))\end{aligned}$$

satisfying the properties of Theorem 15. So now what we have to check is that, if we have already constructed such a ψ over $W = U \cap (\mathrm{pr}_1)^{-1}(V)$, where V is a semialgebraic open subset of X^i, then it is possible to glue ψ_ν with this old ψ to get a new ψ over $W \cup W_\nu$. Let $(\lambda, 1-\lambda)$ be an S^r partition of unity on $V \cup V_\nu$ subordinate to (V, V_ν). Let $G(x,x')$ be the second component of $\eta^{-1}\big(\eta_1(x), \lambda(x)\eta_2(x,x')\big)$. So $G(x,x') = x'$ when $\lambda(x) = 1$ and x when $\lambda(x) = 0$. The new ψ coincides with the old one over $W \setminus W_\nu$, with ψ_ν over $W_\nu \setminus W$, and it sends $(x,y') \in (\pi_2)^{-1}(W \cap W_\nu)$ to

$$\Big(\mathrm{pr}_1 \circ \psi \circ \psi_\nu^{-1}\big(\mathrm{pr}_1(\psi_\nu(x,y')), G(x,x')\big), \pi(y')\Big).$$

This coincides with the old ψ when $\lambda(x) = 1$ and with ψ_ν when $\lambda(x) = 0$, and the new ψ which is obtained so satisfies the properties of Proposition 15. □

6. Approximating by an isotopy which is Nash on each stratum

We have now to finish the proof of Theorem 1, starting from Proposition 11. What is left is to show that we can actually have a trivialization h which is a Nash diffeomorphism in restriction to each stratum, and not only an $\mathcal{S}^r$ diffeomorphism with r finite. Of course we are going to use the approximation Theorem II.4.1 of [S]. We are going to work by induction on the dimension of the strata. So we suppose that our trivialization

$$h = (h_0, p) : X \longrightarrow X(0) \times R^l$$

is such that h is a Nash diffeomorphism from X^j to $X^j(0) \times R^l$ for all $j < i$. We can approximate h_0 on X^i by a Nash mapping $h_0' : X^i \to X^i(0)$. This approximation may be chosen so close that we have an $\mathcal{S}^r$ mapping $H : [0,1] \times X^i \to X^i(0)$, with $H(0,-) = h_0'$ and $H(1,-) = h_0$, and for each $t \in [0,1]$, $(H(t,-), p)$ is an $\mathcal{S}^r$ diffeomorphism from X^i to $X^i(0) \times R^l$ which extends continuously h on $X^{<i}$. Now we have to modify h_0 in a neighborhood of X^i in order to get again a continuous trivialization on the whole of X. So we fix an $\mathcal{S}^r$ tubular neighborhood T of X^i with a retraction π such that $p \circ \pi = p$, and a distance function ρ, which are given by the control data. We have also $\pi \circ h_0 = h_0 \circ \pi$ and $\rho \circ h_0 = \rho$, by Proposition 11. We suppose that the tube T has its radius given by the positive Nash function σ on X^i, and that $h_0(T \cap X)$ is contained in $(T \cap X)(0)$. We introduce an $\mathcal{S}^r$ function $t(s,x)$ on $[0, +\infty[\times X^i$ which is, for $x \in X^i$ fixed, an increasing function of s which is 0 on a neighborhood of 0 and 1 on a neighborhood of $[\sigma(x), +\infty[$.

Now we use our Proposition 15 in the fiber $X(0)$. We have $\psi : T_2 \to T_1$, over a neighborhood U of the diagonal in $X^i(0) \times X^i(0)$. We can suppose that the Nash approximation is so close that for any $y \in X \cap T$ with $\rho(y) < \sigma(\pi(y))/2$ and any $t \in [0,1]$, we have $(h_0(y), H(t, \pi(y))) \in T_1$. Now, for such y, set

$$h_0'(y) = \mathrm{pr}_2 \circ \psi^{-1}\big(h_0(y), H(t(\rho(y), \pi(y)), \pi(y))\big).$$

This coincides with the Nash approximation h_0' on X^i and with h_0 on a neihborhood of $\rho = (\sigma \circ \pi)/2$. So we can set $h_0'(y) = h_0(y)$, for all other $y \in X^{>i}$. This new h_0' gives a continuous trivialisation $h' = (h_0', p) : X^{\geq i} \to X^{\geq i}(0) \times R^l$. And if we have taken the tube T thin enough, this trivialization extends continuously h restricted to $X^{<i}$. This is by Lemma 12. Now the proof of Theorem 1 is complete. □

7. Uniform bounds

Now we prove Theorem 2. The existence of uniform recursive bounds is a rather formal matter, once we have proved Theorem 1 for any real closed field (it is the same as for the effectiveness result Theorem B in [CS]). The only point to check is the following lemma.

Lemma 16. *Given positive integers n, d and k, there is an algorithm to produce a closed formula $\Phi(n,d,k)$ of the theory of real closed fields which says: "For any finite semialgebraic stratification $X = \bigcup_i X^i \subset R^n$ with all the X^i of degree not greater than d, satisfying Whitney's regularity conditions, and for any Nash mapping p from a neighborhood of X to R^l of degree not greater than d, which is a submersion in restriction to each X^i, there is a semialgebraic trivialization h of p on X of degree not greater than k, which is a Nash diffeomorphism in restriction to each X^i."*

We postpone the proof of this lemma. Theorem 1 implies that given n and d, there is a k such that $\Phi(n,d,k)$ is a theorem of the theory of real closed fields. Otherwise there would be counterexamples to $\Phi(n,d,k)$ for each k, and hence, using an ultraproduct construction, a counterexample to Theorem 1. So we have a uniform bound. To show the recursiveness of this bound, it suffices to describe an algorithm to compute k from n and d. Here is one: produce successively $\Phi(n,d,1)$, $\Phi(n,d,2)$,... until the decision algorithm for the theory of real closed field says that the produced sentence is a theorem. This looks not very serious, but at least this proves Theorem 2. □

Proof of Lemma 16. The bounds on the degrees have the effect that the quantifications are actually quantifications over a finite number of variables (those which appear as coefficients of the polynomials in the descriptions). Recall that we know how to say that a semialgebraic subset of given degree is the graph of a Nash function (Lemma 5.1 in [CS]). Then we know how to express the fact that X^i is a Nash manifold of dimension i: any point of X^i has a neighborhood where $n-i$ coordinates are Nash functions of the i others. We have a bound on the degree of the connected components of the X^i's, and we know how to express that a semialgebraic subset of a given degree is a connected component of X^i. To express the properties of the stratification, we have to say that the X^i's are disjoint and that for any connected components V of X^i and W of X^j, with $i < j$, either $V \cap W = \emptyset$ or $V \subset \text{clos}(W)$ and that in this case conditions a and b are satisfied at each point of V. Once we know how to express that a vector is tangent to a Nash manifold at a point, there is no problem in formulating these conditions (see for instance 9.6.3 in [BCR]). To deal with p, we have to realize that l is surely not greater than n. Then the properties of p are easily recognized to be expressible by a formula, and the same for the properties of h. □

References

[B] Bekka, K., Regular stratification of subanalytic sets. Bull. London Math. Soc. 25 (1993), 7–16.

[BCR] Bochnak, J., Coste, M., Roy, M-F., Géométrie algébrique réelle. Ergeb. Math. Grenzgeb. (3) 12, Springer Verlag, Berlin - Heidelberg - New York 1987.

[CS] Coste, M., Shiota, M., Nash triviality in families of Nash manifolds. Invent. Math. 108 (1992), 349–368.

[G-al] Gibson, C.G., Wirthmuller, K., du Plessis, A.A., Looijenga, E.J.N., Topological stability of smooth mappings. Lecture Notes in Math. 552, Springer Verlag, Berlin - Heidelberg - New York 1976.

[H] Hardt, R., Semi-algebraic local-triviality in semi-algebraic mappings. Amer. J. Math 102 (1980), 291–302.

[M] Mather, J., Notes on topological stability. Harvard University 1970.

[S] Shiota, M., Nash manifolds. Lecture Notes in Math. 1269, Springer Verlag, Berlin - Heidelberg - New York 1987.

[SY] Shiota, M., Yokoi, M., Triangulations of subanalytic sets and locally subanalytic manifolds. Trans. Amer. Math. Soc. 286 (1984), 727–750.

Lattice-ordered rings and semialgebraic geometry: I

*Charles N. Delzell and James J. Madden**

Abstract. Lattice-ordered rings provide a useful language for semialgebraic geometry. Constructions such as the free f-ring and various spectra for (partially ordered) rings are closely related to the real spectrum. They can be used to give a new proof of the first author's continuous solution to Hilbert's 17th problem, and they provide a novel approach to various other results. Basic techniques from universal algebra are used repeatedly. We have attempted to make f-rings more meaningful and accessible to real and semialgebraic geometers.

1991 Mathematics Subject Classification: 14P10, 06F25, 13J25.

Introduction

The purpose of this paper is to gather together some elementary facts about f-rings which seem to have special significance for real algebraic and semialgebraic geometry, and to describe (or at least suggest) some of the pretty connections they have with topics such as the real spectrum, the real Nullstellensatz, the Hörmander–Łojasiewicz inequality, Bröcker's ultrafilter theorem, the Pierce–Birkhoff conjecture, Hilbert's 17th Problem, and real closed rings.

The notion of an f-ring was introduced nearly 40 years ago by Birkhoff and Pierce in [BP], the same paper in which the notorious Pierce–Birkhoff conjecture originated. At present, few geometers seem to be aware of these objects, despite the fact that there has been an active and productive group of mathematicians studying them for many years. In the present paper, we hope to alert geometers to the potential that exists here, point out a few guideposts and set up a framework for future work. In a sequel, we shall consider additional topics relating to real closed f-rings and f-rings of abstract piecewise definable functions. At present, we are able only to begin to touch on the many aspects of the theory of f-rings relevant to geometry.

The name "f-ring" is an abbreviation for the term "function ring." Birkhoff and Pierce had in mind rings of functions with values in a totally ordered field or

---

* Both authors supported by NSF-DMS-9104427.

ring. One of the most important observations in their 1956 paper was that certain simple equational laws characterize the class of those lattice-ordered rings which admit an embedding in a product of totally ordered rings. This fact permits the application of all the resources of universal algebra. In semialgebraic geometry, one inevitably deals with rings of functions with natural lattice operations, such as the ring of all piecewise polynomial functions on an algebraic set or the ring of all continuous semialgebraic functions on a pro-constructible set. The formalism of f-rings provides a rich and technically powerful medium of expression. Moreover, the theory of f-rings fits together neatly with the general theory of partially ordered rings, which is already a standard tool in real algebraic geometry.

A set of notes on f-rings and semialgebraic geometry prepared by the second author in the mid 1980's formed the core for the present paper. While the notes were shared freely with all who were interested, they were never prepared for publication. For a long time, the results they contained seemed not much more than a curiosity. This changed, however, when the first author proved his "sup-inf-polynomially varying" solution to Hilbert's 17th Problem [D]. It became clear that the abstract material in the notes provided useful conceptual insights into this work. A preprint of Keimel [Ke] contains a review of recent work on lattice-ordered algebraic systems and their connections with algebra, analysis and geometry.

Here is a summary of the main points covered. In §1, we review background on f-rings and describe the free f-ring over a given ring. This construction plays a pivotal role throughout the present work. In §2, we give some basic facts about radicals of ℓ-ideals, and prove a proposition comparing the category of reduced f-rings with the category of reduced partially ordered rings. (This result is not used further in the present work, so 2.1, 2.2, 2.3, and 2.4 may be skipped at first reading.) §3 describes various spectra for rings, *po*-rings and f-rings, and contains the most useful result of the paper—Proposition 3.3—which says that the real spectrum of a ring A *is* the convex-prime ideal spectrum (or "Brumfiel spectrum") of the free f-ring over A. In §4, we engage in some philosophizing about the Nullstellensatz. The first 5 paragraphs of §4 should be taken in this spirit and, like all philosophizing, should be regarded as tentative. After this, in 4.1 and 4.2, we make explicit some important aspects of the correspondence between the algebraic structure of an f-ring and the topology of its Keimel and Brumfiel spectra. In §5, we treat material related to the first author's recent work on Hilbert's 17th Problem. This used a special Nichtnegativstellensatz for f-rings, a version of which appears here as Theorem 5.4. In §6, we show that the (concrete) Hörmander–Łojasiewicz Inequality is literally a concrete Nullstellensatz for f-rings. From this, and the material in earlier sections, we deduce the ultrafilter theorem. This section closes with some observations about categories of f-algebras, which are used in the last section. In §7, we present a new proof of the first author's "sup-inf-polynomially varying" solution to Hilbert's 17th problem based on results of §5.

1. Categories of partially ordered rings

We shall be concerned with the following categories:

RING: the category of rings and ring morphisms. By assumption, *all rings are commutative and have unity which is preserved by all morphisms.*

PORING: the category of partially ordered rings. The objects are pairs $(A, \mathcal{P})$, where A is a ring and $\mathcal{P}$ is a subset of A satisfying the following conditions:

$$\mathcal{P} + \mathcal{P} \subseteq \mathcal{P}, \quad \mathcal{P} \cdot \mathcal{P} \subseteq \mathcal{P}, \quad \mathcal{P} \cap -\mathcal{P} = \{0\} \qquad (po\text{-}1)$$

$$\forall x \in A, \quad x^2 \in \mathcal{P}\ . \qquad (po\text{-}2)$$

The morphisms from $(A, \mathcal{P})$ to $(B, \mathcal{Q})$ are the ring morphisms $\phi : A \to B$ such that $\phi(\mathcal{P}) \subseteq \mathcal{Q}$. Objects in this category are called *po*-rings. When $\mathcal{P}$ is understood, we sometimes omit explicit reference to it. (Sometimes "*po*-rings" which are not assumed to satisfy (*po*-2) appear in the literature, *e.g.*, [BP]. Such objects are not mentioned in the present work.)

FRING: the equational category of rings equipped with an operation $\vee$ satisfying the following laws: $\vee$ is commutative and associative, $x \vee x = x$,

$$x + (y \vee z) = (x + y) \vee (x + z), \text{ and} \qquad (f\text{-}1)$$

$$x^+(y \vee z) = x^+y \vee x^+z. \qquad (f\text{-}2)$$

Here, $x^+ := x \vee 0$. We call an object of this category an f-ring. The morphisms are the ring morphisms which preserve $\vee$. (Many authors consider f-rings which are not assumed to have unity. For these, stronger defining identities are needed. This is discussed in [BKW], 9.1.14, but is not of concern in the present work.)

For reference and to fix notation, we list some easily proved facts about PORING.

Fact PO1: *Let $(A, \mathcal{P})$ be a po-ring. The relation $\le$ on A defined by $x \le y \Leftrightarrow y - x \in \mathcal{P}$ is a partial order satisfying the following conditions:*

i) *$\forall x, y, z$: $x \le y$ implies $x + z \le y + z$,*
$\forall x, y, z$: $0 \le z$ & $x \le y$ implies $zx \le zy$, and

ii) *$\forall x$: $0 \le x^2$.*

Any partial order relation $\le$ on A satisfying these conditions is called a po-ring order. If $\le$ is a po-ring order, then $(A, \{x \in A \mid 0 \le x\})$ is a po-ring.

Definition. Suppose $(A, \mathcal{P})$ is a *po*-ring and $\le$ the associated partial order on A. An ideal $I \subseteq A$ is said to be *convex* if $x, y \in I$ & $x \le z \le y$ implies $z \in I$.

Fact PO2: *The kernel of any PORING morphism is convex. If $I \subseteq A$ is convex, then $(A/I, \mathcal{P}/I)$ is a po-ring and the canonical map $A \to A/I$ is a PORING morphism.*

Since we expect that many readers will be unfamiliar with the category FRING, we shall list the facts about f-rings which we use repeatedly. Some, but not all, of these are routine to prove. We have supplied references for the harder results.

Fact F1: *Products, subobjects and homomorphic images (under FRING morphisms) of f-rings are f-rings. Also, the free f-ring on any set exists* (see Lemmas 1.2 and 1.3 below).

F1 is an immediate consequence of the fact that FRING is an equational category.

Definition. In any f-ring, one has defined operations $x \wedge y := -(-x \vee -y)$, $x^+ := x \vee 0$, $x^- := (-x) \vee 0$ and $|x| := x \vee (-x)$.

Fact F2: *On any f-ring, there is a canonical po-ring order defined by $x \leq y$ if and only if $x \vee y = y$. Moreover, under $\vee$ and $\wedge$, any f-ring is a distributive lattice (unbounded unless $A = \{0\}$), with $\vee$ giving the supremum of its arguments and $\wedge$ giving the infimum.*

In connection with F2, note that of the defining conditions for a *po*-ring, (*po*-1) is immediate from the f-ring identities. On the other hand, (*po*-2) is non-trivial and is most easily derived from F4, below (whose proof, of course, does not involve (*po*-2)). A nice proof of the distributive law $x \wedge (y \vee z) = (x \wedge y) \vee (x \wedge z)$ is in [Jo], V.4.4. *Caution*: It is possible for (the canonical partial order on) a *po*-ring to be a distributive lattice, but for the supremum operation to fail to satisfy the FRING identity (f-2).

Definition. A ring ideal I in an f-ring A is called an *ℓ-ideal* if it is closed under the operation $\vee$ and is convex with respect to the canonical *po*-ring order on A.

Fact F3: *The kernel of any FRING morphism is an ℓ-ideal. If $I \subseteq A$ is an ℓ-ideal, then A/I is naturally an f-ring, with f-ring operation $x + I \vee y + I := (x \vee y) + I$.*

Definition. An ℓ-ideal I of an f-ring A is said to be *irreducible* if the canonical *po*-ring order on A/I is total.

Fact F4: *In a totally ordered po-ring $(A, \mathcal{P})$ (i.e., $\mathcal{P} \cup -\mathcal{P} = A$), $x \vee y := \max\{x, y\}$ defines an f-ring operation. Hence, by F1, a sub-f-ring (i.e., a subset which is closed under the ring and lattice operations) of a product of totally ordered rings is an f-ring. Conversely, every f-ring is a sub-f-ring of a product of totally ordered f-rings. (Equivalently, in any f-ring the intersection of all irreducible ℓ-ideals is $\{0\}$.)*

F4 says that the class of f-rings coincides with the class of sub-lattice-rings of products of totally ordered rings. This is, perhaps, the single most significant fact about f-rings. To prove it, it suffices to show that an ℓ-ideal maximal with the property of not containing a given non-zero $x \in A$ is irreducible; see [BKW], 9.1.4 and 9.1.1.

Fact F5: *In any f-ring, any convex prime ideal is an ℓ-ideal. In any f-ring, any prime ℓ-ideal is irreducible. A minimal prime ideal in (the underlying ring of) an f-ring is an ℓ-ideal (in the f-ring).*

The first two assertions in F5 follow from the identity

$$(x^+)(x^-) = 0,$$

which is obviously true in any totally ordered f-ring, and hence by F4 in all f-rings. For the first assertion, if I is prime and $x, y \in I$, then either $(x-y)^+ \in I$ or $(x-y)^- \in I$. But $(x-y)^+ + y = x \vee y = (x-y)^- + x$. For the second, the same identity implies that for any $x + I \in A/I$ either $(x+I)^+ = 0$ or $(x+I)^- = 0$, which is equivalent to A/I being totally ordered. The third assertion is much more subtle and at first sight may seem something of a miracle. It provides a useful link between the arithmetic structure of an f-ring and its order. For a proof, see [BKW], 9.3.3.

Fact F6: *If A is an f-ring and E is a subset of A, then the sub-f-ring of A generated by E consists of the elements of A of the form*

$$\bigvee_{i=1}^{m} \bigwedge_{j=1}^{n_i} f_{ij},$$

where f_{ij} is in the subring of A generated by E.

F6 is due to Henriksen and Isbell (see [HI], page 542). Their proof uses F4. The following fact is proved in [HI]. Our 6.5, below, will extend this slightly.

Fact F7: *Let K be any totally ordered field. Let $F(n)$ denote the free f-ring on n generators, and let $\mathcal{N}$ be the ℓ-ideal of nilpotent elements of $F(n)$ (see §2). Then $F(n)/\mathcal{N}$ is isomorphic (as an f-ring) to the sub-f-ring of the f-ring of all functions from K^n to K generated by the projections.*

There are obvious forgetful functors from PORING and FRING to RING. From Fact F2 it follows easily that there is a forgetful functor from FRING to PORING. We shall be quite relaxed about notation for these functors, and shall *sometimes use the same symbol to denote an f-ring and its underlying ring or po-ring.* The text will always contain enough information to resolve any ambiguities that arise from this. Lemmas 1.1 and 1.2 below describe left adjoints for all three of these forgetful functors.

Definition. For any ring A, let $\mathcal{I}_{po}(A)$ (respectively, $\mathcal{I}_f(A)$) denote the intersection of all the ideals of A which occur as the kernel of some ring map $\phi : A \to B$, where B is the underlying ring of a (not necessarily fixed) po-ring (respectively, f-ring).

Lemma 1.1. *Let A be a ring, let $\mathcal{I} = \mathcal{I}_{po}(A)$ and let $\mathcal{Q}$ denote the set of sums of squares in $A/\mathcal{I}$. Then $(A/\mathcal{I}, \mathcal{Q})$ is a po-ring, and if $(B, \mathcal{P})$ is any po-ring and $\phi : A \to B$ is any ring morphism, then there is a unique po-ring morphism $\overline{\phi}$: $(A/\mathcal{I}, \mathcal{Q}) \to (B, \mathcal{P})$ such that $\overline{\phi}(a + \mathcal{I}) = \phi(a)$ for all $a \in A$.* □

The proof of 1.1 is entirely routine and is left to the reader.

Lemma 1.2. (a) *For any ring A, there are an f-ring $F(A)$ and a ring morphism $\eta_A : A \to F(A)$ with the following universal property: for any f-ring H and any ring morphism $\phi : A \to H$, there is a unique f-ring morphism $\overline{\phi} : F(A) \to H$ such that $\overline{\phi} \circ \eta_A = \phi$.*

(b) *For each partially ordered ring $(A, \mathcal{P})$, there are an f-ring $F_{po}(A, \mathcal{P})$ and a PORING morphism $\eta_{(A,\mathcal{P})} : (A, \mathcal{P}) \to F_{po}(A, \mathcal{P})$ with the following universal* *property: for any f-ring H and any PORING morphism $\phi : (A, \mathcal{P}) \to H$, there is a unique f-ring morphism $\overline{\phi} : F_{po}(A, \mathcal{P}) \to H$ such that $\overline{\phi} \circ \eta_{(A,\mathcal{P})} = \phi$.* □

We omit the details of the proof, since this is standard universal algebra—but here is a sketch. For $F(A)$, one takes the free f-ring on the underlying set of A modulo the smallest congruence containing all the ring-theoretic relations of A. Explicitly, for each $x \in A$ let $\jmath(x)$ be a distinct symbol. Let F be the free f-ring on $\{\jmath(x) \mid x \in A\}$. Then $F(A) \cong F/I$, where I is the ℓ-ideal generated by the following elements of F:

$$\begin{array}{ll} \jmath(0_A), \qquad 1_F - \jmath(1_A), & \\ \jmath(x) + \jmath(-x), & x \in A, \\ \jmath(x) + \jmath(y) - \jmath(x+y), & x, y \in A \quad \text{and} \\ \jmath(xy) - \jmath(x)\jmath(y), & x, y \in A. \end{array}$$

For $F_{po}(A, \mathcal{P})$, one takes the free f-ring on the underlying set of A modulo all ring- *and* order-theoretic relations of $(A, \mathcal{P})$. Explicitly, $F_{po}(A, \mathcal{P}) \cong F(A)/I$, where I is the ℓ-ideal generated by all elements $|\eta_A(x)| - \eta_A(x)$ where $x \in \mathcal{P}$. (This is the smallest ℓ-ideal I such that $\eta_A(x) + I \geq 0$ for all $x \in \mathcal{P}$.) We call $F(A)$ (resp., $F_{po}(A, \mathcal{P})$) *the free f-ring on A* (resp., *the free f-ring on* $(A, \mathcal{P})$).

Note that for a ring A, $F(A) \cong F_{po}(A/\mathcal{I}, \mathcal{Q})$, the latter expression referring to the construction in Lemma 1.1.

It is possible to give a more concrete representation of $F(A)$ and $F_{po}(A, \mathcal{P})$, as the next lemma shows.

Lemma 1.3. (a) *Let A be any ring. For each pair $\xi := (I, \mathcal{Q})$, where I is an ideal of A and $\mathcal{Q}$ is a total po-ring order on A/I, let $A[\xi] := (A/I, \mathcal{Q})$, and for each $a \in A$ let $a(\xi)$ denote the image of a in $A[\xi]$. Let $\Delta : A \to \prod_\xi A[\xi]$ be the diagonal map (i.e., $(\Delta(a))_\xi = a(\xi)$), and let $S(A)$ be the sub-f-ring of $\prod_\xi A[\xi]$ generated by $\Delta(A)$. Then the canonical map $\overline{\Delta} : F(A) \to \prod_\xi A[\xi]$ is an isomorphism onto $S(A)$.*

(b) *Let $(A, \mathcal{P})$ be a partially ordered ring. For each pair $\xi := (I, \mathcal{Q})$, where I is a convex ideal of $(A, \mathcal{P})$ and $\mathcal{Q}$ is a total order on A/I extending $\mathcal{P}/I$, let $(A, \mathcal{P})[\xi] := (A/I, \mathcal{Q})$. Let $\Delta : A \to \prod_\xi (A, \mathcal{P})[\xi]$ be the diagonal map, and let $S(A, \mathcal{P})$ be the sub-f-ring of $\prod_\xi (A, \mathcal{P})[\xi]$ generated by $\Delta(A)$. Then the canonical map $\overline{\Delta} : F_{po}(A, \mathcal{P}) \to \prod_\xi (A, \mathcal{P})[\xi]$ is an isomorphism onto $S(A, \mathcal{P})$.*

Proof. We prove (a); the proof of (b) is similar. The image of $\overline{\Delta}$ contains generators for $S(A)$ and must be contained in $S(A)$ by the universal mapping property, so it suffices to show that $\overline{\Delta}$ is injective. If $0 \neq x \in F(A)$, then by F4 there is an irreducible ℓ-ideal $J \subset F(A)$ which does not contain x. Let $q : F(A) \to F(A)/J$ be the canonical (f-ring) morphism. Note that $q \circ \eta_A$ is a *ring* morphism with totally ordered codomain. Moreover, $q \circ \eta_A$ is surjective, because $F(A)/J$ is totally ordered, and hence the operation $\vee$, when applied to elements in $q \circ \eta_A(A)$, produces no elements not already in $q \circ \eta_A(A)$. Now take $\xi = (I, \mathcal{Q})$, where I is the kernel of

$q \circ \eta_A$ and $\mathcal{Q}$ is the canonical order on $F(A)/J$ $(= A/I)$. Then $(\overline{\Delta}(x))_\xi \neq 0$, so $\overline{\Delta}(x) \neq 0$. □

It will be important later to appreciate what happens when the free-f-ring construction is iterated. To describe this, it is useful for the moment to introduce explicit notation for the forgetful functors. Let A be an f-ring. Then $G(A)$ will denote the underlying ring of A, and $(G(A), \mathcal{P}(A))$ will denote the underlying *po*-ring.

Since the operation $\vee$ is completely determined by the partial order, it follows that for any f-ring A:

$$F_{po}(G(A), \mathcal{P}(A)) \cong A.$$

In contrast, it is generally the case that $A \ncong FG(A) := F(G(A))$. For example, suppose that K is a field which admits several different total orders, and let K_0 be the f-ring obtained by choosing one of them. Suppose that $x \in K$ is positive in some but not all of the orders on K. Then obviously neither $(\Delta(x))^+$ nor $(\Delta(x))^-$ equals 0, yet their product vanishes. Thus $FG(K_0)$ is not an integral domain. Indeed, it is possible for all of the f-rings A, $FG(A)$, $FGFG(A), \ldots$ to be distinct.

2. Reduced rings and real rings

We shall present some basic results related to *po*-rings and f-rings with no nilpotent elements. A basic fact about integral domains in PORING is

Fact R1: *If $(A, \mathcal{P})$ is a po-ring and A has no zero-divisors, then there is a totally ordered field $(K, \mathcal{Q})$ with $A \subseteq K$ and $\mathcal{P} \subseteq \mathcal{Q}$* ([B], 1.6.1).

For any ring A, let $\mathcal{N}(A)$ denote the nil-radical of A (= the set of all nilpotent elements of A = the intersection of all prime ideals of A). We call a ring with no non-zero nilpotents *reduced*. It is well-known that $A/\mathcal{N}(A)$ is reduced. The nil-radical behaves very nicely in a *po*-ring or an f-ring, as seen in the following fact R2. The second assertion in R2 is a consequence of F5, but is actually more elementary than F5.

Fact R2: *For any po-ring A, $\mathcal{N}(A)$ is convex and is equal to the intersection of all convex prime ideals of A* ([B], 2.3.6). *For any f-ring A, $\mathcal{N}(A)$ is an ℓ-ideal and is equal to the intersection of all prime ℓ-ideals of A* ([BKW], 9.2.6).

Let A be a ring and $I \subseteq A$ an ideal. The *radical* of I is $\sqrt{I} = \{\, b \in A \mid \exists n \in \mathbb{N} \;\; b^n \in I \,\}$. If $I = \sqrt{I}$, then I is called a *radical ideal*. From R2 and the remark just preceding it, we deduce immediately that:

I an ideal in a ring	$\Rightarrow$	$\sqrt{I} = \bigcap\{\text{primes containing } I\}$
I a convex ideal in a *po*-ring	$\Rightarrow$	$\sqrt{I} = \bigcap\{$*convex* primes containing $I\}$
I an ℓ-ideal in an f-ring	$\Rightarrow$	$\sqrt{I} = \bigcap\{$prime *ℓ-ideals* containing $I\}$.

Clearly, the radical of a convex (resp. ℓ-)ideal is a convex (resp. ℓ-)ideal.

The full subcategories of RING, PORING and FRING whose objects are reduced are denoted respectively RING/N, PORING/N and FRING/N. One obtains a functor from RING to RING/N by "modding out" nil-radicals. The same is true for PORING and FRING. We leave the details to the reader. R1 and R2 together imply that any reduced *po*-ring (f-ring) is a sub-*po*-ring (sub-f-ring) of a product of totally ordered fields.

We shall be concerned with the functor that is obtained by composing the mod-$\mathcal{N}$ functor with the free f-ring functor F_{po}. It is important to note that even when a *po*-ring $(A, \mathcal{P})$ is reduced, $F_{po}(A, \mathcal{P})$ need not be (see [HI]).

Lemma 2.1. *Let $(A, \mathcal{P})$ be a reduced po-ring, let $\mathcal{N} := \mathcal{N}(F_{po}(A, \mathcal{P}))$ and let $\epsilon_{(A,\mathcal{P})}$ denote the composite morphism*

$$(A, \mathcal{P}) \xrightarrow{\eta(A,\mathcal{P})} F_{po}(A, \mathcal{P}) \to F_{po}(A, \mathcal{P})/\mathcal{N}.$$

Then $\epsilon_{(A,\mathcal{P})}$ is injective. Moreover, for any PORING morphism $\phi : (A, \mathcal{P}) \to R$, where R is a reduced f-ring, there is a unique FRING morphism

$$\overline{\phi} : F_{po}(A, \mathcal{P})/\mathcal{N} \to R$$

such that $\overline{\phi} \circ \epsilon_{(A,\mathcal{P})} = \phi$.

Proof. By R1 and R2 a reduced *po*-ring admits a PORING embedding into a product of totally ordered fields, and therefore admits a PORING embedding into an f-ring. The second assertion is an immediate consequence of Lemma 1.2(b). □

Lemma 2.2. *Any PORING morphism between reduced f-rings is an FRING morphism.*

Proof. Let ϕ be such a morphism. It is enough to show $\phi(|x|) = |\phi(x)|$. (A map which preserves $|\ |$ must preserve $\vee$, since $2(x \vee y) = (x + y + |x - y|)$ is an f-ring identity, by F4.) First we show that the squares of these elements are equal: $(\phi(|x|))^2 = \phi(|x|^2) = \phi(x^2) = (\phi(x))^2 = |\phi(x)|^2$. Second, we note that they are both positive (or 0). Third, we note that in any object of FRING/N, positive square roots—when they exist—are unique, since every object of FRING/N is a sub-f-ring of a product of totally ordered *domains*, by R2. □

In category-theoretic language, the preceding lemmas can be restated as follows:

Proposition 2.3. *FRING/N is full and mono-reflective in PORING/N.* □

Since many important properties of a category can be detected in a full, mono-reflective subcategory, Proposition 2.3 helps to explain why f-rings are useful in semialgebraic geometry. This proposition also has several interesting formal consequences, as described, *e.g.*, in [HS]. We give one example. Recall that a morphism in a category C is called an *epimorphism* if $\phi_1 \circ \epsilon = \phi_2 \circ \epsilon$ implies $\phi_1 = \phi_2$ for all morphisms ϕ_1, ϕ_2 in C. For instance, the embedding of the integers in the rationals is a RING epimorphism. The following is a special case of a general fact about reflections; see [M], p. 90, exercise 3.

Corollary 2.4. *For any reduced po-ring $(A,\mathcal{P})$, $\epsilon_{(A,\mathcal{P})}$ is an epimorphism in PORING/N.* □

Interestingly, the mono-reflection of Proposition 2.3 is *not* "essential" (see [HM]). In other words, $F_{po}(A,\mathcal{P})$ may (and generally does) have convex prime ideals (hence, by 2.2 and F3, ℓ-ideals) which have trivial intersection with the image of $\epsilon_{(A,\mathcal{P})}$. For example, if K is a field admitting several total orders and $\mathcal{Q}$ is the set of sums of squares in K, then $F_{po}(K,\mathcal{Q})$, as pointed out at the end of §1, is not a field, but it admits a surjective f-ring morphism onto $(K,\mathcal{P})$ for any total order $\mathcal{P}$. The kernel of any such morphism misses the image of $\epsilon_{(A,\mathcal{P})}$. (From this point of view, the image of $\epsilon_{(A,\mathcal{P})}$ is rather sparse, though 2.4 suggests the opposite.)

In real geometry, the category of reduced rings is not as important as the category of real rings. For any ring A, let $\mathcal{R}(A)$ denote the *real radical* of A, *i.e.*,

$$\mathcal{R}(A) := \{\, a \in A \mid \exists\, m \in \mathbb{N} \;\; \exists\, b_1,\dots,b_p \in A \quad a^{2m} + b_1^2 + \dots + b_p^2 = 0 \,\}.$$

A ring is called *real* if its real radical is $\{0\}$. The full subcategory of RING whose objects are real is denoted RING/R.

Fact R3: *For any ring A, $\mathcal{R}(A) = \{a \in A \mid \phi(a) = 0$ for all ring morphisms $\phi : A \to k$ with k a totally ordered field$\}$* ([BCR], 4.1.7).

From R3 it follows immediately that one obtains a functor from RING to RING/R by "modding out" real radicals. It also follows that the underlying ring of any reduced *po*-ring or f-ring is real, and that $\mathcal{R}(A) = \sqrt{\mathcal{I}_{po}(A)} = \sqrt{\mathcal{I}_f(A)}$.

FRING/N is a monoreflective subcategory of RING/R, but since FRING/N is *not* full in RING/R this does not have many interesting consequences beyond what obviously follows from the results in §1. We leave it to the reader to formulate and prove the analogue of Lemma 2.1 with RING/R in place of PORING/N.

3. Spectra

One may view the Zariski spectrum as a functor from RING to topology. Similar functors from PORING and from FRING to topology have been defined and studied. For a unified exposition, the reader may consult chapter 5 of Peter Johnstone's book [Jo]. In the present section, we point out that the various spectra are related by the functors described in §1 and §2. We begin by listing the spectra to be discussed, together with their definitions and some associated terminology. The names "Brumfiel spectrum" and "Keimel spectrum" were coined by Johnstone.

The *Zariski spectrum* of a ring A is the set of all prime ideals of A endowed with the topology generated by the "cozero sets," these being the sets of the form $\operatorname{coz}(a) := \{p \mid a \notin p\}$, where $a \in A$. This topological space is denoted $\operatorname{Spec}_Z A$. Observe that the cozero sets actually form a basis for the topology, since the intersection of two cozero sets is a cozero set: $\operatorname{coz}(a) \cap \operatorname{coz}(b) = \operatorname{coz}(ab)$. (We shall use the same symbol, coz(a), for sets in several different spectra. No confusion can occur as long as the context is observed.)

The *Brumfiel spectrum* of a *po*-ring $(A, \mathcal{P})$ is the subspace of the Zariski spectrum of A whose points are the *convex* primes. This is denoted $\mathrm{Spec}_B(A, \mathcal{P})$, or $\mathrm{Spec}_B A$ if the partial order is understood.

The *Keimel spectrum* of an f-ring A is the set of all irreducible ℓ-ideals of A endowed with the topology generated by the subsets $\mathrm{coz}(a) = \{ I \mid a \notin I \}$, $a \in A$. This is denoted $\mathrm{Spec}_K A$. Note that $\mathrm{coz}(a) = \mathrm{coz}(|a|)$. The sets $\mathrm{coz}(a)$ form an open basis for the topology of $\mathrm{Spec}_K A$, because $\mathrm{coz}(a) \cap \mathrm{coz}(b) = \mathrm{coz}(|a| \wedge |b|)$. Moreover, finite unions of cozero sets are cozero sets, because $\mathrm{coz}(a) \cup \mathrm{coz}(b) = \mathrm{coz}(|a| \vee |b|)$.

The *real spectrum* of a ring A, denoted $\mathrm{Spec}_R A$, is the set of pairs $(p, \mathcal{T})$ where p is a prime ideal of A and $\mathcal{T}$ is a total order on A/p. Before describing the topology, we introduce some notation which will be used repeatedly. If $\xi := (p, \mathcal{T}) \in \mathrm{Spec}_R A$, then $A[\xi]$ denotes the *po*-ring $(A/p, \mathcal{T})$, and for any $a \in A$, $a(\xi)$ denotes $a + p \in A/p$; $a(\xi) \geq b(\xi)$ means $a + p \geq b + p$ in the order $\mathcal{T}$. Any function f defined on $\mathrm{Spec}_R A$ with $f(\xi)$ in the real closure of the fraction field of $A[\xi]$ is called an *abstract function*. In other words, the abstract functions are the elements of $\prod_\xi \overline{A[\xi]}$, where $\overline{A[\xi]}$ denotes the real closure of the (ordered) fraction field of $A[\xi]$. The map $\xi \mapsto a(\xi)$ is called the abstract function induced by a. We define $\mathrm{coz}(a) := \{ \xi \mid a(\xi) \neq 0 \}$. The topology on $\mathrm{Spec}_R A$ is that generated by the subsets $\mathrm{pos}(a) := \{ \xi \mid a(\xi) > 0 \}$, $a \in A$, which we call *positivity sets*. Note that in contrast to the Zariski and Keimel spectra, the canonical generators for this topology (that is, the positivity sets $\mathrm{pos}(a)$, $a \in A$) do not necessarily form a basis: there may not be any $x \in A$ such that $\mathrm{pos}(a) \cap \mathrm{pos}(b) = \mathrm{pos}(x)$.

We define the real spectrum of a *po*-ring $(A, \mathcal{P})$ to be the subspace of $\mathrm{Spec}_R A$ consisting of those pairs $(p, \mathcal{T})$ where p is a *convex* prime ideal of $(A, \mathcal{P})$ and $\mathcal{T} \supseteq \mathcal{P}/p$. In order to avoid ambiguity, we denote this $\mathrm{Spec}_{Rpo}(A, \mathcal{P})$, or $\mathrm{Spec}_{Rpo} A$, if the partial order on A is understood.

Lemma 3.1. *Let A be a ring. $F(A)/\mathcal{N}(F(A))$ is isomorphic to the sub-f-ring of the f-ring of all abstract functions on* $\mathrm{Spec}_R A$ *generated by (the images of) A.*

Proof. This is immediate from 1.3(a) and R2. The latter shows that $\mathcal{N}(F(A))$ is the kernel of the composite map:

$$F(A) \to \prod_{\xi \in X} A[\xi] \to \prod_{\xi \in \mathrm{Spec}_R A} A[\xi],$$

where X is the set of all $\xi := (I, \mathcal{Q})$ as in 1.3(a), and the first map is $\overline{\Delta}$. □

Proposition 3.2. *If A is an f-ring, then* $\mathrm{Spec}_B A$ *is the subspace of* $\mathrm{Spec}_K A$ *whose points are the prime ℓ-ideals. Moreover, the map $(p, \mathcal{T}) \mapsto p$ from* $\mathrm{Spec}_{Rpo} A$ *to* $\mathrm{Spec}_B A$ *is a homeomorphism.*

Proof. The first assertion is obvious from the definitions above and F5. It is also clear that the map in the second assertion is surjective. By F5, again, the canonical partial order on A/p for any convex prime p is total, so the map is injective. The fact that $\mathrm{coz}(a) = \mathrm{coz}(|a|) = \mathrm{pos}(|a|)$ (in $\mathrm{Spec}_{Rpo} A$) shows that the map is a homeomorphism. □

The following proposition was noticed by the second author in 1983; we know of no other references to it.

Proposition 3.3. *Let A be any ring. For each $\xi = (p,T) \in \mathrm{Spec}_R A$, let $\overline{\phi_\xi} : F(A) \to A[\xi]$ be the f-ring morphism determined as in* 1.2(a) *by the ring map $\phi_\xi : a \mapsto a + p$ from A to the totally ordered domain $A[\xi]$. Then the map*

$$q_A : \xi \mapsto \ker \overline{\phi_\xi}$$

is a homeomorphism of $\mathrm{Spec}_R A$ *with* $\mathrm{Spec}_B F(A)$.

Proof. There is no loss of generality in assuming that A is real, since $\mathrm{Spec}_R A$ and $\mathrm{Spec}_B F(A)$ are unaffected by reducing A mod its real radical. In this case, η_A is injective. We shall identify A with its image in $F(A)$. To see that q_A is injective, suppose $\xi, \zeta \in \mathrm{Spec}_R A$ and $\xi \neq \zeta$. Switching the names of these points if necessary, there is $a \in A$ such that $a(\xi) > 0$ and $a(\zeta) \leq 0$. Therefore $a \vee 0 \notin \ker \overline{\phi_\xi}$, but $a \vee 0 \in \ker \overline{\phi_\zeta}$. That q_A is surjective can be seen by considering an arbitrary prime ℓ-ideal $I \subseteq F(A)$. Let $\pi : F(A) \to F(A)/I$ be the canonical map. The restriction of π to A is a ring surjection onto a totally domain (arguing as in the proof of 1.3). It determines a point $\xi \in \mathrm{Spec}_R A$ for which $\overline{\phi_\xi} = \pi$. Now we show that both q_A and its inverse carry open sets to open sets. Pick any $w \in F(A)$. There is a finite subset $\{a_{ij}\} \subseteq A$ such that $|w| = \bigvee_i \bigwedge_j a_{ij} = \bigvee_i \bigwedge_j (a_{ij} \vee 0)$, so $\mathrm{coz}(w) = \mathrm{coz}(|w|) = \bigcup_i \bigcap_j \mathrm{coz}(a_{ij} \vee 0)$ in $\mathrm{Spec}_B F(A)$. Thus the basic opens of the form $\mathrm{coz}(a \vee 0)$, $a \in A$, are a subbasis for the topology on $\mathrm{Spec}_B F(A)$. Since $q_A(\mathrm{pos}(a)) = \mathrm{coz}(a \vee 0)$ for $\mathrm{pos}(a) \subseteq \mathrm{Spec}_R A$, q_A induces a bijection of subbases. □

All the spectra we listed above are functors. There is a unified way of treating them, which can be sketched as follows. Suppose $\mathrm{Spec}_X A$ is one of the spectra above. Each point of $\mathrm{Spec}_X A$ corresponds in an obvious way to a morphism $\phi : A \to Q$, where Q is a ring which may carry some additional structure, *e.g.*, a total order. Now, if $f : A' \to A$ is a morphism, then $\phi \circ f$ (and its codomain) carry whatever structure is necessary to determine a point of $\mathrm{Spec}_X A'$. It is routine to verify that the map $\mathrm{Spec}_X f : \mathrm{Spec}_X A \to \mathrm{Spec}_X A'$ defined this way is continuous. We leave the details to the reader.

In fact, the map q_A in Proposition 3.3 is a natural equivalence between Spec_R and the composite $\mathrm{Spec}_B \circ F$. In other words, in addition to what is asserted in 3.3, it is also the case that for any RING morphism $f : A \to A'$, the following diagram commutes:

$$\begin{array}{ccc} \mathrm{Spec}_R A' & \xrightarrow{q_{A'}} & \mathrm{Spec}_B F(A') \\ {\scriptstyle \mathrm{Spec}_R f}\downarrow & & \downarrow{\scriptstyle \mathrm{Spec}_B F(f)} \\ \mathrm{Spec}_R A & \xrightarrow{q_A} & \mathrm{Spec}_B F(A), \end{array}$$

where $F(f) := \overline{\eta_{A'} \circ f}$ (the bar notation being that from Lemma 1.2(a)).

There are several other natural transformations which deserve mention. The maps in Proposition 3.2 give rise to a natural transformation

$$\mathrm{Spec}_B|_{\mathrm{FRING}} \xrightarrow{\subseteq} \mathrm{Spec}_K$$

and to a natural equivalence

$$\mathrm{Spec}_{Rpo}|_{\mathrm{FRING}} \xrightarrow{\cong} \mathrm{Spec}_B|_{\mathrm{FRING}} .$$

Also, there is an obvious version of 3.3 for PORING in place of RING, which gives a natural equivalence

$$\mathrm{Spec}_{Rpo} \xrightarrow{\cong} \mathrm{Spec}_B \circ F_{po}.$$

All the details are routine.

Actually, all the spectra described in this section are functors not merely to topology but to the category of spectral spaces and spectral maps. It is a classical theorem—originating in work of M. Stone—that this category is equivalent to the category of distributive lattices. In [Jo], chapter V, Johnstone describes the distributive lattices corresponding to all these spectra directly in terms of generators and relations. This elegant approach to spectra seems to have originated with Joyal in the early 1970's. It is essentially more powerful than treating spectra as spaces of prime ideals because it generalizes to contexts in which the existence of sufficiently many prime ideals may not be verifiable. The space available does not permit a complete discussion, but even if it did, we could never write a better treatment than that which appears already in [Jo]. The discussion of principal-radical ideals in an f-ring immediately following 4.2 (below), however, shows some of the ideas involved.

Propositions 3.2 and 3.3 show that for any ring A, the real spectrum of A is contained in the Keimel spectrum of the free f-ring over A. Using the universal mapping property of $F(A)$, it is not difficult to show that $\mathrm{Spec}_K F(A)$ can be constructed directly from A as the set of all pairs $(I, \mathcal{T})$, where I is an ideal of A (not necessarily prime) and $\mathcal{T}$ is a total *po*-ring order on A/I, with the topology defined as for $\mathrm{Spec}_R A$. Since $\mathrm{Spec}_R A$ has a well-understood geometric meaning, one is led to ask if the additional points present in $\mathrm{Spec}_K F(A)$ have any geometric significance. We have room only for a very brief remark. If $(I, \mathcal{T}) \in \mathrm{Spec}_K F(A)$ is in the closure of a point $(p, \mathcal{Q}) \in \mathrm{Spec}_R A \subseteq \mathrm{Spec}_K F(A)$ (*i.e.*, I contains a real prime ideal p, and A/p supports an order $\mathcal{Q}$, with $\mathcal{T} = \mathcal{Q}/I$), then I/p is a valuation ideal in A/p. In [Z], Zariski demonstrates that under some strong hypotheses, valuation ideals which are simple (*i.e.*, have no non-trivial factorization) correspond in a precise way to infinitely near points. Alvis, Johnston and Madden [AJM] have applied Zariski's theory to give a very clear geometric interpretation of the points of $\mathrm{Spec}_K F(A)$ supported at simple valuation ideals in the case that A is a two-dimensional regular ring. As for the points of $\mathrm{Spec}_K F(A)$ which do not specialize points of $\mathrm{Spec}_R A$, no geometric interpretation is known. Isbell's work [I] gives some idea of where points of the latter type arise.

4. Abstract "Stellensätze" for f-rings

Hilbert's Nullstellensatz establishes a correspondence between geometric objects—algebraic subsets of K^n (K an algebraically closed field)—and algebraic objects—radical ideals of $K[X_1, \ldots, X_n]$. Various "concrete" and "abstract" real Nullstellensätze, "Positivstellensätze" and "Nichtnegativstellensätze" have appeared in the literature. Generally, the concrete versions state a correspondence between semialgebraic subsets of R^n (R a real closed field) and certain ideals, sub-semirings or multiplicative subsets of $R[X_1, \ldots, X_n]$. In the abstract versions, the real spectrum replaces R^n as the stage for the construction of geometric objects. The spectra which we looked at in the previous section permit us to formulate and prove geometric-algebraic correspondences involving f-rings which are similar in spirit to these Stellensätze.

If we analyze proofs of the Nullstellensatz with an eye to making generalizations, we find three elements which account for the character of the theorem. They are:

(1) the facts that a) the cozero sets form a basis for the topology on the Zariski spectrum of any ring, and b) an ideal is radical if and only if it is an intersection of primes;

(2) the fact that each prime ideal of $K[X_1, \ldots, X_n]$ is an intersection of K-points, *i.e.*, ideals at which the residue ring is isomorphic to K;

(3) the Noetherian property of $K[X_1, \ldots, X_n]$.

We shall discuss these briefly in order to understand the variations that can occur in generalizations. The first element by itself is a kind of "abstract Nullstellensatz." Just by unravelling definitions, one can show that for an arbitrary ring A there is an order-preserving bijection between the *po*-set of all those subsets of $\mathrm{Spec}_Z A$ that can be obtained as unions of cozero sets and the *po*-set of all those ideals of A that can be obtained by intersecting prime ideals. Combining this with (1), one gets an order-preserving bijection:

$$\begin{array}{rcl} \text{Open subsets of } \mathrm{Spec}_Z A & \cong & \text{Radical ideals of } A \\ U & \longmapsto & \mathcal{I}(U) := \{\, x \in A \mid \mathrm{coz}(x) \subseteq U \,\} \\ \bigcup\{\, \mathrm{coz}(x) \mid x \in I \,\} =: \mathcal{U}(I) & \longleftarrow & I. \end{array}$$

The syntactic description of the radical of an ideal, *i.e.*, $\sqrt{I} = \{\, x \in A \mid \exists n \ x^n \in I \,\}$, yields the following abstract Nullstellensatz: *For $a \in A$ and $X \subseteq A$, $\mathrm{coz}(a) \subseteq \bigcup\{\, \mathrm{coz}(x) \mid x \in X \,\}$ if and only if there are $n \in \mathbb{N}$, $b_i \in A$ and $x_i \in X$ such that $a^n = \sum_i b_i x_i$.* To prove this, one just notes that each of the conditions is equivalent the condition that a belongs to the radical of the ideal generated by X.

The second element means that no two open sets in $\mathrm{Spec}_Z K[X_1, \ldots, X_n]$ contain exactly the same points of K^n. (Here, we are identifying K^n with a subset of $\mathrm{Spec}_Z K[X_1, \ldots, X_n]$ in the usual way.) This element is the deepest and most difficult to prove, and there are many possible approaches. One way is via elimination of quantifiers for algebraically closed fields, and it is not unusual in these times to hear the opinion that elimination is the ground and origin of the Nullstellensatz. (Not long ago, elimination was somewhat out of favor; see, *e.g.*, A. Weil,

[W], footnote on page 31. What Weil had in mind, however, is probably somewhat different from what is meant by elimination today.)

Element (3) simply assures that every radical ideal (being finitely generated) corresponds to a set with a *finite* description. Without using (3), a correspondence between "algebraic sets" in K^n and radical ideals of $K[X_1, \ldots, X_n]$ can be proved *provided* that by "algebraic set" we understand a subset of K^n defined by the vanishing of some (not necessarily finite) set of polynomials. A demonstration of such a correspondence without any use of (3) can be found in the textbook [Ka] §1-3 and exercise 12 on page 20. Actually, (3) and (2) have some subtle relationships; see [Ka], Theorems 30 and 147.

In 6.3, we have a concrete Nullstellensatz for f-rings of semialgebraic functions. For these rings, (1) holds, (3) fails, and only a weak analogue of (2) is valid. This example involves R-algebras with f-ring structure (R being a real closed field) in which there are only enough R-points to distinguish between cozero sets in the Brumfiel spectrum, not between arbitrary open sets. The most general theorem possible in this case is obtained by considering the traces which *ideals in the lattice of cozero sets* leave on R^n, rather than the traces left by the *open subsets* of $\mathrm{Spec}_K A$.

Let us formulate what might be called an abstract Nullstellensatz for an f-ring and its Keimel spectrum. Let A be an f-ring. It follows from F4 that each ℓ-ideal in A is equal to the intersection of all the irreducible ℓ-ideals that contain it. Also, as mentioned in §3, the cozero sets form a basis for the topology, and so by unravelling definitions, we get an order-preserving bijection between the *po*-set of all open subsets of $\mathrm{Spec}_K A$ and the *po*-set of all ℓ-ideals of A. A syntactic description of ℓ-ideals may be given as follows. If $X \subseteq A$, let

$$\langle X \rangle := \left\{ a \in A \;\middle|\; \exists b_1, \ldots, b_n \in A \quad \exists x_1, \ldots, x_n \in X \quad |a| \leq \sum_{i=1}^{n} b_i |x_i| \right\}.$$

One checks that $\langle X \rangle$ is an ℓ-ideal and is contained in every ℓ-ideal containing X. From these observations, we can deduce the following Nullstellensatz for $\mathrm{Spec}_K A$. (The reader is reminded that the symbol $\mathrm{coz}(x)$ is used with reference to the context—in this case, the Keimel spectrum.)

Proposition 4.1. *Let A be an f-ring. There is an order-preserving bijection*

$$\begin{array}{ccc} \textit{Open subsets of } \mathrm{Spec}_K A & \cong & \ell\textit{-ideals of } A \\ U & \longmapsto & \mathcal{I}(U) := \{ x \in A \mid \mathrm{coz}(x) \subseteq U \} \\ \bigcup\{ \mathrm{coz}(x) \mid x \in I \} =: \mathcal{U}(I) & \longleftarrow & I. \end{array}$$

Moreover, if $a \in A$ and $X \subseteq A$, then $\mathrm{coz}(a) \subseteq \cup\{ \mathrm{coz}(x) \mid x \in X \}$ in $\mathrm{Spec}_K A$ if and only if $|a| \leq \sum_{i=1}^{n} b_i |x_i|$ for some $b_1, \ldots, b_n \in A$ and $x_1, \ldots, x_n \in X$. □

It is clear that the cozero sets in the Keimel spectrum correspond exactly to principal ℓ-ideals. In fact, the cozero sets can be recognized from the topology of $\mathrm{Spec}_K A$: they are precisely the quasi-compact open sets. As mentioned in §3 just after the definition of the Keimel spectrum, the cozero sets form a lattice. In fact,

Spec_K can be constructed directly from the lattice of cozero sets without any other reference to A. The ideas involved in this are elaborated in connection with the Brumfiel spectrum of an f-ring, immediately below. See also [Jo].

Now we shift attention to the Brumfiel spectrum. By the remarks following R2, §2, a radical ℓ-ideal in an f-ring is the intersection of the prime ℓ-ideals containing it. Observing that

$$\sqrt{\langle X\rangle} = \left\{ a \in A \;\middle|\; \exists m \in \mathbb{N}\, \exists b_1,\ldots,b_n \in A\, \exists x_1,\ldots,x_n \in X \quad |a|^m \le \sum_{i=1}^{n} b_i|x_i| \right\},$$

we get the following Nullstellensatz:

Proposition 4.2. *Let A be an f-ring. There is an order-preserving bijection:*

$$\textit{Open subsets of } \mathrm{Spec}_B A \quad\cong\quad \textit{Radical } \ell\textit{-ideals of } A,$$

given by the operators $\mathcal{I}$ and $\mathcal{U}$ (to be precise, the—obvious—versions of these operators for $\mathrm{Spec}_B A$). Moreover, if $a \in A$ and $X \subseteq A$, then $\mathrm{coz}(a) \subseteq \bigcup\{\mathrm{coz}(x) \mid x \in X\}$ in $\mathrm{Spec}_B A$ if and only if $|a|^m \le \sum_{i=1}^{n} b_i|x_i|$ for some $m \in \mathbb{N}$, $b_1,\ldots,b_n \in A$ and $x_1,\ldots,x_n \in X$. □

We shall elaborate a bit on the correspondence between open sets in $\mathrm{Spec}_B A$ and radical ℓ-ideals. These remarks will be used in §6. Let $a \in A$. We call the ideal

$$\langle\langle a\rangle\rangle := \sqrt{\langle a\rangle} = \{\, b \in A \mid \exists c \in A\, \exists\, m \in \mathbb{N} \quad |b|^n \le c|a| \,\}$$

the *principal-radical ℓ-ideal generated by a*. As $\langle\langle a\rangle\rangle = \langle\langle |a|\rangle\rangle$, every principal-radical ℓ-ideal is generated by a non-negative element of A. The following identities, which are quite easy to prove, show that the principal-radical ideals form a sublattice of the lattice of all radical ℓ-ideals:

$$\begin{aligned} \langle\langle a \vee b\rangle\rangle &= \langle\langle a\rangle\rangle \vee \langle\langle b\rangle\rangle \qquad a,b \ge 0 \\ \langle\langle a \wedge b\rangle\rangle &= \langle\langle a\rangle\rangle \wedge \langle\langle b\rangle\rangle \qquad a,b \ge 0. \end{aligned}$$

(The lattice operations on the right are those in the lattice of all radical ℓ-ideals: $I \wedge J = I \cap J$; $I \vee J$ is the intersection of all prime ℓ-ideals containing both I and J.) From the first equation, it is clear that any ℓ-ideal which is finitely generated as a radical ideal is principal-radical.

The lattice of principal-radical ℓ-ideals of A is denoted $\langle\langle A\rangle\rangle$. Since the lattice operations in A are distributive, $\langle\langle A\rangle\rangle$ is a distributive lattice (with least element $\langle\langle 0\rangle\rangle$ and greatest element $\langle\langle 1\rangle\rangle$). The correspondence in 4.2 restricts to a lattice isomorphism

$$\text{Cozero subsets of } \mathrm{Spec}_B A \quad\cong\quad \langle\langle A\rangle\rangle. \tag{4.2.1}$$

In fact, it is possible to construct $\mathrm{Spec}_B A$ directly from $\langle\langle A\rangle\rangle$. We need to recall the construction of the spectrum of a distributive lattice. An *ideal* in a distributive lattice L is a subset $I \subseteq L$ such that: (i) I is non-empty and closed under $\vee$, and (ii) $b \in I$ whenever $b \le a \in I$. The set of all ideals of L (ordered by set-theoretic inclusion) forms a complete lattice, which is denoted $\mathrm{Idl}(L)$. An ideal I is called *prime* if it satisfies: (iii) either $a \in I$ or $b \in I$ whenever $a \wedge b \in I$. The set of all prime ideals of L endowed with the topology generated by the sets of the form $\{ I \mid a \notin I \}$ $(a \in L)$ is called the *(lattice) spectrum of* L, and is denoted spec$L$. It is possible to show that there is an order-preserving bijection between the open sets in spec$L$ and Idl(L) and that spec$L$ is completely determined by Idl(L); see [Jo].

Using (4.2.1), it is routine to show that there is an order-preserving bijection,

$$\text{Radical } \ell\text{-ideals of } A \quad \cong \quad \mathrm{Idl}(\langle\langle A \rangle\rangle). \tag{4.2.2}$$

Because $\mathrm{Spec}_B A$ is completely determined by its topology, which is (isomorphic as a lattice to) the lattice of radical ℓ-ideals of A—and similarly for spec$\langle\langle A \rangle\rangle$—it follows from (4.2.2) that for any f-ring A,

$$\mathrm{Spec}_B A \cong \mathrm{spec}\langle\langle A \rangle\rangle. \tag{4.2.3}$$

To finish this section, we give an application of 4.2. Let G denote an arbitrary ring. By applying 4.2 and 3.3 to $F(G)$, the free f-ring over G, we obtain the following version of the abstract Hörmander–Łojasiewicz inequality. This theorem concerns the behavior of abstract functions on the real spectrum of G.

Corollary 4.3. *Let G be a ring. Suppose $g \in G$ and $X \subseteq G$. If $\mathrm{coz}(g) \subseteq \cup\{ \mathrm{coz}(x) \mid x \in X \}$ in $\mathrm{Spec}_R G$, then for some $m \in \mathbb{N}$, $c_1, \ldots, c_n \in G$ and $x_1, \ldots, x_n \in X$, we have*

$$\forall \xi \in \mathrm{Spec}_R G \qquad |g(\xi)|^m \leq \sum_{i=1}^{n} c_i(\xi)|x_i(\xi)|.$$

Proof. From 3.3 and 4.2, we get $b_1, \ldots, b_n \in F(G)$ such that $|g|^n \leq \sum_{i=1}^{n} b_i|x_i|$ (in $F(G)$). It is only necessary to show that the coefficients can be chosen in G, and this is obvious from F6 and the following f-ring inequality (which follows from the simpler inequality $|y| \leq 1 + y^2$ valid in any totally ordered ring, hence in any f-ring):

$$\bigvee_i \bigwedge_j y_{ij} \leq \sum_{ij} 1 + y_{ij}^2. \qquad \square$$

The abstract Hörmander–Łojasiewicz inequality first appeared in [ABR] in a stronger form than 4.3. The version we have presented can be relativized to closed subsets of $\mathrm{Spec}_R G$ defined by partial orders on G by using Spec_{Rpo} and F_{po} in place of Spec_R and F, as the reader may verify without trouble. In this way, the full strength of the original version is attained.

5. Nichtnegativstellensatz for sup-inf definable functions

The abstract Nichtnegativstellensatz concerns the real spectrum. Let A be a ring. For $S \subseteq A$, we define

$$\mathcal{W}(S) := \{\xi \in \mathrm{Spec}_R A \mid \forall s \in S \;\; s(\xi) \geq 0\},$$

and for $X \subseteq \mathrm{Spec}_R A$, we define

$$\mathcal{P}(X) := \{a \in A \mid \forall \xi \in X \;\; a(\xi) \geq 0\}.$$

We call $\mathcal{WP}(X)$ the *nn-hull of* X, and if $X = \mathcal{WP}(X)$ we say X is *nn-convex* ("*nn*" means "non-negative"). Note: *nn*-convexity and the *nn*-hull of X depend on A and not only on $\mathrm{Spec}_R A$. For example, although $\mathrm{Spec}_R R(A)$ and $\mathrm{Spec}_R A$ are homeomorphic, where $R(A)$ is the real closure of A, every closed subset is *nn*-convex relative to $R(A)$ but not necessarily relative to A—see the example following 5.1. We call $\mathcal{PW}(S)$ the *nn-saturation of* S, and if $S = \mathcal{PW}(S)$ we say S is a *saturated cone*. It is trivial that $\mathcal{WPW}(S) = \mathcal{W}(S)$ and $\mathcal{PWP}(X) = \mathcal{P}(X)$. Thus $\mathcal{W}$ and $\mathcal{P}$ give an order-reversing correspondence between the *po*-set of *nn*-convex sets in $\mathrm{Spec}_R A$ and the *po*-set of saturated cones in A. Since

$$\mathcal{W}(S) = \mathrm{Spec}_R A \setminus \bigcup_{s \in S} \mathrm{pos}(-s),$$

there is an order-preserving correspondence between the saturated cones and the sets of the form $\cup\{\mathrm{pos}(-s) \mid s \in S\}$. This makes the analogy with the description of the Nullstellensatz in §4 clear.

The so-called Nichtnegativstellensatz is the following syntactic description of the saturated cone generated by a given subset of A.

Lemma 5.1. *Let X be a subset of A, and let $S \subseteq A$ be the subsemiring of A generated by $X \cup A^2$. Then*

$$\mathcal{P}(X) = \{f \in A \mid \exists s_1, s_2 \in S \; \exists k \in \mathbb{N} \quad (f^{2k} + s_1)f = f^{2k} + s_2\}.$$

Proof. See [L], 7.4(2). □

In the Zariski spectrum, the sets of the form $\cup\{\mathrm{coz}(s) \mid s \in S\}$ are exactly *the open sets*. The Nichtnegativstellensatz appears to have a shortcoming in that we lack a similar nice description of the *nn*-convex sets. In the real spectrum, the intersection of two positivity sets (*i.e.*, sets of the form $\mathrm{pos}(a)$, $a \in A$) need not be a positivity set, and so not every open set need be a union of positivity sets. As a result, for an arbitrary ring the *nn*-convex sets are difficult to characterize.

Example. Let $\mathbb{R}$ be the field of real numbers and let $A = \mathbb{R}[x, y]$, where x and y are indeterminates. In $\mathrm{Spec}_R A$, the set $C := \{\xi \mid x(\xi) \leq 0 \text{ or } y(\xi) \leq 0\}$ is a closed constructible set which is not nn-convex. To see this, argue as follows. Suppose the complement $C' := \{\xi \mid x(\xi) > 0 \text{ and } y(\xi) > 0\}$ were a union of positivity sets. Since C' is quasicompact, finitely many would suffice, say $C' = \text{pos}(a_1) \cup \cdots \cup \text{pos}(a_n)$. Then at least one of the a_i would change sign on an interval in the positive x-axis. By the Transversal Zeros Theorem this element would also change sign at points on the negative x-axis, so $\text{pos}(a_i)$ would contain points not in C'. This would be a contradiction. One can show that the nn-hull of C is C together with all $\xi \in \text{Spec}_R A$ having a specialization supported on one of the primes (x) or (y).

The fact that some open sets are missing from the correspondence may or may not be seen as a defect, according to taste. Recent work on Hilbet's 17th Problem, however, has uncovered a need arising out of solid technical reasons going beyond mere taste for a correspondence comprehending closed sets which are not nn-convex; see [D]. In this work, an association is made between arbitrary closed sets in $\text{Spec}_R A$ and syntactic data in $F(A)$, the free f-ring over A. To understand this, it is important to grasp the relationship between $\text{Spec}_R A$ and $\text{Spec}_R F(A)$ (the latter expression, as always, referring to the real spectrum of the *underlying ring of* $F(A)$). Below, we continue to suppress notation for forgetful functors. For example, we write $F(F(A))$ for the free f-ring on the underlying ring of $F(A)$.

Lemma 5.2. $\text{Spec}_R A$ *is a retract of* $\text{Spec}_R F(A)$. *I.e., there are continuous maps* $s : \text{Spec}_R A \to \text{Spec}_R F(A)$ *and* $r : \text{Spec}_R F(A) \to \text{Spec}_R A$ *such that* $r \circ s$ *is the identity on* $\text{Spec}_R A$.

Proof. Let $\iota_{F(A)}$ denote the identity on $F(A)$. Then

$$\overline{\iota_{F(A)}} \circ \overline{\eta_{F(A)} \circ \eta_A} \circ \eta_A = \overline{\iota_{F(A)}} \circ \eta_{F(A)} \circ \eta_A = \iota_{F(A)} \circ \eta_A,$$

and so by the uniqueness clause of 1.2(a),

$$\overline{\iota_{F(A)}} \circ \overline{\eta_{F(A)} \circ \eta_A} = \iota_{F(A)}.$$

This shows that $F(A)$ is a retract of $F(F(A))$. Applying the functor Spec_B, one gets that $\text{Spec}_B F(A)$ is a retract of $\text{Spec}_B F(F(A))$ (retraction is self-dual). The lemma follows by applying 3.3. □

Remarks. Note that the image of $\overline{\eta_{F(A)} \circ \eta_A}$ is simply the sub-f-ring of $F(F(A))$ generated by $\eta_{F(A)} \circ \eta_A(A)$. The proof gives $s = q_{F(A)}^{-1} \circ \text{Spec}_B(\overline{\iota_{F(A)}}) \circ q_A$ and $r = q_A^{-1} \circ \text{Spec}_B(\overline{\eta_{F(A)} \circ \eta_A}) \circ q_{F(A)}$. The real spectra of $F(A)$ and of certain rings containing $F(A)/\mathcal{N}$ have been examined in detail by Schwartz in [S]. Our 5.2 may be extracted from the material in §1 and §2 of his paper. Schwartz's methods are quite different from those we employ, depending on the theory of real closed rings rather than universal algebra. His paper therefore is an important reference for anyone desiring to pursue the topics we are discussing here.

By 5.2, $\text{Spec}_R A$ is homeomorphic with the image of s endowed with the subspace topology. From now on, we treat $\text{Spec}_R A$ as a subspace of $\text{Spec}_R F(A)$. There is no loss of generality in assuming that $\eta_A : A \to F(A)$ is injective, since the kernel of this morphism is contained in the real radical of A. We shall identify A with its image in $F(A)$. *In the remainder of this section,* $\mathcal{W}(S)$ *refers to a subset of* $\mathrm{Spec}_R F(A)$, *and* $\mathrm{pos}_A(f) := \{\xi \in \mathrm{Spec}_R A \mid f(\xi) > 0\}$*—even when* $f \in F(A)$.

For any f-ring F, we define $|F| := \{\,|f| \mid f \in F\,\}$.

Lemma 5.3. *In* $\mathrm{Spec}_R F(A)$, $\mathrm{Spec}_R A = \mathcal{W}(|F(A)|)$. *Every closed subset of* $\mathrm{Spec}_R A$ *is nn-convex in* $\mathrm{Spec}_R F(A)$ *(relative to* $F(A)$*).*

Proof. Let $\xi \in \mathrm{Spec}_R F(A)$. Then,

$$\begin{aligned}\xi \in \mathrm{Spec}_R A &\iff F(A) \ni f \mapsto f(\xi) \in F(A)[\xi] \text{ is an } f\text{-ring morphism}\\ &\iff f \geq 0 \text{ (in } F(A)) \text{ implies } f(\xi) \geq 0\\ &\iff \text{for all } f \in F(A),\ |f|(\xi) \geq 0.\end{aligned}$$

(The first equivalence is by the same idea used in the proof of 1.3. In the last line, $|f|(\xi)$ denotes the image of $|f|$ in $F(A)[\xi]$, which need not be nonnegative in the order on $F(A)[\xi]$.) Thus,

$$\mathrm{Spec}_R A = \bigcap_{f \in F(A)} \{\xi \in \mathrm{Spec}_R F(A) \mid |f|(\xi) \geq 0\} = \mathcal{W}(|F(A)|)$$

(*i.e.*, the complement of $\mathrm{Spec}_R A$ is a union of positivity sets). It remains to see that *every* closed subset of $\mathrm{Spec}_R A$ is nn-convex in $\mathrm{Spec}_R F(A)$ (relative to $F(A)$). From 3.1, it follows that for any $f_1, \ldots, f_n \in F(A)$,

$$\mathrm{pos}_A\Big(\bigwedge_{i=1}^{n} f_i\Big) = \bigcap_{i=1}^{n} \mathrm{pos}_A(f_i)\,.$$

This shows that a finite intersection of sets of the form $\mathrm{pos}_A(f)$, $f \in A$, is the intersection with $\mathrm{Spec}_R A$ of a positivity set in $\mathrm{Spec}_R F(A)$. Hence, the complement in $\mathrm{Spec}_R F(A)$ of any closed subset of $\mathrm{Spec}_R A$ is a union of positivity sets, which is equivalent to the claim. □

From 5.1 and 5.3, we immediately get the following Nichtnegativstellensatz for sup-inf-definable abstract functions, a version of which was described by the first author at the Trento Meeting.

Theorem 5.4. *Let A be a ring and let $F(A)$ be the free f-ring over A.*
a) *There is an order-preserving bijection:*

$$\begin{array}{ccc} \textit{Open subsets of } \mathrm{Spec}_R A & \cong & \begin{array}{c}\textit{Saturated cones in } F(A)\\ \textit{containing } |F(A)|\end{array}\\ U & \longmapsto & \{f \in F(A) \mid \mathrm{pos}_A(-f) \subseteq U\}\\ \bigcup\{\mathrm{pos}_A(-f) \mid f \in P\} & \longleftarrow & P \end{array}$$

b) *Moreover, if* $f \in F(A)$ *and* $X \subseteq F(A)$*, then*

$$\mathrm{pos}_A(-f) \subseteq \bigcup_{x \in X} \mathrm{pos}_A(-x)$$

if and only if $(f^{2k} + s_1)f = f^{2k} + s_2$ *for some* $k \in \mathbb{N}$, *and some* s_1, s_2 *in the subsemiring of* $F(A)$ *generated by* $X \cup (F(A))^2 \cup |F(A)|$. □

There is a version of 5.4 for *po*-rings obtained by replacing A with $(A, \mathcal{P})$, F with F_{po} and Spec_R with Spec_{Rpo} in the above. We leave it to the reader to prove this by verifying the (obvious) analogues 5.2 and 5.3. (In doing this, it is useful to note that $\eta_{(A,\mathcal{P})}(\mathcal{P}) \subseteq |F_{po}(A, \mathcal{P})|$.)

6. Semialgebraic and SIPD functions

Throughout this section, R *denotes a real closed field,* X *denotes a closed semialgebraic subset of* R^n, *and* A *is an* f*-ring of* R*-valued continuous semialgebraic functions on* X *containing the restrictions to* X *of the polynomials.*

We give a correspondence between the closed semialgebraic subsets of X and the principal-radical ℓ-ideals of A. In a way reminiscent of the classical Nullstellensatz, this result is closely connected with elimination of quantifiers for the base field. It is possible to obtain the correspondence by combining the abstract theorems of §4 with an elimination argument. Our purpose, however, is to give readers a sense of the connections between f-rings and geometry, not to write a formal development from minimum assumptions. Therefore, we take a different approach. In 4.3, we derived the abstract Hörmander–Łojasiewicz inequality from an abstract Nullstellensatz for f-rings. In this section, we turn the tables and take the concrete version of the inequality as the starting point. The only other result from geometry we call on is the Finiteness Theorem. As a by-product of what we prove here, we get a version of Bröcker's ultrafilter theorem.

At the end of this section, we include a brief discussion of partially ordered algebras over partially ordered rings, which is used in §7, but is of independent interest.

HŁI (Hörmander–Łojasiewicz Inequality). *Let* $f, g : X \to R$ *be continuous semialgebraic functions. Suppose that for each* $\varepsilon > 0$ *in* R, $\{x \in X \mid |g(x)| \geq \varepsilon\}$ *is closed and bounded, and* $Z(f) \subseteq Z(g)$. *Then, there are* $c \in R$ *and a positive integer* m *such that* $|g|^m \leq c|f|$ *on* X.

Proof. See [C], p. 118–119. □

Lemma 6.1. *Suppose* $I \subseteq A$ *is a radical* l*-ideal. If* $f \in I$, $h \in A$ *and* $Z(f) \subseteq Z(h)$, *then* $h \in I$.

Proof. Let $H := |h| \vee 1$ and $D := (1 + ||x||^2)^{-1}$. By HŁI, there are $c_0 \in R$ and $m_0 \in \mathbb{N}$ such that

$$D^{m_0} \leq c_0/H.$$

Thus

$$g := |h|D^{m_0+1} \leq HD^{m_0+1} \leq c_0 D.$$

Now f and g satisfy the hypothesis of HŁI, so there are $c \in R$ and $m \in \mathbb{N}$ such that $g^m \leq c|f|$, *i.e.*,

$$|h|^m \leq c(1 + ||x||^2)^{m(m_0+1)}|f|.$$

The right side belongs to I because $f \in I$. Since I is convex and radical, $h \in I$. □

If $a \in A$, let $\mathrm{COZ}(a) := \{ x \in X \mid a(x) \neq 0 \}$. This is an open semialgebraic subset of X.

Proposition 6.2. *The map*

$$\langle\langle |a| \rangle\rangle \mapsto \mathrm{COZ}(a)$$

is a lattice isomorphism from $\langle\langle A \rangle\rangle$*, the lattice of principal-radical ℓ-ideals of A (see §4), to the lattice of open semialgebraic subsets of X.*

Proof. First, the mapping is well defined, since if $\langle\langle |a| \rangle\rangle = \langle\langle |b| \rangle\rangle$, then $|a|^m \leq t|b|$ and $|b|^{m'} \leq t'|a|$ (for some $m, m' \in \mathbb{N}$ and $t, t' \in A$), so $\mathrm{COZ}(a) = \mathrm{COZ}(b)$. The map is surjective, for if $C \subseteq X$ is any closed semialgebraic set, then by the Finiteness Theorem (*e.g.*, [C]) there are finitely many polynomials f_{ij} such that $C = \bigcup_i \bigcap_j \{ x \in X \mid f_{ij}(x) \geq 0 \}$. Then $C = \{ x \in X \mid 0 = \bigvee_i \bigwedge_j (f_{ij} \wedge 0) \}$, which shows that every open semialgebraic subset of X is $\mathrm{COZ}(a)$ for some $a \in A$. The map is injective, for suppose $\mathrm{COZ}(a) = \mathrm{COZ}(b)$. By the lemma, $a \in \langle\langle |b| \rangle\rangle$ and $b \in \langle\langle |a| \rangle\rangle$, so $\langle\langle |a| \rangle\rangle = \langle\langle |b| \rangle\rangle$. It is easy to see that both the map and its inverse are order-preserving. □

Corollary 6.3. *Let L be the lattice of open semialgebraic subsets of X. Then*

$$\mathrm{Spec}_B A \cong \mathrm{spec} L.$$

Proof. This is immediate from the homeomorphism in equation (4.2.3). □

By an argument similar to that used in [HI] to establish F7 of §1, the reduced free R-f-algebra $F(R[Z_1, \dots, Z_m])/\mathcal{N}$, where the Z_i are indeterminates, is the smallest sub-f-ring of the f-ring of all R-valued functions on R^m which contains the polynomials. (See 6.5 for an even more general fact.) Using this, and applying 3.3, we immediately get:

Corollary 6.4. $\mathrm{Spec}_R R[Z_1, \dots, Z_m]$ *is homeomorphic to the lattice spectrum of the lattice of all open semialgebraic subsets of R^m. In particular, each point of* $\mathrm{Spec}_R R[Z_1, \dots, Z_m]$ *corresponds to a prime ideal of open semialgebraic subsets of R^m.* □

This corollary can, of course, be rephrased in terms of prime filters in the lattice of all closed semialgebraic subsets of R^m, or in terms of ultrafilters in the Boolean reflection of this lattice. This is just a matter of notation. Thus the corollary is Bröcker's ultrafilter theorem; see [BCR], 7.2.4.

We now develop some basic facts about algebras with f-ring structure. We intend to treat this topic in more detail in a future publication.

Let K be a ring. As usual, a *K-algebra* is a ring map $K \to A$. If K and A are *po*-rings and $K \to A$ is a *po*-ring morphism, we have what we call a *po-K-algebra*.

If, in addition, A is an f-ring, we speak of an *f-K-algebra.* (Note that the *po*-ring action $K \to A$ extends to an action of f-rings $F_{po}(K) \to A$. For this reason, there seems to be no reason to consider separately the case in which K is an f-ring.) An f-K-algebra morphism is a commuting triangle

$$\begin{array}{ccc} & K & \\ \swarrow & & \searrow \\ A & \longrightarrow & B \end{array}$$

in which $A \to B$ is an f-ring morphism. The action of K may be considered as a collection of unary operations. The category of f-K-algebras is obviously equational.

If $a : K \to A$ is a *po*-K-algebra, then $F_{po}(A)$ is naturally an f-K-algebra via the composite *po*-ring morphism

$$K \xrightarrow{a} A \xrightarrow{\eta_A} F_{po}(A).$$

Moreover, any *po*-ring morphism $A \to B$ which is also a *po*-K-algebra morphism to an f-K-algebra B lifts uniquely to an f-K-algebra morphism $F_{po}(A) \to B$. Thus $F_{po}(A)$ may be identified with the free f-K-algebra over A. (If K has a unique order and A carries the weakest order for which $K \to A$ is a *po*-ring morphism, then $F_{po}(A) \cong F(A)$.)

Definition. Let A be an f-K-algebra. Suppose $Y \subseteq A^m$. Then $\mathrm{SIPD}_K(Y)$ denotes the sub-f-K-algebra of the f-K-algebra of all functions $Y \to A$ generated by the restrictions to Y of the canonical projections.

By F6, every element of $\mathrm{SIPD}_K(Y)$ is the restriction to Y of some function of the form

$$\bigvee_{i=1}^{m} \bigwedge_{j=1}^{n_i} f_{ij},$$

where the f_{ij} are polynomial functions with coefficients in A. (By a polynomial function, we mean a function obtained by substituting the canonical projections for the variables in an element of $A[Z_1, \ldots, Z_m]$.)

For any f-K-algebra A, we have the subcategory (of the category of all f-K-algebras) $\mathbf{HSP}(A)$, the smallest subcategory containing A and closed under homomorphic images, subalgebras and products. By Birkhoff's theorem (stated, for example, in [Ja], Theorem 2.15), $\mathbf{HSP}(A)$ consists of all f-K-algebras which satisfy all f-K-algebra identities valid in A. By standard universal algebra (as in [HI]), $\mathrm{SIPD}_K(A^m)$ is the free $\mathbf{HSP}(A)$-object on m generators.

From now until the end of this section, assume that R is a real closed field and K is a subdomain with the inherited total order. Both K and R are f-K-algebras via the identity map on K. In general the categories $\mathbf{HSP}(K)$ and $\mathbf{HSP}(R)$ are different. For instance, $Z(Z-1) \wedge 0 = 0$ is a valid identity in the integers, but not in any totally ordered field. In fact, there are examples of (non-archimedean) *fields* K for which there are polynomials $f \in K[Z]$ such that $f(a) \geq 0$ for all $a \in K$, but $f(r) < 0$ for some $r \in R$; see p. 99 of [AS]. If f is such a polynomial, then $f \wedge 0 = 0$

is an equation valid in K but not in R. (What we are saying is *stronger* than just that the restriction map $\mathrm{SIPD}_K(R^m) \to \mathrm{SIPD}_K(K^m)$ need not be injective. R need not even be embeddable in any homomorphic image of any sub-f-K-algebra of any power of K.)

Lemma 6.5. *Regard $K[Z_1, \ldots, Z_m]$ as a po-ring, by endowing it with the weakest partial order inducing the given order on K. Let $F = F_{po}(K[Z_1, \ldots, Z_m])$ and $\mathcal{N} = \mathcal{N}(F)$. Then $F/\mathcal{N} \cong \mathrm{SIPD}_K(R^m)$.*

Proof. This is very much like the proof of 4.5 in [HI]. By Tarski's transfer principle, it is immediate that any f-K-algebra law violated in a totally ordered f-K-algebra without zero-divisors is violated in R. Every totally ordered reduced f-K-algebra, therefore, belongs to $\mathbf{HSP}(R)$, by Birkhoff's theorem. Hence, any reduced f-K-algebra—being a sub-algebra of a product of totally ordered reduced f-K-algebras, by R2 and F4—is in $\mathbf{HSP}(R)$. The kernel of the canonical surjection $F \to \mathrm{SIPD}_K(R^m)$ contains $\mathcal{N}$ since $\mathrm{SIPD}_K(R^m)$ is reduced, and the kernel is no larger than $\mathcal{N}$ since $F/\mathcal{N}$ belongs to $\mathbf{HSP}(R)$. (Reminder: $\mathrm{SIPD}_K(R^m)$ is the free $\mathbf{HSP}(R)$-object on m generators, and therefore maps onto $F/\mathcal{N}$.) □

7. Application to Hilbert's 17th problem

The abstract Nichtnegativstellensatz 5.1 easily implies Stengle's "concrete" Nichtnegativstellensatz for polynomials, either by Tarski's elimination of quantifiers, or by its special case, Lang's homomorphism theorem; see [L], §8. In a similar way, the abstract Nichtnegativstellensatz for "sup-inf-A functions" 5.4(b) easily implies the following concrete Nichtnegativstellensatz.

Theorem 7.1. *Assume that R is real closed field and K is a subfield with the inherited total order. Suppose $f, g_1, \ldots, g_t \in \mathrm{SIPD}_K(R^m)$. Then $f(z) \geq 0$ for all $z \in R^m$ at which each $g_l(z) \geq 0$ if and only if*

$$(f^{2e} + s_1)f = f^{2e} + s_2,$$

for some $e \in \mathbb{N}$ and some s_1, s_2 in the subsemiring of $\mathrm{SIPD}_K(R^m)$ generated by the perfect squares, the g_l, and $|\mathrm{SIPD}_K(R^m)|$.

Proof. The "if" direction is obvious. For the "only if" direction, suppose that no such identity exists. Regard $A := K[Z_1, \ldots, Z_m]$ as a *po*-ring, by endowing it with the weakest partial order inducing the given order on K. By 6.5, $F_{po}(A)/\mathcal{N} \cong \mathrm{SIPD}_K(R^m)$. Thus, by 5.4(b) (or, to be precise, its analogue for *po*-rings) there exists

$$\alpha \in \mathrm{Spec}_R R[Z_1, \ldots, Z_m] \cong \mathrm{Spec}_{Rpo} K[Z_1, \ldots, Z_m]$$

such that $f(\alpha) < 0$ and each $g_l(\alpha) \geq 0$. Let $A[\alpha]$ be the "ordered residue domain at α," as described in §3. Write $f = \bigvee_i \bigwedge_j f_{ij}$ and $g_l = \bigvee_i \bigwedge_j g_{lij}$, for finitely many $f_{ij}, g_{lij} \in K[Z_1, \ldots, Z_m]$. Consider the following first-order statement in the language of ordered rings with constants from K:

$$\exists y_1, \ldots, y_m \ \big[\forall i \ \exists j \ f_{ij}(y_1, \ldots, y_m) < 0 \ \& \ \forall l \ \exists i \ \forall j \ g_{lij}(y_1, \ldots, y_m) \geq 0\big].$$

(The apparent quantification over indices i, j and l is used to abbreviate a long formula, built up from finitely many conjunctions and disjunctions.)

Taking $y_r = Z_r(\alpha) \in A[\alpha]$ shows that this sentence is satisfied in $A[\alpha]$ and *a fortiori* in the real closure of the ordered fraction field of $A[\alpha]$. Tarski's elimination procedure allows us to replace this real closed field by any other real closed field which contains K and induces the same ordering on K. We choose R, obtaining a refutation of the hypothesis that $f \geq 0$ wherever all the $g_l \geq 0$ in R^m. □

In 1988 [D] the first author had derived 7.1 from 5.1 rather than from 5.4(b) as above; in 1991 González-Vega and Lombardi [GL] re-discovered 7.1, deriving it from Stengle's concrete Nichtnegativstellensatz for polynomials, rather than from 5.1 or 5.4(b).

In this paper we do not state in detail the many refinements and variations of 5.1, Stengle's theorem, 5.4(b), or 7.1. For an example of such a refinement, instead of starting with the single set $S \subseteq A$ (or $\subseteq F(A)$), one may start with 3 sets $S_>, S_\geq, S_= \subseteq A$ (or even a fourth set, $S_{\neq}$); then, instead of defining $\mathcal{W}(S)$ (§5), one considers the set $\mathcal{V}$ of points where all functions in $S_>, S_\geq, S_=, S_{\neq}$ are > 0, ≥ 0, $= 0$, and $\neq 0$, respectively. For variations, one may give sum-of-squares identities similar to those in 5.1 or 5.4(b) which are equivalent to f being *positive* on $\mathcal{V}$, or *zero* on $\mathcal{V}$, instead of nonnegative there. Another variation is to omit explicit mention of any distinguished function f, and give a sum-of-squares identity equivalent to the *emptiness* of $\mathcal{V}$; perhaps this should be called the "Keinstellensatz," as opposed to the Nullstellensatz, Nichtnegativstellensatz, and Positivstellensatz. Anyway, each of these Stellensätze is easily derived from any of the others.

Also in 1988, the first author applied 7.1 to Hilbert's 17th problem, obtaining a continuously (even sup-inf-polynomially) varying version of the latter. To state it, let $X = \{X_1, \ldots, X_n\}$ be a set of indeterminants. For any multi-index $\theta = (\theta_1, \ldots, \theta_n) \in \mathbb{N}^n$, let X^θ denote the monomial $X_1^{\theta_1} \cdots X_n^{\theta_n}$ and let $|\theta| = \sum_i \theta_i$. Let d be a positive integer, which will remain fixed throughout. The number of monomials of degree $\leq d$ is $D := \binom{n+d}{n}$. For each θ with $|\theta| \leq d$, let C_θ be a distinct indeterminate, and let $C := \{\, C_\theta \mid |\theta| \leq d \,\}$. The general polynomial of degree d is

$$f(C; X) := \sum_{|\theta| \leq d} C_\theta X^\theta \in \mathbb{Z}[C; X].$$

Let

$$P_{nd} = \{\, c \in R^D \mid \forall x \in R^n \quad f(c; x) \geq 0 \,\}.$$

By Tarski's elimination of quantifiers, P_{nd} is semialgebraic, and in fact definable by inequalities involving only polynomials in $\mathbb{Z}[C]$. It is easy to see that P_{nd} is closed in the usual topology on R^D. By the finiteness theorem for closed semialgebraic sets (e.g., [BCR] or [C]), there are finitely many $h_{kl} \in \mathbb{Z}[C]$ such that

$$P_{nd} = \bigcup_k \bigcap_l \{ c \in R^D \mid h_{kl}(c) \geq 0 \}.$$

Let $h := \bigvee_k \bigwedge_l h_{kl} \in \mathrm{SIPD}_{\mathbf{Z}}(R^D)$. Then

$$P_{nd} = \{ c \in R^D \mid h(c) \geq 0 \}.$$

Corollary 7.2 (SIPD Hilbert 17). *There is an identity*

$$f(c;X) = \sum_i p_i(c) \left(\frac{q_i(c;X)}{s(c;X)} \right)^2, \tag{7.2.1}$$

where:

a) $p_i \in \mathrm{SIPD}_{\mathbf{Z}}(R^D)$ *and* $p_i \geq 0$ *on* P_{nd}, *and*

b) $q_i, s \in \mathrm{SIPD}_{\mathbf{Z}}(R^D)[X]$, *and* $\exists e \in \mathbb{N}$ *such that*

$$s(c;X) = f(c;X)^{2e} + \sum_j t_j(c) g_j(c;X)^2,$$

where $g_j \in \mathrm{SIPD}_K(R^D)[X]$, $t_j \in \mathrm{SIPD}_{\mathbf{Z}}(R^D)$ *and* $t_j \geq 0$ *on* P_{nd}.

Moreover, each $p_i(q_i/s)^2$ *extends (by 0) to a semialgebraic, locally uniformly (even locally Lipschitz) continuous function* $P_{nd} \times R^n \to R$.

Proof. Use the notation of 7.1, taking $m = D + n$, $C = \{Z_1, \ldots, Z_D\}$, $X = \{Z_{D+1}, \ldots, Z_m\}$, and $\{ g_1, \ldots, g_t \} = \{h\}$. (Regard $\mathrm{SIPD}_K(R^D)$ as a sub-f-ring of $\mathrm{SIPD}_K(R^m)$ in the obvious way, via the projection $R^m \to R^D$.) By the definition of P_{nd}, $f \geq 0$ on $P_{nd} \times R^n$. By 7.1, $(f^{2e} + v_1)f = f^{2e} + v_2$, for some $e \in \mathbb{N}$ and v_1, v_2 in the subsemiring of $\mathrm{SIPD}_K(R^m)$ generated by the perfect squares, h, and $|\mathrm{SIPD}_K(R^m)|$. (Actually, from the last set, we need only the finitely many elements $h - \bigwedge_l h_{kl}$ and $h_{kl} - \bigwedge_l h_{kl}$, which lie even in $\mathrm{SIPD}_{\mathbf{Z}}(R^D)$). Thus

$$f = \frac{f^{2e} + v_2}{f^{2e} + v_1} = \frac{f^{2e} + v_2}{f^{2e} + v_1} \cdot \frac{f^{2e} + v_1}{f^{2e} + v_1} = \frac{f^{4e} + v_3}{(f^{2e} + v_1)^2} = \sum_i p_i(c) \left(\frac{q_i(c;X)}{s(c;X)} \right)^2,$$

where v_3 has the same form as v_1 and v_2, $s = f^{2e} + v_1$, and the p_i have the form stated.

To prove the last statement, extend $p_i(q_i/s)^2$ to R^m by declaring the value of the extension to be 0 wherever $s = 0$. The semialgebraicity of the extension is obvious. The only points $(c;x) \in P_{nd} \times R^n$ at which the (locally Lipschitz) continuity of the extension is in question are those where $s = 0$; but provided that $(c;x) \in P_{nd} \times R^n$, $s(c;x) = 0$ only if $f(c;x) = 0$. By 7.2.1, each $p_i(q_i/s)^2$ tends to 0 near such $(c;x)$. This shows pointwise continuity; locally Lipschitz continuity follows from the corresponding property for f. □

A derivation of 7.2 from 7.1 similar to the above will appear in [GL]. In [D] and [DGL], however, 7.2 was derived without explicit use of 7.1 (though this was only a minor variation of the above proof).

References

[AJM] D. Alvis, B. Johnston, J. Madden, Complete ideals defined by sign conditions and the real spectrum of a two-dimensional local ring. Submitted.

[ABR] C. Andradas, L. Bröcker, J. Ruiz, Minimal generation of basic open semianalytic sets. Invent. Math. 92 (1988), 409–430.

[AS] E. Artin, O. Schreier, Algebraische Konstruktionen reeller Körper. Abhandl. Math. Sem. Hamburg 5 (1926), 85–99.

[BKW] A. Bigard, K. Keimel, S. Wolfenstein, Groupes et anneaux réticulés. Lecture Notes in Math. 608, Springer-Verlag, Berlin-Heidelberg-New York 1977.

[BP] G. Birkhoff, R.S. Pierce, Lattice ordered rings. Anais Acad. Bras. Ci. 28 (1956), 41–69.

[BCR] J. Bochnak, M. Coste, M.-F. Roy, Géométrie algébrique réelle. Ergebnisse der Mathematik und ihrer Grenzgebiete, 3. Folge, Band 12, Springer-Verlag, Berlin-Heidelberg-New York 1987.

[B] G. Brumfiel, Partially Ordered Rings and Semi-Algebraic Geometry. London Math. Soc. Lecture Note Series 37, Cambridge 1979.

[C] M. Coste, Ensembles semi-algébriques. In: Géométrie Algébrique Réelle et Formes Quadratiques (ed. by J.-L. Colliot-Thélène, M. Coste, L. Mahé, M.-F. Roy), 109–138. Lecture Notes in Math. 959, Springer-Verlag, Berlin-New York 1982.

[D] C. Delzell, Continuous, piecewise-polynomial functions which solve Hilbert's 17th problem. J. reine angew. Math. 440 (1993), 157–173. Abstract in AMS Abstracts 10, Number 3, Issue 63 (April 1989), #849-14-160, 208–9.

[DGL] C. Delzell, L. González-Vega, H. Lombardi. A continuous and rational solution to Hilbert's 17th problem, and several cases of the Positivstellensatz. In: Computational Algebraic Geometry (ed. by F. Eyssette and A. Galligo), Progress in Math. 109, Birkhäuser (1993), 61–75.

[GL] L. González-Vega, H. Lombardi, A real Nullstellensatz and Positivstellensatz for the semipolynomials over an ordered field. J. Pure & Appl. Algebra 90 (1993), in press.

[HM] A.W. Hager, J.J. Madden, Essential reflections versus minimal embeddings. J. Pure Appl. Algebra 37 (1985), 27–32.

[HI] M. Henriksen, J.R. Isbell, Lattice ordered rings and function rings. Pacific J. Math. 12 (1962), 533–566.

[HS] H. Herrlich, G. Strecker, Category Theory: An Introduction. Allyn & Bacon, Boston 1973.

[I] J. Isbell, Notes on ordered rings. Algebra Universalis 1 (1972), 393–399.

[Ja] N. Jacobson, Basic Algebra II. W.H. Freeman, San Francisco 1980.

[Jo] P.T. Johnstone, Stone Spaces. Cambridge Univ. Press 1983.

[Ka] I. Kaplansky, Commutative Rings, Revised edition. Univ. Chicago Press 1974.

[Ke] K. Keimel, Trends in lattice-ordered groups and rings. In: Proc. Symp. on Lattice Theory and its Applications (in Honor of the 80th Birthday of Garrett Birkhoff), Darmstadt 1991, in press.

[L] T.-Y. Lam, Introduction to real algebra. Rocky Mountain J. Math. 14(4) (1984), 767–814.

[M] S. MacLane, Categories for the Working Mathematician. Graduate Texts in Math. 5, Springer-Verlag, Berlin-Heidelberg-New York 1971.

[S] N. Schwartz, Piecewise polynomial functions. In: Ordered Algebraic Structures, The 1991 Conrad Conference (ed. by J. Martinez, C. Holland). Kluwer Academic Publishers, Dordrecht 1993.

[W] A. Weil, Foundations of Algebraic Geometry, revised and enlarged edition. A.M.S. Colloquium Publications, vol. 29, 1962.

[Z] O. Zariski, Polynomial ideals defined by infinitely near points. Amer. J. Math. 60 (1938), 151–204.

Bounds for the number of connected components of a family of differentiable manifolds

Jean-Pierre Françoise

Abstract. We give an integral formula to bound the number of connected components of a manifold. We present then one of the motivations related to Dynamical Systems.

1991 Mathematics Subject Classification: 58A05, 58A07, 34C05.

Introduction

The index formula due to Kronecker was modified by E. Picard in 1892 ([P]). We use E. Picard's formula here to write a bound for the number of connected components of a manifold of class C^3 with an integral formula.

Next, we consider a family U_λ ($\lambda \in \Lambda$) of C^3 manifolds which depends C^1 of a parameter λ. It is useful for the purpose of our study to define a controlled family U_λ as such that there is a uniform bound for the number of critical points of the distance function (to a fixed point) restricted to U_λ.

This notion appears to be useful in several subjects related to control theory and dynamical systems. We give here an example about the periodic orbits of polynomial vector fields of the plane.

I. An integral formula to bound the number of connected components of a differentiable manifold

Let M be a closed ball in $\mathbb{R}^n$. Let $f : M \to \mathbb{R}$ be a function of class C^3 such that $f^{-1}(0) \cap \partial M = \emptyset$. Let $x = (x_1, \ldots, x_n) \in M$ and let x^0 be a fixed point of $M \setminus \partial M$. We denote by $\sum = f^{-1}(0)$.

Definition I.1. The Picard mapping associated to (x^0, f) is defined as follows:

$$M \times \mathbb{R} \times [-\varepsilon, +\varepsilon] \xrightarrow{\phi} \mathbb{R}^{n+2}$$

$$\phi_i(x_1,\ldots,x_n;x_{n+1},x_{n+2}) = x_i - x_i^0 - x_{n+1}\frac{\partial f}{\partial x_i}\ (i=1,\ldots,n)$$

$$\Phi_{n+1}(x_1,\ldots,x_n;x_{n+1},x_{n+2}) = f(x_1,\ldots,x_n)$$

$$\Phi_{n+2}(x_1,\ldots,x_n;x_{n+1},x_{n+2}) = x_{n+2}\det\left(\frac{\partial\phi_i}{\partial x_j};\ i,j=1,\ldots,n+1\right)$$

Remarks. - ϕ is of class C^1.
- We call the $n+1^{th}$-variable x_{n+1} the Lagrange multiplier.

Let Ω be the standard volume form on $\mathbb{R}^{n+2}$,

$$\Omega = dy_1 \wedge \ldots \wedge dy_{n+2}$$

and let X be the vector fields

$$X = \sum_{i=1}^{n+2} \frac{(-1)^i y_i}{(y_1^2+\ldots+y_{n+2}^2)^{n+3)/2}}\frac{\partial}{\partial y_i}.$$

Definition I.2. The Picard residue $P_A(x^0,f)$ is

$$\int_{\partial M\times[-A,+A]\times[-\varepsilon,+\varepsilon]} \phi^*(\iota_X\Omega).$$

Remarks. - The singular locus of $\phi^*(\iota_X\Omega)$ is exactly given by $\phi_1^2+\ldots+\phi_{n+2}^2=0$. As we assume that $\phi_{n+1}(x)=f(x)$ does not vanish on ∂M, the integral of $\phi^*(\iota_X\Omega)$ over $\partial M\times[-A,+A]\times[-\varepsilon,+\varepsilon]$ exists.
- These definitions although not classic are inspired by the article [P].

Theorem I.3. *There is a set U, which is a countable intersection of open and dense sets in M, and a real number A such that if $x^0\in U$, the number of connected components of $f^{-1}(0)$ is less than*

$$\frac{1}{c_n}P_A(x^0,f)$$

where c_n is the volume of the unit ball in $\mathbb{R}^{n+2}$.

Proof. (1) We first construct U by Sard's lemma.

Let $\varphi:\mathbb{R}^{n+1}\to\mathbb{R}^{n+1}$ be the mapping

$$\underline{x}\to\begin{cases}\varphi_i(\underline{x}) = x_i - x_i^0 - x_{n+1}\dfrac{\partial f}{\partial x_i} & i=1,\ldots,n\\ \varphi_{n+1}(\underline{x}) = f(\underline{x})\end{cases}$$

The set U is given by the points $\underline{x}^0$ such that

$$\det\left(\frac{\partial\varphi_i}{\partial x_j}\right) = \left[\begin{array}{c|c} \mathrm{I} - x_{n+1}\dfrac{\partial^2 f}{\partial x_i \partial x_j} & \begin{matrix} -\partial f/\partial x_1 \\ \vdots \\ -\partial f/\partial x_n \end{matrix} \\ \hline \dfrac{\partial f}{\partial x_1} \cdots \dfrac{\partial f}{\partial x_n} & 0 \end{array}\right] \neq 0$$

on $\varphi^{-1}(0)$.

Let us consider the mapping

$$S : \sum \times \mathbb{R} \to \mathbb{R}^n$$

$$(\underline{x}, \lambda) \mapsto \left(x_1 - \lambda\frac{\partial f}{\partial x_1}, x_2 - \lambda\frac{\partial f}{\partial x_2}, \ldots, x_n - \lambda\frac{\partial f}{\partial x_n}\right).$$

The critical points S are given by the equation

$$\det\left(\frac{\partial\varphi_i}{\partial x_j}\right) = 0.$$

The set U is the complement in $\sum \times\mathbb{R}$ of the set of critical values of S; by Sard's lemma, it is a countable intersection of open and dense sets.

(2) We have furthermore the two following properties:

(i) $d(\iota_X\Omega) = 0$ (X is volume-preserving);

(ii) $\int_{\partial B(0,\varepsilon)} \iota_X\Omega = \int_{\partial B(0,1)} \iota_X\Omega = C_n$ (independent of ε).

We consider the set $\phi^{-1}(0)$. We have $\det \mathrm{Jac}\,\phi = (\det \mathrm{Jac}\,\varphi)^2 > 0$ on $\phi^{-1}(0)$. We denote by $N = \sharp(\phi^{-1}(0))$. Let $\varepsilon > 0$ be small enough so that $\phi^{-1}(B(0,\varepsilon))$ is a disjoint union of N open sets V_j ($j = 1, \ldots, N$) such that $\phi : V_j \to \phi(V_j)$ is a diffeomorphism.

(3) On a connected component of $f^{-1}(0)$, the function

$$\frac{1}{2}\sum_{i=1}^{n}(x_i - x_i^0)^2$$

has at least two critical points given by the equations

$$(*) \qquad \begin{cases} x_i - x_i^0 = x_{n+1}\dfrac{\partial f}{\partial x_i} & i = 1, \ldots, n \\ f(x) = 0. \end{cases}$$

Consequently, we obtain that the number of connected components of $f^{-1}(0)$ is less than the number of solutions of (*) which is N.

Let $m > 0$ be the minimum of the function $\sum_{i=1}^{n}(\frac{\partial f}{\partial x_i})^2$ on the compact set $\sum$.

The Lagrange multiplier x_{n+1} which is a solution of (*) necessarily satisfies

$$|x_{n+1}| \leq \Big(\sum_{i=1}^{n}(x_i - x_i^0)\frac{\partial f}{\partial x_i}\Big)/\sum_{i=1}^{n}(\frac{\partial f}{\partial x_i})^2 \leq \frac{1}{m}\,\mathrm{Diam}(M)\max_{x\in M}|\partial f/\partial x_i| = A.$$

(4) We apply then Stokes theorem and we get the result

$$\int_{\partial M\times[-A,+A]\times[\epsilon,+\epsilon]}\phi^*(L_\times\Omega) = \sum_{j=1}^{N}\int_{\partial V_j}\phi^*(L_\times\Omega) = \sum_{j=1}^{N}\int_{\partial B(0,\epsilon)}L_\times\Omega = C_n\cdot N.$$

Remark. If f would have critical points, it should be impossible to get a bound on the number of singular components by using this method. D. Pecker suggested to me to consider the so-called Napoléon hat as a possible example.

In this case the system (*) may have no solution and, nevertheless, the distance function may have critical points. (We can think in these examples, that the Lagrange multiplier becomes unbounded.)

We know that there are at least two critical points on each connected component. And so, the Picard residue may be used also to decide if the set $f^{-1}(0)$ is empty or not. We have already used this idea in a previous article in the algebraic case ([F-R]).

II. The uniform control of the number of connected components of $\sum_\lambda = f_\lambda^{-1}(0)$; an example of a controlled manifold

Let $f_\lambda : M \to \mathbb{R}$ be a family of functions of class C^3 which depend C^1 of a parameter λ ($\lambda \in \Lambda, \Lambda$ is a compact set).

Definition II.1. The family f_λ is controlled if there is a number N such that for all $\lambda \in \Lambda$, the number of critical points of $\delta = \sum(x_i - x_i^0)^2$ restricted to a connected component of $f_\lambda^{-1}(0)$ is less than N.

A manifold S is said to be controlled over the set Λ if there is a differentiable submersion $p : S \to \Lambda$ and an integer N such that the number of the critical points of the distance function δ restricted to a connected component of $p^{-1}(\lambda)$ is less than N.

Proposition II.2. *Given a controlled family f_λ (resp. a controlled manifold S), the number of connected components of $f_\lambda^{-1}(0)$ (resp. of $p^{-1}(\lambda)$) is uniformly bounded.*

II.3. Application to the study of the periodic orbits of planar vector fields.

Let $X = f(x,y)\frac{\partial}{\partial x} + g(x,y)\frac{\partial}{\partial y}$ be a planar vector where $f, g : \mathbb{R}^2 \to \mathbb{R}$ are two polynomials of degree less than n. We assume that f, g are normalized so that

$$f = \sum_{i+j\leq n} f_{ij}x^i y^j, \quad g = \sum_{i+j\leq n} g_{ij}x^i y^j, \quad \sum_{i+j\leq n} f_{ij}^2 + g_{ij}^2 = 1.$$

We denote by Λ the set of the coefficients. We proved in [F.P.] that the set of periodic orbits whose period is less than T (T can be any real number fixed in advance) is a sub-analytic set S.

Proposition II.3. *The manifold S is controlled over Λ.*

Proof. Let (x^0, y^0) be a point of the plane $\mathbb{R}^2$. A critical point (x, y) of the distance function restricted to a periodic orbit is a point where the vector field $(x-x^0, y-y^0)$ is orthogonal to the vector field X. Consequently, the critical points are on the algebraic curve $\gamma : (x - x^0)f(x,y) + (y - y^0)g(x,y) = 0$.

Next, we observe that the number of points where γ is tangent to the periodic orbit can be easily bounded because these points are solutions of the system

$$\begin{cases} \varphi(x,y) = (x - x^0)f + (y - y^0)g = 0 \\ f\dfrac{\partial\varphi}{\partial x} + g\dfrac{\partial\varphi}{\partial y} = 0 \end{cases}$$

Then, we need to bound uniformly the number of points where γ crosses transversally the periodic orbit. Between two such points there is a point where the angle $(x - x^0, y - y^0; X)$ has an extremum. The locus of the points where this angle has an extremum is again an algebraic curve $\tilde{\gamma}$.

Now the number of points where γ intersects transversally the periodic orbit is bounded by the number of intersection points of γ and $\tilde{\gamma}$ and hence uniformly bounded and we conclude the result.

Remark. This final argument is rather classical in the subject for instance to show that the limit cycles of a quadratic vector fields are convex.

References

[F-P] J.P. Françoise, C.C. Pugh, Keeping track of limit cycles. J. Differential Equations 65 (1986), 139–157.

[F-R] J.P. Françoise, F. Ronga, The decidability of real algebraic sets by the index formula. In: Real algebraic geometry, M. Coste, L. Mahé, M.-F. Roy (eds.), pp. 235–239, Lecture Notes in Math. 1524, Springer-Verlag 1992.

[P] E. Picard, Sur le nombre des racines communes à plusieurs équations simultanées. J. Math. Pures Appl. (1892), 5–24.

Voronoi computability in $SE(3)$

M.J. González-López and T. Recio**

Abstract. We discuss Riemannian distances in the Lie groups $SO(3)$ and $SE(3)$, in relation with proximity problems in robotics. We study the problem of the computability of this kind of metrics, showing that any left-invariant metric over $SO(3)$ cannot be a semialgebraic function and that the same applies for $SE(3)$. Moreover, we prove that the Voronoi diagram of a finite set of points for such distances in $SO(3)$ is a semialgebraic set that coincides with the diagram computed under the Euclidean distance. However, in $SE(3)$, the Voronoi diagram with the Riemannian distance is not computable in any of the real models of computation, which implies, in particular, that it is not a semialgebraic set.

1991 Mathematics Subject Classification: 68U05, 70B15, 14P99, 68Q05.

1. Introduction

The Voronoi diagram of a collection of points $P = \{p_1, p_2, \dots, p_n\}$, over a metric space G, under a distance d, is the set of points $x \in M$ such that $d(x, p_i) = d(x, p_j)$ for some two points p_i and p_j; and $d(x, p_i) \leq d(x, p_k)$ for all other points p_k. In some sense, the Voronoi diagram is the safest way (with respect to the given distance) to "travel" among P over G, avoiding equally to get closer to one of the points in P than to any other. For this reason the Voronoi diagram is of interest in Computational Geometry and in Robotics. It is well known that such diagrams are efficiently computed over the Euclidean plane with the usual distance. Many generalizations and methods of computation appear, for instance, in F. Aurenhammer [1].

On the other hand, a usual procedure in Robotics considers the positions of a rigid body (or a collection of bodies) in Euclidean three-dimensional space as a point in $\mathbb{R}^{12}$ (or a point in a cartesian product of several copies of $\mathbb{R}^{12}$), fixing two reference frames (one universal or exterior and one attached to the body) and associating with each position of the rigid body the matrix of the affine isometry that gives the deplacement of the body's frame with respect to the universal frame. In this way the set of positions of the robot is a (in general, semialgebraic) subset of $SE(3)$, which is the cartesian product of the algebraic smooth variety

* Partially supported by C.I.C.Y.T PB89/0379/C02/01 and ESPRIT-BRA 6846 "POSSO".

$SO(3)$ of proper orthogonal 3×3 matrices (as included in $\mathbb{R}^9$ by considering the concatenation of the three columns) and the translation part $\mathbb{R}^3$. Thus $SE(3)$ is also an algebraic smooth variety of dimension 6 in $\mathbb{R}^{12}$.

Any obstacles to the robot's movements in physical three-dimensional space will correspond, in the above described setting, with subsets of prohibited positions in $SE(3)$ for the rigid body. If the obstacles have a semialgebraic description in $\mathbb{R}^3$, then the corresponding subsets of $SE(3)$ will be also semialgebraic. Both $SO(3)$ and $SE(3)$ can be also considered as metric (Riemannian) spaces, with a metric that measures the length of the shortest path over the manifold, between two given points, i.e. the "amount" of deplacement needed to pass from one position of a rigid body to another. In principle, it seems that a "natural" measure of the danger of a given placement of a robot body with respect to an obstacle could be the Riemannian distance of the corresponding point in $SE(3)$ to the corresponding subset of prohibited positions induced in $SE(3)$ by the obstacle. Other distances for pairs of positions of rigid bodies are considered, for example in J. C. Latombe [7]. They all deal with the body as situated in the physical space in the two given positions and measure just some extreme Euclidean distance between points of the body when placed in the two locations. This seems unsatisfactory, in theory, as it is unclear for us, (provided that there is a relation between moving a robot and time needed to move it), whether positions which are closer in time would be necessarily closer under these distances. Moreover, it implies that the usual theoretical frame for working in motion planning (the consideration of $SE(3)$ as working space) has to be left aside to study collision proximity.

From the several possible Riemannian metrics in $SO(3)$ and $SE(3)$, one should consider those which are left-invariant, in the sense that the distance between two points of $SE(3)$ regarded as two positions of the same body under a fixed universal frame is equal to the distance between two other points that represent the same positions after making a change of the universal frame (we recall that we are considering, by convention, the columns of a matrix as the image of the basis vectors; if we adopt the row convention, then we should speak, instead, of right invariant metrics): in fact, a position of matrix M viewed from reference frame F, changes to have a matrix $H^{-1}M$ when viewed from a frame F' that has position H with respect to F. For $SO(3)$ a(ny) left invariant metric is induced by considering a (essentially unique) scalar product over the tangent space over the identity and then translating this product to every tangent space by left translations: it agrees with the induced scalar product of the ambient space $\mathbb{R}^9$ and therefore is also right-invariant (bi-invariant). Now, being $SO(3)$ compact, by the theorem of Hopf-Rinow ([10]), the distance of a position g to the identity can be computed considering a geodesic from g to the identity e; and being bi-invariant, such geodesic is an exponential mapping; moreover, also because of the bi-invariance, the computation of the distance can be made changing g to its canonical form $\exp(\theta S)$, where θ is an angle in the interval $[0, \pi]$ and S is a skew-symmetric matrix. It yields that the distance $d(g, e)$ is then equal to the angle θ or, equivalently, to the arc cosinus in $[0, \pi]$ of $\frac{\text{trace}(g)-1}{2}$. As the arc cos is not a semialgebraic function, we will show (Proposition 3.1.) that a left invariant metric over $SO(3)$ is not a semialgebraic function. But, considering in $SO(3)$ the Euclidean distance d_e induced from the embedding in $\mathbb{R}^9$, the following relation holds (c.f. Section 3) for the distance between two elements $g, h \in SO(3)$

$$d_e(g,h) = 2\sqrt{1 - \cos(d(g,h)}.$$

Therefore, as cosinus is, in the chosen interval, a monotone function, we have (Proposition 3.3.) that for any finite set of positions $P = \{p_1, p_2, \ldots, p_n\}$ in $SO(3)$, its Voronoi diagram under distance d is equal to the Voronoi diagram in R^9 under the Euclidean distance, intersected with $SO(3)$. In particular the Voronoi diagram is a semialgebraic set, result that can be also reached following another arguments (Proposition 3.2.).

In the case of $SE(3)$ it can be shown that there is not bi-invariant metric (see [9]), but still a (essentially unique) left invariant metric can be computed by considering the presentation of $SE(3)$ as $SO(3) \times \mathbb{R}^3$; then for a given position g of matrix with a rotational part of angle θ and a translational part of vector of module m, the distance to the identity matrix is $\sqrt{\theta^2 + m^2}$. As it is well known, under a change of orthogonal coordinates, the module m is not conserved and that is why the distance is not bi-invariant. One needs information about the specific universal frame to compute the distance and not merely (as in the case of $SO(3)$) some geometricl information such as the angle of rotation. What is worst, we can prove (Proposition 4.2.) that the Voronoi diagram over $SE(3)$ is not computable. Here, by "computable" we can understand any of the models of computation that accept the real numbers as alphabet for the space of admissible inputs: for instance, the algebraic computation or decision tree model of Ben-Or [2] or the real RAM machine used by Preparata and Shamos [11] for the analysis of algorithms in Computational Geometry; it applies also to the real Turing machine of Blum-Shub-Smale (BSS) [3] or the class of machines analized in Cucker-Montaña-Pardo [5] as subclasses of the BSS model. In other words, it seems that under no one of these models is the Voronoi diagram a computable object. Our example could be related to a recent result of Heintz-Krick-Slissenko-Solernó [6] showing that in three-dimensional space, under the Euclidean distance, the smallest path joining two points in a connected semialgebraic set is not locally semialgebraic. In some sense, both results represent the failure of the semialgebraic framework for dealing with metric aspects.

Our negative result has motivated some questions on the topological shape of the Voronoi diagram, demanding, in particular, how rigid is the shape of a Voronoi diagram under different distances (this kind of results could be used to determine the Voronoi diagram, as we could compute exactly its topological shape and then aproximate it by numerical methods). We can find some conclusions on this topic in the paper ([4]) "On the topological shape of planar Voronoi diagrams", by A. Corbalán, M. Mazón, T. Recio and F. Santos, essentially showing that, at least in the simplest case of normed distances in the Euclidean plane, all inducing the same topology as the usual one, the conservation of the shape of Voronoi diagrams under different distances is highly demanding (i.e., the distances must be basically the same).

2. Notions and notation

Throughout this paper (G, d) denotes a metric space.

Definition 2.1. Let $g_1, g_2 \in G$. The *Voronoi region of g_1 with respect to g_2* is the subset of G

$$H(g_1, g_2) := \{g \in G \mid d(g, g_1) < d(g, g_2)\}.$$

If $P := \{g_1, \ldots, g_s\} \subset G$, the *Voronoi region of g_i with respect to P* is the subset of G

$$R(g_i, P) := \cap_{i \neq j} H(g_i, g_j).$$

The *Voronoi diagram of P in G with the distance d* is the complement in G of the set

$$\bigcup_{i=1}^{s} R(g_i, P).$$

It will be denoted $\mathrm{Vor}(P, d)$, or $\mathrm{Vor}(P)$ if d is clear. See [11] for more details on Voronoi diagrams. We shall use freely results concerning the Lie and metric structure of $SO(3)$ and $SE(3)$; basic references are [8], [9] and [10].

3. Riemannian bi-invariant metric and computability of the Voronoi diagram in $SO(3)$

The set of three-dimensional skew symetric matrices SS is the tangent space to the identity in $SO(3)$, and if we fix a scalar product in SS and translate it to every tangent space by left translations, we have that it agrees with the induced scalar product of the ambient space $\mathbb{R}^9$ and therefore it is bi-invariant. The theorem of Hopf-Rinow (c.f. for example, [10]) can be applied to this situation, as $SO(3)$ is compact, therefore, the distance of a position $g \in SO(3)$ to the identity is the length of a geodesic from g to the identity e; as this metric is bi-invariant, the geodesics are the exponential mappings and the computation of the distance can be made changing g to its canonical form, i.e., the matrix

$$\exp(\theta\tilde{u}) := \exp\left\{\theta\begin{pmatrix} 0 & -u_3 & u_2 \\ u_3 & 0 & -u_1 \\ -u_2 & u_1 & 0 \end{pmatrix}\right\}$$

where $\theta \in [0, \pi]$ is the angle of g and $u := (u_1, u_2, u_3) \in \mathbb{R}^3$ is the unitary oriented axis of g.

Now, if $g \in SO(3)$ and s is a proper orthogonal matrix representing g, there is a geodesic $\gamma(t) = \exp t \cdot s$ such that $\gamma(1) = g$; besides, in a certain orthogonal basis in the Euclidean tree-dimensional space we have that $s = \theta\tilde{u}$. We compute its length:

$$d_R(e,g) = \text{length}(\gamma) = \int_0^1 \sqrt{\langle \gamma'(t), \gamma'(t)\rangle_{\gamma(t)}}\ dt =$$

$$= \int_0^1 \sqrt{\langle \theta\tilde{u}, \theta\tilde{u}\rangle_e}\ dt = \sqrt{\langle \theta\tilde{u}, \theta\tilde{u}\rangle_e} = \theta$$

where $\langle , \rangle_q$ denotes the scalar product in the tangent space to the element q. Remark that if we take another orthogonal basis in the Euclidean space, the expression of $g \in SO(3)$ changes to $h \circ g \circ h^{-1}$, where h represents the basis change, and, as the metric is bi-invariant, we have that $d_R(g_1, g_2) = d_R(h \circ g_1 \circ h^{-1}, h \circ g_2 \circ h^{-1})$; i.e., the distance d_R is an intrinsic concept in $SO(3)$.

Any other Riemannian bi-invariant metric in $SO(3)$ is equal to the proposed here up to a constant, as all these metrics have the same geodesics. Thus, the general expression for the Riemannian bi-invariant distance between an element $g \in SO(3)$ and the identity is

$$\theta\|u\|$$

where (θ, u) is the associated pair to g, and the norm $\|u\|$ depends on the considered scalar product.

If we are given the matrix M_g representing $g \in SO(3)$ in a fixed orthogonal basis (not necessarily the canonical expression), we have that θ is the unique solution in $[0, \pi]$ of the equation

$$\cos(\theta) = \frac{\text{trace}(M_g) - 1}{2}.$$

Therefore, we can write

$$d_R(e,g) = \text{arc-cosinus}_{[0,\pi]}\Big(\frac{\text{trace}(M_g) - 1}{2}\Big).$$

A consequence of this expression is:

Proposition 3.1. *The Riemannian distance d_R in $SO(3)$ is not a semialgebraic function.*

Proof. Considering the semialgebraic mapping

$$\begin{array}{cccc} H: & SO(3) \subset \mathbb{R}^9 & \longrightarrow & \mathbb{R} \\ & M_g & \longmapsto & \dfrac{\text{trace}(M_g) - 1}{2} \end{array}$$

we have that $d_R = \text{arc-cosinus}_{[0,\pi]} \circ H$. Besides, we define the semialgebraic subset of $SO(3)$

$$T := \{M = (a_{ij})_{ij} \in SO(3) \mid a_{11} = 1,\ a_{23} > 0\}$$

where the restriction $H_{|T} : T \to [-1, 1]$ is a semialgebraic bijective mapping. If d_R were a semialgebraic mapping, then the same could be said of the restriction $d_{R|T}$, which in this case coincides with $\text{arc-cosinus}_{[0,\pi]} \circ (H_{|T})$, as $\text{Im}(H) = \text{Im}(H_{|T})$.

However, composing to the right with the semialgebraic mapping $(H_{|T})^{-1}$, we would conclude that arc-cosinus$_{[0,\pi]}$ is a semialgebraic mapping, which is a contradiction.

□

Moreover, following the same argument as in the proof of 3.1 we can conclude that d_R is not even locally a semialgebraic function, as if this were the case, we would have (following the notation in 3.1) that the restriction $d_{R|T} =$ arc-cosinus$_{[0,\pi]} \circ (H_{|T})$ is locally a semialgebraic function, and composing to the right with the semialgebraic mapping $(H_{|T})^{-1}$ we would conclude that the mapping arc-cosinus$_{[0,\pi]}$ is locally semialgebraic, which is impossible, as shown in [6]. We can also prove that the graph of d_R is not a countably union of semialgebraic sets and therefore that it is not computable under the BSS model [5]. In fact, if $\text{Graph}(d_R) = \{(M,r) \in SO(3) \times \mathbb{R} \mid d_R(e,M) = r\}$ were a countably union of semialgebraic sets, then the same could be said of the set $W := \{(M,r) \in SO(3) \times \mathbb{R} \mid d_R(e,M) = r, M \in T\}$. Remark that matrices in T have the shape

$$M = \begin{pmatrix} 1 & 0 & 0 \\ 0 & a & -b \\ 0 & b & a \end{pmatrix}$$

with $a^2 + b^2 = 1$ and $b < 0$. Therefore, $d_R(e,M) = \text{arc-cosinus}_{[0,\pi]}(a)$, and we can write

$$W = \left\{ \left(\begin{pmatrix} 1 & 0 & 0 \\ 0 & a & -b \\ 0 & b & a \end{pmatrix}, \quad \text{arc-cosinus}_{[0,\pi]}(a) \right) \middle| \; a^2 + b^2 = 1, \; b < 0 \right\}.$$

Projecting this set on the variables corresponding to a and arc-cosinus$_{[0,\pi]}(a)$ we would have that the set

$$W' := \{(a, \text{arc-cosinus}_{[0,\pi]}(a)) \mid a \in [-1,1]\}$$

is a countably union of semialgebraic sets, but it is the graph of the arc-cosinus$_{[0,\pi]}$ function, which is a one dimensional set; therefore, at least one of the sets of the countably union would have dimension one; this implies that there would be a neighbourhood of a point $(x_0, y_0) \in W'$ in which the graph of arc-cosinus$_{[0,\pi]}$ is a semialgebraic set, which is impossible as in [6] it is shown that this mapping is not locally semialgebraic in any point.

However, the conclusion in 3.1. is not an obstruction to make equality tests among these expressions; in fact we have that:

Proposition 3.2. *The Voronoi diagram in $SO(3)$ with the distance d_R is a semialgebraic set.*

Proof. Given a finite set of points P, $\text{Vor}(P, d_R)$ is a semialgebraic set if we find an algebraic tree (c.f. [5]) to decide if a given point $g \in SO(3)$ belongs to $\text{Vor}(P, d_R)$, as in this case $\text{Vor}(P, d_R)$ is described as the set of affirmative branches of the tree.

For the Voronoi diagram we have a "natural" algebraic tree, which for every input (P, g) makes the polynomial operations

$$\mathrm{trace}(M_g^{-1} M_{g_i})$$

for all $g_i \in P$ and makes equality and inequality tests among these expressions, because, as arc-cosinus is a monotone function in $[0, \pi]$ we have that

$$d_R(g, g_1) < d_R(g, g_2) \Longleftrightarrow \mathrm{trace}(M_g^{-1} M_{g_1}) < \mathrm{trace}(M_g^{-1} M_{g_2}). \qquad \square$$

We remark that the above proof shows, in fact, that there exists a time bounded real Turing machine computing Voronoi diagrams in $SO(3)$ (c.f. [5]), as it is presented an uniform algorithm for its computation whose complexity depends only on the number of points.

Fix an ortonormal basis in E, and let $g, h \in SO(3)$ be elements whose matrices in this basis will be denoted by $M_g = (g_{ij})_{i,j}$ and $M_h = (h_{ij})_{i,j}$, respectively. The Euclidean distance between g and h is

$$d_E(g, h) = \|M_g - M_h\| = \sqrt{\sum_{ij} (g_{ij} - h_{ij})^2} = \sqrt{2(3 - \sum_{ij} g_{ij} h_{ij})}.$$

The Riemannian distance is:

$$d_R(g, h) = d_R(e, g^{-1} \circ h) = \mathrm{angle}(g^{-1} \circ h) =$$

$$= \mathrm{arcos}_{[0,\pi]} \left(\frac{\mathrm{trace}(M_g^{-1} M_h) - 1}{2} \right) = \mathrm{arcos}_{[0,\pi]} \left(\frac{\sum_{ij} g_{ij} h_{ij} - 1}{2} \right).$$

Therefore

$$d_E(g, h) = 2\sqrt{1 - \cos(d_R(g, h))}$$

and we obtain:

Proposition 3.3. *Given $P = \{g_1, \ldots, g_s\} \in SO(3)$ we have that*

$$\mathrm{Vor}(P, d_R) = \mathrm{Vor}(P, d_E) \cap SO(3).$$

Proof. For every $g \in SO(3)$ and for all $i, j \in \{1, \ldots, s\}$ the following equivalence holds

$$d_R(g, g_i) < d_R(g, g_j) \Longleftrightarrow d_E(g, g_i) < d_E(g, g_j).$$

Likewise if we change "$<$" by "$>$" or "$=$". $\square$

4. Riemannian left-invariant metric and computability of the Voronoi diagram in $SE(3)$

Considering the elements of $SE(3)$ as affine isometries of the Euclidean space, and fixing a reference frame, we can identify $SE(3)$ with $SO(3) \times \mathbb{R}^3$ and the tangent space at the identity, $T_e(SE(3))$, with $SS \times \mathbb{R}^3$. Thus we will put

$$g \equiv \begin{pmatrix} 1 & 0 \\ b & R \end{pmatrix} \in SE(3), \qquad s \equiv \begin{pmatrix} 0 & 0 \\ v & \tilde{w} \end{pmatrix} \in T_e(SE(3))$$

where $b, v, w \in \mathbb{R}^3$ and R is a proper orthogonal matrix. The metric induced in every tangent space $T_g(SE(3))$ by $\mathbb{R}^{12}$ is equal to the metric obtained translating to the left the scalar product in $T_e(SE(3))$. In fact, considering the scalar product in $\mathbb{R}^{12}$ we have

$$\|g \circ s\| = \left\| \begin{pmatrix} 1 & 0 \\ b & R \end{pmatrix} \cdot \begin{pmatrix} 0 & 0 \\ v & \tilde{w} \end{pmatrix} \right\| = \left\| \begin{pmatrix} 0 & 0 \\ R \cdot v & R \cdot \tilde{w} \end{pmatrix} \right\| = \left\| \begin{pmatrix} 0 & 0 \\ v & \tilde{w} \end{pmatrix} \right\| = \|s\|.$$

But, making these computations to the right we get

$$\|s \circ g\| = \left\| \begin{pmatrix} 0 & 0 \\ v & \tilde{w} \end{pmatrix} \cdot \begin{pmatrix} 1 & 0 \\ b & R \end{pmatrix} \right\| = \left\| \begin{pmatrix} 0 & 0 \\ v + \tilde{w} \cdot b & \tilde{w} \cdot R \end{pmatrix} \right\| \neq$$

$$\neq \left\| \begin{pmatrix} 0 & 0 \\ v & \tilde{w} \end{pmatrix} \right\| = \|s\|.$$

Therefore, it does not exist a bi-invariant metric in $SE(3)$ induced by the Euclidean metric in $\mathbb{R}^{12}$. As choosing a scalar product in $T_e(SE(3))$ is determined up to a constant and it is equal to the scalar product in $\mathbb{R}^{12}$, we conclude that a left-invariant metric in $SE(3)$ cannot be bi-invariant; in fact, the distance depends on the fixed reference frame in the Euclidean space.

Next, we will state the explicit expression of the Riemannian left-invariant distance d_R in $SE(3)$ given by the canonical scalar product in $\mathbb{R}^{12}$; remark that it suffices to determine $d_R(e, g)$, $g \in SE(3)$, as $d_R(g_1, g_2) = d_R(e, g_1^{-1} \circ g_2)$. The process is the following:

We fix an orthogonal reference frame in $\mathbb{R}^3$, which becomes identified with $SO(3) \times \mathbb{R}^3$. $SE(3)$ is complete and connected (as product of complete and connected spaces), therefore we can apply Hopf-Rinow theorem and the distance between the elements e and g is given by the length of a geodesic connecting them. The mapping

$$\begin{array}{rccc} \gamma: & [0,1] & \longrightarrow & SO(3) \times \mathbb{R}^3 \\ & t & \mapsto & (\gamma_1(t), \gamma_2(t)) \end{array}$$

is a geodesic if and only if γ_1 and γ_2 are geodesics in their respective spaces. Denote

$$e = \begin{pmatrix} 1 & 0 \\ 0 & e \end{pmatrix} \quad \text{and} \quad g = \begin{pmatrix} 1 & 0 \\ b & R \end{pmatrix}$$

In $\mathbb{R}^3$ the geodesics are the lines, therefore the geodesic connecting 0 and b is, up to reparametrization, $\gamma_2(t) := t \cdot b$, with $\|\gamma_2'(t)\| = \|b\|$. With respect to $SO(3)$, the geodesic connecting e and R is, as shown before, $\gamma_1(t) := e^{t \cdot \theta \cdot \tilde{u}}$, where $\theta \in [0, \pi]$ is the angle of R and u its unitary oriented axis. Thus, considering the scalar product in the product space, we have

$$d_R(e, g) = \int_0^1 \sqrt{\langle \gamma'(t), \gamma'(t) \rangle_{\gamma(t)}}\ dt =$$

$$= \int_0^1 \sqrt{\langle \gamma_1'(t), \gamma_1'(t) \rangle_{\gamma_1(t)} + \langle \gamma_2'(t), \gamma_2'(t) \rangle_{\gamma_2(t)}}\ dt =$$

$$= \int_0^1 \sqrt{\langle \theta\tilde{u}, \theta\tilde{u} \rangle_e + \langle b, b \rangle}\ dt = \sqrt{\langle \theta\tilde{u}, \theta\tilde{u} \rangle_e + \langle b, b \rangle} = \sqrt{\theta^2 + \|b\|^2}$$

where the factor $\|b\|$ indicates the dependency of this expression on the chosen reference frame: in particular, on the origin of such frame.

If we consider $g, h \in SE(3)$, with $g = (b, R)$ and $h = (c, P)$, then

$$g^{-1}h = \begin{pmatrix} 1 & 0 \\ R^t(c-b) & R^tP \end{pmatrix},$$

therefore

$$d_R(g, h) = \sqrt{\theta^2 + \|R^t(c-b)\|^2} = \sqrt{\theta^2 + \|c-b\|^2}$$

where $\theta \in [0, \pi]$ is the angle corresponding to the rotation R^tP. We remark that this is not a right-invariant metric, as, for example, $d_R(g, h) \neq d_R(e, h \circ g^{-1})$.

Let us fix an orthonormal reference frame and let $g, h \in SE(3)$ elements whose matrices in the fixed frame are denoted, respectively, by

$$M_g = \begin{pmatrix} 1 & 0 \\ (b_i)_i & (g_{ij})_{i,j} \end{pmatrix} \quad \text{and} \quad M_h = \begin{pmatrix} 1 & 0 \\ (c_i)_i & (h_{ij})_{i,j} \end{pmatrix}.$$

The Euclidean distance between g and h is

$$d_E(g, h) = \|M_g - M_h\| = \sqrt{2(3 - \sum_{ij} g_{ij}h_{ij}) + \|b-c\|^2}$$

The Riemannian distance is

$$d_R(g, h) = d_R(e, g^{-1} \circ h) = \sqrt{\text{angle}(g^{-1} \circ h) + \|b-c\|^2}$$

Therefore, we have the identity

$$d_E(g, h)^2 = \|b-c\|^2 + 4(1 - \cos(\sqrt{d_R(g, h)^2 - \|b-c\|^2})\ ),$$

which in this case is not useful to identify the Voronoi diagrams computed with the two distances (as we have made in the case of $SO(3)$). In fact, the Voronoi diagram with the distance d_R changes if we change the reference frame as we show in the following example: let us consider $g_1, g_2 \in SE(3)$ such that angle(g_1) = angle(g_2)

and suppose that $d_R(g_1, e) = d_R(g_2, e)$. Let us fix two reference frames χ and χ'. The matricial expression of g_1 and g_2 in these frames is, respectively,

$$M_i^{\chi} = \begin{pmatrix} 1 & 0 \\ v_i^{\chi} & R \end{pmatrix}; \qquad M_i^{\chi'} = \begin{pmatrix} 1 & 0 \\ v_i^{\chi'} & R \end{pmatrix}$$

where R is a proper orthogonal matrix and $v_i^{\chi}, v_i^{\chi'} \in \mathbb{R}^3$. As the angles of g_1 and g_2 are equal, in order to keep the equality $d_R(g_1, e) = d_R(g_2, e)$ in the two frames, we must have

$$\|v_1^{\chi}\| = \|v_2^{\chi}\| \iff \|v_1^{\chi'}\| = \|v_2^{\chi'}\|.$$

If we denote g the afinne isometry such that $\chi' = \chi \circ g$, we have that $M_i^{\chi'} = Q^{-1}.M_i^{\chi'}.Q$, where $Q = (v, P)$ is the matricial expression of g, therefore $v_i^{\chi'} = P^t(v_i^{\chi} - v + Rv)$. As the rotation P^t keeps the norm, we must have

$$\|v_1^{\chi}\| = \|v_2^{\chi}\| \iff \|v_1^{\chi} - v + Rv\| = \|v_2^{\chi} - v + Rv\|$$

which is, in general, false. The conclusion is that, when computing in χ we have that $e \in \mathrm{Vor}(\{g_1, g_2\}, d_R)$, but if we make the computations in χ', then $e \notin \mathrm{Vor}(\{g_1, g_2\}, d_R)$. Thus, it makes no sense to speak of a Voronoi diagram in $SE(3)$ but in the chosen immersion of this group in $\mathbb{R}^{12}$.

Proposition 4.1. *The Riemannian distance d_R in $SE(3)$ is not a semialgebraic function.*

Proof. It suffices to remark that for some particular cases (in which there is not translation) the computation of d_R in $SE(3)$ is equal to the computation of d_R in $SO(3)$, and then use 3.1. □

Proposition 4.2. *For any given immersion of $SO(3)$ in $\mathbb{R}^{12}$, the Voronoi diagram with the Riemannian distance is not computable in the real Turing machine model of Blum-Shub-Smale* [3].

Proof. Fix a reference frame in the Euclidean space and consider $g_1, g_2 \in SE(3)$ whose expressions with respect to this frame are

$$g_1 \equiv \begin{pmatrix} 1 & 0 \\ c_1 & M_{\theta_1} \end{pmatrix}, \qquad g_2 \equiv \begin{pmatrix} 1 & 0 \\ c_2 & M_{\theta_2} \end{pmatrix}$$

where $c_1 \neq c_2$ are vectors in $\mathbb{R}^3$, $\theta_1 \neq \theta_2$ are angles in $[0, \pi]$, and we use the notation $M_{\theta_i} = \begin{pmatrix} 1 & 0 & 0 \\ 0 & \cos\theta_i & -\sin\theta_i \\ 0 & \sin\theta_i & \cos\theta_i \end{pmatrix}$, $i = 1, 2$. We will prove that $\mathrm{Vor}(\{g_1, g_2\}, d_R)$ is not a countably union of semialgebraic sets. In fact, if it were a countably union of semialgebraic sets, then its intersection with the semialgebraic set

$$\left\{ \begin{pmatrix} 1 & 0 \\ c & M_{(x,y)} \end{pmatrix} := \begin{pmatrix} 1 & 0 & 0 & 0 \\ c_{11} & 1 & 0 & 0 \\ c_{12} & 0 & x & -y \\ c_{13} & 0 & y & x \end{pmatrix} \middle| \, y \geq 0, \; x^2 + y^2 = 1 \right\}$$

would be also a countably union of semialgebraic sets and the same could be said for the set

$$A := \Big\{(c, x, y) \in \mathbb{R}^5 \mid y \geq 0, \; x^2 + y^2 = 1,$$

$$(\text{angle}(M_{(x,y)}^{-1} M_{\theta_1}))^2 + \|c - c_1\|^2 = (\text{angle}(M_{(x,y)}^{-1} M_{\theta_2}))^2 + \|c - c_2\|^2 \Big\}.$$

We remark that, for $i = 1, 2$ we have:

$$\text{angle}(M_{(x,y)}^{-1} M_{\theta_i}) = \text{arc-cosinus}_{[0,\pi]}(x \cos\theta_i + y \sin\theta_i) =$$

$$= \begin{cases} \theta_i - \text{arc-cosinus}_{[0,\pi]}(x), & \text{if } \theta_i \geq \text{arc-cosinus}_{[0,\pi]}(x), \\ \text{arc-cosinus}_{[0,\pi]}(x) - \theta_i, & \text{otherwise.} \end{cases}$$

Moreover, $\|c - c_i\|^2 = \|c\|^2 + \|c_i\|^2 - 2\langle c, c_i \rangle$. Therefore, we have that

$$A = \Bigg\{ (c, x, y) \mid y \geq 0, \; x^2 + y^2 = 1,$$

$$\text{arc-cosinus}_{[0,\pi]}(x) = \frac{\theta_1^2 - \theta_2^2 - \|c_2\|^2 + \|c_1\|^2 - 2\langle c, c_1 - c_2 \rangle}{2(\theta_1 - \theta_2)} \Bigg\}$$

will be a countable union of semialgebraic sets. Denote by f the semialgebraic function $\mathbb{R}^3 \to \mathbb{R}$ given by $f(c) := \dfrac{\theta_1^2 - \theta_2^2 - \|c_2\|^2 + \|c_1\|^2 - 2\langle c, c_1 - c_2 \rangle}{2(\theta_1 - \theta_2)}$. Remark that f is a surjective mapping. The projection Π of A on the variables (c, x) must be also a countably union of semialgebraic sets

$$\Pi(A) = \{(c, x) \mid x \in [-1, 1], \; \text{arc-cosinus}_{[0,\pi]}(x) = f(c)\}$$

and also the image of this set by the semialgebraic mapping (f, Id), where Id denotes the identity mapping,

$$(f, \text{Id}) \circ \Pi(A) = \{(z, x) \in \mathbb{R}^2 \mid x \in [-1, 1], \; \text{arc-cosinus}_{[0,\pi]}(x) = z\}$$

must be a countably union of semialgebraic sets, which is a contradiction, as we noted in the comments to proposition 3.1.

Finally, we remark that if Voronoi diagrams could be computed by a real Turing machine, it would recognize for a specific size of the input, countably unions of semialgebraic sets [5]. Therefore, for the inputs g_1, g_2 we would have that $\text{Vor}(\{g_1, g_2\}, d_R)$ would also be of the same kind, and this is impossible as shown above. □

Acknowledgements. We are very grateful to professors J. Lafuente and A.M. Amores, from the University Complutense of Madrid, and to the Cantabria Complexity group, for many valuable ideas on the subject of this paper.

References

[1] F. Aurenhammer, Voronoi Diagrams: A Survey of a Fundamental Geometric Data Structure. ACM Comp. Surveys 23, No. 3, (1991).
[2] M. Ben-Or, Lower bounds for algebraic computation trees. A.C.M. 15th Symp. on Theory of Computing (1993), 80–86.
[3] L. Blum, M. Shub, S. Smale, On a theory of computation and complexity over the real numbers: NP-completeness, recursive functions and universal machines. Bull. Amer. Math. Soc. 21 (1989), 1–46.
[4] A. Corbalán, M. Mazón, T. Recio, F. Santos, On the topological shape of planar Voronoi diagrams. ACM Conference on Computational Geometry, San Diego 1993.
[5] F. Cucker, J.L. Montaña, L.M. Pardo, Time bounded computations over the reals. Internat. J. Algebra Comput. 2 (1992), 395–408.
[6] J. Heintz, T. Krick, A.O. Slissenko, P. Solernó, Searching for a shortest path around semialgebraic obstructions in the plane. Zapiski Nauchnick Seminarov LOMI, Leningrad Branch of the Steklov Math. Inst., Vol. 192 (1991), 163–173.
[7] J.C. Latombe, Robot Motion Planning. Kluwer Academic Publishers 1991.
[8] M. Le Borgne, Quaternions et controle sur l'espace des rotations. Tech. Rep. No. 377, INRIA Rennes (France) 1987.
[9] J. Loncaric, Geometrical analysis of compliant mechanisms in robotics. Ph. D. Dissertation, Division of Applied Sciences, Harvard University 1985.
[10] B. O'Neill, Semi-Riemannian geometry. With applications to relativity. Pure and Applied Mathematics 103, Academic Press 1983.
[11] F.P. Preparata, M.I. Shamos, Computational Geometry. Texts Monographs Comput. Sci., Springer-Verlag, New York 1985.

On the Gabrielov theorem for subpfaffian sets

Zbigniew Hajto

Abstract. For subpfaffian sets we obtain an analog of the Gabrielov theorem about the complement of subanalytic set. Previously we set up basic notions of the theory of subpfaffian sets.

1991 Mathematics Subject Classification: 32B20, 53C12, 34C30.

0. Introduction

In this paper, we give an introduction to the theory of subpfaffian sets. The motivation for developing that kind of transcendental analogy of subanalytic geometry is the following question posed by R. Moussu and M. Shiota: "What can we get adding to subanalytic sets the solutions of Pfaffian equations?".

Briefly, the theory is composed of three parts.

A) Semipfaffian geometry, where basic objects are defined. The building blocks of our geometry are the intersections of leaves of Pfaffian foliations with the strata of normal decompositions of S. Lojasiewicz [9]. In fact, this part of the theory was suggested in several papers of R. Moussu, C. Roche and J-M. Lion [6], [10], [11], [12]. In the first section we are giving a summary of semipfaffian geometry selecting various elementary objects, i.e. semipfaffian, basic semipfaffian and Pfaffian sets (cf. Definitions 1.11, 1.11.1).

B) Subpfaffian geometry, which includes the analog of Gabrielov theorem about the complement from subanalytic geometry [3], [4]. The main obstruction in building a self-contained theory of semipfaffian sets is the lack of the theorem about the closure of a semipfaffian set ([1], [6]). In the fourth section, using the theorem about the complement of a subpfaffian set, we prove that the closure of a subpfaffian set is subpfaffian. This part of the theory is based on the conditional assumption that the boundary of a subpfaffian set E is contained in a closed subpfaffian set of dimension smaller than the dimension of E. In the following section we prove that our assumption is not restrictive, i.e. that any basic semipfaffian set is subregular (cf. Definition 4.1.). It should be mentioned that in the subanalytic case the theorem on the closure is used in the proof of Gabrielov's Theorem [3], [4], [7].

C) The third part is in fact the proof that any basic semipfaffian set is subregular by reducing the problem to the hypersurface case. The theorem about the complement and its consequence theorem about the closure together with our version of finiteness theorem (cf. Theorem 2.5.) are the fundamentals of the basic subpfaffian geometry. They form a good starting point for further study of the solutions of systems of Pfaffian equations. In this context, it seems to be valuable (and rather simple) to reprove the triangulation theorem from subanalytic geometry and to establish a kind of equisingularity theory for subpfaffian sets. Finally, an interesting connection with the works of Lou van den Dries should be mentioned as well [13], [14].

1. Preliminaries

In this section, we shall gather together some necessary facts from the semipfaffian geometry and some theorems about foliations (cf. [6], [10], [11]) which we shall use later.

1.1. Let N be a real analytic manifold of dimension n.

Definition. A *Pfaffian hypersurface* of N is a triplet $(V, \mathfrak{F}, M)$ where
a) $M \subset N$ is a semianalytic open subset of N,
b) $\mathfrak{F}$ is a foliation of codimension one (with singularities) defined in an orientable neighbourhood of the closure $\overline{M}$ of M (cf. [12]),
c) V is a leaf of $\mathfrak{F}_{|M}$,
d) $\operatorname{Sing} \mathfrak{F} \cap M = \emptyset$.

1.2 Definition. Let $H = (V, \mathfrak{F}, M)$ be a Pfaffian hypersurface in N. We say that H is *separant* if and only if $M \setminus V$ has exactly two connected components and V is the common frontier (in M) of both.

We say that H has the *Rolle property* (i.e. is Rollian) if and only if each analytic path $\gamma : [0,1] \to M$ such that $\gamma(0), \gamma(1)$ are points in V has at least one point $\gamma(t)$ such that the tangent vector $\gamma'(t)$ is tangent to the foliation $\mathfrak{F}$.

1.3 Remark. If M is orientable, then separant implies Rollian. Moreover, if in addition M is simple connected, then any Pfaffian hypersurface $H = (V, \mathfrak{F}, M)$ is separant and therefore Rollian [11].

1.4. One of the most important theorems in the semipfaffian geometry is the Finiteness Theorem of Moussu and Roche [10], [11]:

Theorem. *Let M be an open semianalytic subset of N, $X \subset M$ semianalytic and relatively compact in N. For a finite collection $(V_1, \mathfrak{F}_1, M), \ldots, (V_p, \mathfrak{F}_p, M)$ of Pfaffian hypersurfaces, which have the Rolle property for the paths in X, there* *exists an integer* $b_0 \in \mathbb{N}$ *(depending only on* M, X, *and* $\mathfrak{F}_1, \ldots, \mathfrak{F}_p$*) such that the number of connected components of* $X \cap V_1 \cap \cdots \cap V_p$ *is smaller than* b_0.

1.5 Finiteness Theorem (relative version). *Let* X, M *and* $\{V_i, \mathfrak{F}_i, M\}$ *be like in 1.4 and let* $N = \mathbb{R}^p \times \mathbb{R}^r, \pi : \mathbb{R}^p \times \mathbb{R}^r \to \mathbb{R}^r$ *the natural projection. There exists an integer* $b \in \mathbb{N}$ *(depending only on* M, X *and* $\{\mathfrak{F}_i\}_1^p$*) such that for every* $z \in \mathbb{R}^r$ *the number of connected components of* $X \cap V_1 \cap \cdots \cap V_p \cap \pi^{-1}(z)$ *is smaller than* b.

1.6. Recall some useful notions from the theory developed by Moussu, Roche and Lion [6], [10], [11]. Let $\omega_k, k = 1, \ldots, q$ be the differential 1-forms given in a neighbourhood of $c \in \mathbb{R}^n$ by the foliations of codimension one $\mathfrak{F}_k, k = 1, \ldots, q$. In some neighbourhood U of c, $\omega_k = \sum_{i=1}^n b_i^k dx_i$, where b_i^k are real analytic functions in U. Let us define $\Omega = \{\omega_k, k = 1, \ldots, q\}$. For a subset $J \subset \{1, \ldots, q\}$ we denote by Ω_J the collection $\{\omega_j\}_{j \in J}$, by $\Omega_J(x)$ the subspace of $(\mathbb{R}^n)^*$ (linear forms in $\mathbb{R}^n$) generated by $\{\omega_j(x)\}_{j \in J}$ and put $\operatorname{Ker} \Omega_J(x) = \bigcap_{j \in J} \operatorname{Ker} \omega_j(x) \subset \mathbb{R}^n$.

1.6.1 Definition. For an analytic submanifold $Y \subset \mathbb{R}^n$, we say that Ω_J is *tranverse* to Y if and only if for each point $x \in Y$

$$\dim(\operatorname{Ker} \Omega_J(x) \cap T_x(Y)) = \dim Y - \sharp J.$$

If moreover for each $x \in Y$, $\Omega(x) + T_x^0(Y) = \Omega_J(x) + T_x^0(Y)$ (i.e. $\operatorname{Ker} \Omega(x) \cap T_x(Y) = \operatorname{Ker} \Omega_J(x) \cap T_x(Y)$), we say that Ω_J is a *basis for* Ω *along* Y, where $T_x^0(Y) \subset (\mathbb{R}^n)^*$ is the space of vanishing forms on $T_x(Y)$.

1.6.2 Definition. Let $\mathfrak{N}$ be a normal stratification of some neighbourhood U of $c \in \mathbb{R}^n$ [10] and let $\Omega = \{\omega_k, k = 1, \ldots, q\}$. We say that $\mathfrak{N}$ is *adapted to the family* Ω if and only if there exists a mapping J of $\mathfrak{N}$ in the family of subsets of $\{1, \ldots, q\}$ such that for every $Y \in \mathfrak{N}$, $\Omega_{J(Y)}$ is a basis of Ω along Y. If additionally $\mathfrak{N}$ is compatible with a finite family of semianalytic sets $\{E_\nu\}_{\nu \in I}$ we say that $\mathfrak{N}$ is *adapted to* Ω *and to* $\{E_\nu\}_{\nu \in I}$.

1.7 Proposition [6]. *Let* $\{M_i\}$ *be a finite family of open semianalytic sets in* N, Ω *a finite family of differential* 1*-forms analytic in a neighbourhood of* $c \in N$ *and* $c \in \overline{M}_i$ *for each* i. *Let* $\{E_\nu\}$ *be a finite family of semianalytic sets in* N.

There exists a basis of neighbourhoods $\mathfrak{B}$ *of* c *such that for every* $U \in \mathfrak{B}$, *there exists a normal stratification* $\mathfrak{N}$ *of* U *adapted to* $\{M_i\}, \{E_\nu\}$ *and any subfamily* $\Omega' \subset \Omega$.

Moreover, if the forms in the family Ω *are integrable and the Pfaffian hypersurfaces* $\{(V_i, \omega_i, M_i)\}$, $\omega_i \in \Omega'$ *have the Rolle property for the paths in a leaf* $Y \in \mathfrak{N}$, *then for a basis* $\Omega'_J = \{\omega_{i_1}, \ldots, \omega_{i_p}\}$ *of* Ω' *along* Y, *the intersection* $\bigcap_{\omega_i \in \Omega'}(V_i \cap Y)$ *is composed of connected components of* $\bigcap_{j \in J}(V_j \cap Y)$. (cf. Theorem 1.4.)

1.8 Definition. We say that a normal stratification $\mathfrak{N}$ is *strongly adapted* to a finite family of Pfaffian hypersurfaces $\mathfrak{H} = \{(V_i, \omega_i, M_i)\}_{\omega_i \in \Omega}$ if it is adapted to $\{M_i\}$ and any subfamily $\Omega' \subset \Omega$.

1.9 Proposition. *Let $\mathfrak{N}$ be a normal stratification of a neighbourhood U of $c \in N$, which is strongly adapted to the finite family of Pfaffian hypersurfaces $\mathfrak{H} = \{(V_i, \omega_i, M_i)\}_{\omega_i \in \Omega}$. Let Y be a leaf of $\mathfrak{N}$ such that the family $\mathfrak{H}$ has the Rolle property for the paths in Y. Then the collection $\{\bigcap_{\omega_i \in \Omega'} V_i \cap Y\}_{\Omega' \subset \Omega}$ forms a finite family $\mathfrak{K}_Y$ of analytic submanifolds of Y with normal crossings in Y. Moreover, if the neighbourhood U is sufficiently small, then, for each $\Gamma \in \mathfrak{K}_Y$, we have $0 \in \overline{\Gamma}$.*

Proof. Directly from Proposition 1.7.

1.9.1 Definition. For $k = 0, 1, \dots, p = \dim Y$, we define the *skeleton varieties*

$$L_k = \bigcup_{\substack{N_i \in \mathfrak{K}_Y \\ \dim N_i \leq k}} N_i.$$

1.9.2 Corollary. *The filtration $Y = L_p \supset \cdots \supset L_0$ gives a finite stratification $\{Y_\nu^i\}$ of Y into analytic submanifolds, where $\{Y_\nu^k\}$ is a family of connected components of $L_k \setminus L_{k-1}$ (we assume that $L_{-1} = \emptyset$).*

1.9.3 Definition. We say that the above stratification $\{Y_\nu^i\}$ is the stratification *induced on Y* by the family $\mathfrak{H}$.

1.10 Definition. An analytic submanifold Γ in a neighbourhood U of $c \in N$ is a *semipfaffian leaf* if and only if there exists a normal stratification $\mathfrak{N}$ of U and a finite family $\mathfrak{H}$ of Pfaffian hypersurfaces such that Γ is an element of the induced stratification of some $Y \in \mathfrak{N}$. If $\Gamma \in \mathfrak{K}_Y$ for some $Y \in \mathfrak{N}$, we call it a *Pfaffian leaf.*

1.11 Definition. A subset $E \subset N$ is *semipfaffian* (resp. *basic semipfaffian*) if and only if for every $c \in N$, there exist
1) a normal neighbourhood U of c,
2) a finite family of pfaffian hypersurfaces $\mathfrak{H}$ given by 1-forms defined in a neighbourhood of c,
3) a normal stratification $\mathfrak{N}$ of U strongly adapted to $\mathfrak{H}$ such that $E \cap U = \bigcup_{i,\nu} \Gamma_\nu^i$, where $\{\Gamma_\nu^i\}$ is a finite collection of semipfaffian leaves (resp. elements of $\mathfrak{K}_{Y_j}$ where $\{Y_j\} \subset \mathfrak{N}$).

1.11.1 Definition. Basic semipfaffian sets of the special type $V_1 \cap \cdots \cap V_k \cap X$, where $\{V_i\}$ are Pfaffian hypersurfaces and X is a semianalytic subset in N are called *Pfaffian sets.*

1.12 Remark. Let us observe that both categories of basic semipfaffian sets and semipfaffian sets are closed on locally finite unions and intersections and the category of semipfaffian sets is in addition closed on taking the complementary set.

1.13 Remark. The cartesian product of finitely many semipfaffian (resp. basic semipfaffian) sets is semipfaffian (resp. basic semipfaffian).

1.14 Theorem. *Any connected component of a semipfaffian (resp. basic semipfaffian) set is semipfaffian (resp. basic semipfaffian).*

1.15 Remark. The family of connected components of a semipfaffian (resp. basic semipfaffian) set is locally finite.

1.16 Corollary. *Any open and closed subset of a semipfaffian (resp. basic semipfaffian) set is semipfaffian (resp. basic semipfaffian).*

1.17. Using the result of C. Roche [12, p.6], i.e. interpretation of S. Lojasiewicz's theorem on the closure of partially semialgebraic set [8], we can formulate the Tangent Mapping Theorem in the following way:

Let Y be a semianalytic leaf in $\mathbb{R}^n$, i.e., an analytic submanifold which is a semianalytic subset as well, $\Omega = \{\omega_1, \dots, \omega_p\}$ a family of 1-differential forms which are transverse to Y. Let $\dim Y = k + p$ and $\Gamma = (V_1 \cap \dots \cap V_p) \cap Y$, where (V_i, ω_i, M_i) is a family of Pfaffian hypersurfaces which have the Rolle property in Y, then:

1) *The graph of the tangent mapping*

$$\tau_\Gamma = \{(x, T_x\Gamma) \in \Gamma \times \mathbb{G}_k(\mathbb{R}^n)\}$$

is a Pfaffian subset of $\mathbb{R}^n \times \mathbb{G}_k(\mathbb{R}^n)$.

2) *For a semialgebraic $E \subset \mathbb{G}_k(\mathbb{R}^n)$, the inverse image $\tau^{-1}(E)$ by the mapping*

$$\tau : Y \ni x \longmapsto \ker \omega_1(x) \cap \dots \cap \operatorname{Ker} \omega_p(x) \cap T_x Y \in \mathbb{G}_k(\mathbb{R})$$

is semianalitic in $\mathbb{R}^n$.

3) *For a semialgebraic $E \subset \mathbb{G}_k(\mathbb{R}^n)$ the inverse image $\tau_\Gamma^{-1}(E)$ is a Pfaffian subset in $\mathbb{R}^n$.*

2. Subpfaffian sets

In this section, we shall define subpfaffian sets and state the properties which follow directly from the definition.

2.1 Definition. Let N be a real analytic manifold. We say that a subset $E \subset N$ is *subpfaffian* (in N) if for each $c \in N$ there exists a neighbourhood U such that $E \cap U$ is an image of a basic semipfaffian and relatively compact subset $A \subset N \times \mathbb{R}^k$ by the natural projection $\pi : N \times \mathbb{R}^k \to N$.

2.2 Proposition. *Let $E \subset N = \bigcup_{i \in I} G_i$, where $\{G_i\}$ is an open covering of N. Then E is subpfaffian in $N \Leftrightarrow$ for every $i \in I$, $E \cap G_i$ is subpfaffian in G_i.*

2.3. The union and the intersection of any locally finite family of subpfaffian sets are subpfaffian.

2.4. If $E \subset M \times N$ is a subpfaffian set N-relatively compact, then $\pi(E)$ is subpfaffian in M.

Especially, if $f : M \to N$ is a proper analytic mapping then $f(F)$ is a subpfaffian set, for $F \subset M$ subpfaffian.

2.5 Theorem. *Connected components of subpfaffian sets are subpfaffian and the family of connected components of a subpfaffian set is locally finite.*

Proof. If S is a connected component of E and $U \cap E = \pi(A)$, then $U \cap S$ is a union of $\pi(S_i)$, for S_i the connected components of A.

We shall need some elementary facts from the dimension theory.

2.6. The dimension of a subpfaffian set $E \subset N$ is by definition the maximum of dimensions of analytic submanifolds which are contained in E.

Directly, we have the following properties.

2.6.1. If $E \subset F$, then $\dim E \leq \dim F$.

2.6.2. $\dim E = \dim N \Leftrightarrow \operatorname{int} E \neq \emptyset$.

2.6.3. If $F \subset M \times N$ is subpfaffian and $\pi(F)$ is contained in a subpfaffian set $E \subset M$, where $\pi : M \times N \to M$ is the natural projection, then $\dim F \leq \dim E + \max_{x \in E} \dim \pi_F^{-1}(x)$.

3. Basic lemmas

The technical tools for studying projections of basic semipfaffian sets are the Lemmas A and B which are counterparts of analogical lemmas in subanalytic geometry ([2], [7]).

3.1. Let $\pi : \mathbb{R}^n \times \mathbb{R}^p \to \mathbb{R}^n$ be the natural projection. Let us consider semialgebraic sets

$$A_{k,r} = \{S : \dim \pi(S) = r\} \subset \mathbb{G}_k(R^{n+p})$$

where $k - p \leq r \leq k \leq n + p$.

For the pair (k, r), we assume the existence of a finite covering of $A_{k,r}$ by semialgebraic sets $G_\nu^{k,r}$.

Lemma A. *Every basic semipfaffian bounded subset $A \subset \mathbb{R}^{n+p}$ can be presented as a finite union of Pfaffian leaves Γ_j^k such that*

1) *the rank of $\pi_{\Gamma_j^k} : \Gamma_j^k \to \mathbb{R}^n$ is constant,*
2) *for a fixed Γ_j^k, the subset $\{T_x\Gamma_j^k : x \in \Gamma_j^k\}$ is contained in some $G_\nu^{k,r}$, where $\dim \Gamma_j^k = k$, $r = \operatorname{rg} \pi_{\Gamma_j^k}$.*

3.2 Remark. Let $V = 0 \times \mathbb{R}^p \subset \mathbb{R}^n \times \mathbb{R}^p$. There exists a finite covering of $A_{k,r}$ by semialgebraic $G_\nu^{k,r}$ such that for each ν subspaces $S \cap V$ (where $S \in G_\nu^{k,r}$) have a common suplement in V and all the subspaces $\pi(S)$ (where $S \in G_\nu^{k,r}$) have a common suplement in $\mathbb{R}^n$.

In fact, for (k,r) as in 3.1, it is enough to have a covering by $G_{\nu,\mu} = \{S \in A_{k,r} : S \cap V_\nu = 0, \pi(S) \cap W_\mu = 0\}$, where V_ν are elements of $\mathbb{G}_{p-k+r}(V)$ generated by the vectors of the canonical basis in $\mathbb{R}^p$ and W_μ are elements of $\mathbb{G}_{n-r}(\mathbb{R}^n)$ generated by vectors of the canonical basis in $\mathbb{R}^n$.

Proof. We proceed in the analogical manner like in the subanalytic case using induction on $k = \dim A$ and our version of the Tangent Mapping Theorem (1.17)(cf. [2]).

Lemma B. *Every basic semipfaffian and bounded set $A \subset \mathbb{R}^{n+p}$ contains a finite union of Pfaffian leaves Γ_j^k such that $\pi(A) = \bigcup_{k,j} \pi(\Gamma_j^k)$ and such that*

1) $\pi_{\Gamma_j^k} : \Gamma_j^k \to \mathbb{R}^n$ *is an immersion,*
2) *each subset $\{T_x\Gamma_j^k : x \in \Gamma_j^k\}$ is contained in some $G_\nu^{k,k}$.*

Proof. Let us observe that any Pfaffian leaf Γ has the natural carpeting function g, which is a restriction of the carpeting function of the semianalytic leaf of the normal decomposition (cf. 1.9). More precisely, there exists a function g, analytic in a neighbourhood of $\overline{\Gamma}$, such that $g_\Gamma > 0$ and $g_{\partial\Gamma} \equiv 0$.

After this observation, we proceed like in the subanalytic case (cf. [2]).

4. Complement of a subpfaffian set

In this section, we shall prove that the complement of a subpfaffian set is subpfaffian, under the assumption that its boundary is contained in a closed subpfaffian set of smaller dimension. This is the analog of Gabrielov theorem from subanalytic geometry. As a consequence of our result, we prove that the closure of a subpfaffian set is subpfaffian. Let us observe that in the subanalytic case the theorem about the closure is used in the proof of Gabrielov Theorem [3], [4], [7].

In the next section, we shall prove that in fact our assumption is not restrictive, i.e. that any semipfaffian set is subregular (cf. Definition 4.1).

4.1 Definition. Let E be a semipfaffian set in $\mathbb{R}^n$. We say that E is *subregular* if there exists a closed subpfaffian set $dE \subset \mathbb{R}^n$ such that $\partial E \subset dE$ and $\dim dE < \dim E$.

4.1.1 Remark. Notice that such a dE is not unique. If E is bounded, we shall assume that dE is taken compact. Finally, let us define $\widetilde{E} = E \cup dE$.

We shall need the following lemma which is a direct corollary of Lemma B.

4.2 Lemma. *Let A be a relatively compact basic semipfaffian set in $\mathbb{R}^n \times \mathbb{R}^p$. Then there exist finitely many Pfaffian leaves $\Gamma_j \subset A$ such that $\pi_{\Gamma_j} : \Gamma_j \to \mathbb{R}^n$ is an immersion and such that, for each j, the spaces $\pi(T_x\Gamma_j)$, $x \in \Gamma_j$, have a common suplement in $\mathbb{R}^n$ and $\pi(A) = \cup\pi(\Gamma_j)$, with $\dim\Gamma_j \le \dim\pi(A)$.*

4.3 Lemma. *Let $p : \mathbb{R}^m \to \mathbb{R}^k$ be the natural projection. Let us suppose that the theorem about complement is valid in $\mathbb{R}^k$. If $\Gamma \subset \mathbb{R}^m$ is a Pfaffian leaf of dimension k such that $p_\Gamma : \Gamma \to \mathbb{R}^k$ is a local diffeomorphism then the fibres $\Gamma_u = p_\Gamma^{-1}(u)$, $u \in \mathbb{R}^k$ are finite and $\{\sharp\Gamma_u : u \in \mathbb{R}^k\}$ is bounded.*

Proof. The function $\nu : \mathbb{R}^k \ni u \longmapsto \sharp\Gamma_u$ is finite and lower semicontinuous (cf. [3]). The subset $\Lambda = p(d\Gamma)$ is subpfaffian, compact and nowhere dense in $\mathbb{R}^k$. Therefore it is enough to prove that ν is bounded in $\mathbb{R}^k \setminus \Lambda$. The mapping $p_\Delta : \Delta \to \mathbb{R}^k \setminus \Lambda$ with $\Delta = p_\Gamma^{-1}(\mathbb{R}^k \setminus \Lambda)$ is a local homeomorphism, hence $\nu(u) = \sharp\Gamma_u$ is locally constant. By the hypothesis $\mathbb{R}^k \setminus \Lambda$ is subpfaffian. By Theorem 2.5, $\mathbb{R}^k \setminus \Lambda$ has only finitely many connected components and on each of them ν is constant, therefore ν is bounded in $\mathbb{R}^k$.

4.4 Lemma of Gabrielov. *Let $\pi : \mathbb{R}^n \to \mathbb{R}^k$ be the natural projection. Let us suppose that the theorem about complement is valid in $\mathbb{R}^k$. If E is a subpfaffian bounded set in $\mathbb{R}^n$ such that $\sharp E_u = \pi_E^{-1}(u)$ is bounded when $u \in \mathbb{R}^k$, then the complement $\mathbb{R}^n \setminus E$ is subpfaffian as well.*

Proof. Let us consider the semialgebraic set

$$\Lambda_l = \{(z_1, \dots, z_l) \in (\mathbb{R}^n)^l : \pi(z_i) = \pi(z_j) \text{ and } z_i \neq z_j, \text{ for } i \neq j\}.$$

Let $\nu(u) = \sharp E_u$ for $u \in \mathbb{R}^k$, $\pi_l : (\mathbb{R}^n)^l \ni (z_1, \dots, z_l) \longmapsto z_1 \in \mathbb{R}^n$, $l = 1, 2, \dots$

The subsets $B_l = \{u \in \mathbb{R}^k : \nu(u) \ge l\} = (\pi \circ \pi_l)(E^l \cap \Lambda_l)$ are subpfaffian and bounded in $\mathbb{R}^k$, then $A_l = \{u \in \mathbb{R}^k : \nu(u) = l\} = B_l \setminus B_{l+1}$ by the hypothesis are subpfaffian and bounded (we put $B_0 = \mathbb{R}^k$). But the sets $\pi^{-1}(A_l) \setminus E = \pi^{-1}(A_l) \cap \pi_{l+1}((\mathbb{R}^n \times E^l) \cap \Lambda_{l+1})$ are subpfaffian and finally $\mathbb{R}^n \setminus E = \bigcup_{i=0}^{s}(\pi^{-1}(A_l) \setminus E)$ is subpfaffian.

4.5 Theorem. *The complement of any subpfaffian set is subpfaffian.*

Proof. It is enough to show the theorem for bounded E in $\mathbb{R}^n$. We proceed inductively on n. We can assume that $E = \pi(A)$ where A is basic semipfaffian and bounded.

First, let us assume that $\dim E < n$. From the emma 4.2 we can assume that $E = \pi(\Gamma)$ with Γ a Pfaffian leaf of dimension $k < n$ in $\mathbb{R}^n \times \mathbb{R}^p$ such that π_Γ is an immersion and that $\pi(T_x\Gamma)$ for $x \in \Gamma$ have a common supplement in $\mathbb{R}^n$. Changing coordinates in $\mathbb{R}^n$, we can assume that this supplement is $0 \times \mathbb{R}^{n-k}$. Therefore $p_\Gamma : \Gamma \to \mathbb{R}^k$ (where $p = \widetilde{\pi} \circ \pi$ and $\widetilde{\pi} : \mathbb{R}^n \to \mathbb{R}^k$ is the natural projection is a local diffeomorphism and from the Lemma 4.3, $\{\sharp p_\Gamma^{-1}(u) : u \in \mathbb{R}^k\}$ is bounded. Hence, from the Lemma of Gabrielov, $\mathbb{R}^n \setminus E$ is subpfaffian.

In the case when $\dim E = n$, we can assume that $E = \pi(\Gamma)$ where Γ is a Pfaffian leaf of dimension n such that π_Γ is a local diffeomorphism. Let us take $\widetilde{\Gamma} = \Gamma \cup d\Gamma$ and $\Lambda = d\Gamma$. Then $\pi(\Lambda)$ is a subpfaffian subset in $\mathbb{R}^n$ of dimension $< n$. Therefore $\mathbb{R}^n \setminus \pi(\Lambda)$ is subpfaffian. Moreover $\mathbb{R}^n \setminus \pi(\widetilde{\Gamma})$ is open and closed in $\mathbb{R}^n \setminus \pi(\Lambda)$ (it is because $\mathbb{R}^n \setminus \pi(\widetilde{\Gamma}) = \mathbb{R}^n \setminus \pi(\Lambda) \setminus \pi(\Gamma)$ and $\pi(\Gamma)$ is open in $\mathbb{R}^n$). Hence $\mathbb{R}^n \setminus \pi(\widetilde{\Gamma})$ is a union of connected components of $\mathbb{R}^n \setminus \pi(\Lambda)$ and therefore is subpfaffian. Finally $\mathbb{R}^n \setminus \pi(\Gamma) = (\mathbb{R}^n \setminus \pi(\widetilde{\Gamma})) \cup (\pi(\widetilde{\Gamma}) \setminus \pi(\Gamma)) = \mathbb{R}^n \setminus \pi(\widetilde{\Gamma}) \cup \pi(\Lambda) \setminus (\pi(\Gamma) \cap \pi(\Lambda))$ is subpfaffian because $\dim(\pi(\Gamma) \cap \pi(\Lambda)) < n$. This ends the proof.

4.6 Remark. Let us observe that if any basic semipfaffian set is subregular (cf. Definition 4.1) then the closure of any subpfaffian set is subpfaffian. More precisely, let $E \subset \mathbb{R}^n$ be subpfaffian and relatively compact. We have the following formula for the closure:

$$\overline{E} = \{x \in \mathbb{R}^n : \forall \epsilon > 0 \ \exists x' \in E \ |x - x'|^2 < \epsilon\}.$$

By Theorem 4.5 it is enough to prove that $\mathbb{R}^n \setminus \overline{E}$ is subpfaffian. Now $\mathbb{R}^n \setminus \overline{E} = \pi_1(E_1)$, where

$$E_1 = \{(x, \epsilon) \in \mathbb{R}^n \times (0,1) : \forall x' \in E \ |\mathrm{x} - \mathrm{x}'|^2 \geq \epsilon\}$$

and $\pi_1 : \mathbb{R}^n \times \mathbb{R} \to \mathbb{R}^n$ is the natural projection. But E_1 is $\mathbb{R}$- relatively compact (cf. 2.4) and subpfaffian because

$$\mathbb{R}^{n+1} \setminus E_1 = \pi_2(E_2) \cup (\mathbb{R}^n \times (\mathbb{R} \setminus (0,1)))$$

where

$$E_2 = \{(x, \epsilon, x') \in \mathbb{R}^n \times (0,1) \times \mathbb{R}^n : |x - x'|^2 < \epsilon\} \cap (\mathbb{R}^n \times \mathbb{R} \times E)$$

and $\pi_2 : (\mathbb{R}^n \times \mathbb{R}) \times \mathbb{R}^n \to \mathbb{R}^n \times \mathbb{R}$ is the natural projection.

5. Boundary of a basic semipfaffian set

In this section, we prove that any basic semipfaffian set is subregular (cf. Definition 4.1.). Let $\pi : \mathbb{R}^n \to \mathbb{R}^{n-1}$ be the natural projection.

5.1 Lemma.

1) *Any one-dimensional basic semipfaffian set $E \subset \mathbb{R}^n$ is subregular.*
2) *Any basic semipfaffian set $E \subset \mathbb{R}^n$ of constant dimension $n - 1$ is subregular.*

2a) *If moreover the natural projection π_E has finite fibres, and $\pi^{-1}(0) \cap T_x\Gamma = 0$ for any $x \in \Gamma$, then dE can be chosen in such a way that π_{dE} has finite fibres as well.*

Proof. 1) By Definition 1.11, at any point $c \in \mathbb{R}^n$, E is a finite union of one-dimensional Pfaffian leaves. From the Pfaffian Curve Selection Lemma (cf. [6],[12]), the boundary of a one-dimensional Pfaffian leaf is a point.

2) By the definition of a singular foliation $\mathfrak{F}$, the set Sing $\mathfrak{F}$ is of codimension at least 2 (cf. [12]). From this and the fact that subregularity is a local property, we have subregularity of E.

2a) By the result of Cano, Lion and Moussu [1], we know that for a Pfaffian hypersurface $H = (V, \mathfrak{F}, M)$ with the Rolle property, the boundary ∂V is a subpfaffian set. Now, applying Lemma 4.2, and the Pfaffian curve selection lemma, we can use the Whitney property (see [9] p. 103) in the analogical manner like for semianalytic or subanalytic sets (see [11a] Lemma 2.2 p.3).

5.2 Proposition. *Any basic semipfaffian set is subregular.*

Proof. By Definition 1.11. it is enough to prove that any k-dimensional Pfaffian leaf Γ is subregular. Let $\Gamma = (V_1 \cap \cdots \cap V_p) \cap Y_\nu^{k+p}$, where Y_ν^{k+p} is a $(k+p)$-dimensional leaf of a normal decomposition $\mathfrak{N}$ at $0 \in \mathbb{R}^n$, $\Omega = \{\omega_1, \ldots, \omega_p\}$ a family of 1-differential forms which are transverse to Y_ν^{k+p} and $\{(V_i, \omega_i, M_i)\}_{\omega_i \in \Omega}$ a family of Pfaffian hypersurfaces which have the Rolle property in Y_ν^{k+p}. The proof will proceed by induction with respect to p.

Using Hironaka's rectilinearization theorem [5a] we can assume that Y_ν^{k+p} is an open quadrant of the form $Y_\nu^{k+p} = Y = \{x_1 > 0, \ldots, x_n > 0\}$. From Lemma 5.1, we have the case $p = 1$. Thus let us consider the case $p > 1$.

Recall that, for $e_1, \ldots, e_n$ the canonical basis of $\mathbb{R}^n$, we have a finite atlas $\{\phi_{V_\alpha, V_{\alpha'}}\}_{\alpha_1 < \cdots < \alpha_k}$ of inverse charts on $\mathbb{G}_k(\mathbb{R}^n)$ with $\binom{n}{k}$ elements, where $\alpha' = (\alpha'_1, \ldots, \alpha'_{n-k})$ is the complementary multi-index to α and $V_\alpha = \sum_{i=1}^k \mathbb{R} e_{\alpha_i}$.

Moreover, the images $\widehat{K}_\alpha = \phi_{V_\alpha, V_{\alpha'}}(K_\alpha)$, where

$$\phi_{V_\alpha, V_{\alpha'}} : L(V_\alpha, V_{\alpha'}) \ni f \mapsto \widehat{f} = \{u + f(u)\} \in \mathbb{G}_k(\mathbb{R}^n)$$

and $K_\alpha = \{f \in L(V_\alpha, V_{\alpha'}) : |f| < 2\sqrt{\binom{n}{k}}\}$ form an open finite semialgebraic covering of $\mathbb{G}_k(\mathbb{R}^n)$.

Using the Tangent Mapping Theorem 1.17, for

$$\tau : Y \ni x \mapsto \operatorname{Ker} \omega_1(x) \cap \cdots \cap \operatorname{Ker} \omega_p(x) \in \mathbb{G}_k(\mathbb{R}^n), \quad (k = n - p),$$

we see that Y can be covered by finitely many open connected semianalytic sets $U_1, \ldots, U_r$ such that $p^\alpha_{U_j \cap \Gamma} : U_j \cap \Gamma \to V_\alpha$ is a local isomorphism, where $p^\alpha : \mathbb{R}^n \to V_\alpha$ is the natural projection and U_j is a connected component of $\tau^{-1}(\widehat{K}_\alpha)$

Let us observe that for having Γ subregular, it is enough to know that for any U_j, $(U_j \cap \Gamma)$ is subregular. For any $c \in \Gamma$, we can take a sufficiently small neighbourhood $Q_c = \{|x_i - c_i| < \delta_i\}$ such that $p^\alpha_{Q_c \cap \Gamma} : Q_c \cap \Gamma \to V_\alpha$ is an analytic isomorphism on $Q_c \cap V_\alpha$.

By a permutation of coordinates, we can assume that $Q_c \cap \Gamma$ is a graph of an analytic map

$$\phi = (\phi_{k+1}, \ldots, \phi_n) : \{|x_i - c_i| < \delta_i, i = 1, \ldots, k\} \to \mathbb{R}^{n-k} \simeq V_{\alpha'}.$$

Locally at c, ϕ is a solution of the system of Pfaffian equations

$$(*) \qquad \sum_{i=1}^{n} a_i^s dx_i = 0, \quad s = 1, \dots, p.$$

Hence, taking the scalar product, we have

$$(**) \qquad \langle \sum_{i=1}^{n} a_i^s(x, \phi(x)) e_i, (e_q, \frac{\partial \phi}{\partial x_q}(x)) \rangle = 0, \quad s = 1, \dots, p;\ q = 1, \dots, k$$

and for fixed q and s:

$$-a_q^s = \sum_{i=k+1}^{n} a_i^s \frac{\partial \phi_i}{\partial x_q}.$$

The matrix $A = \{a_i^s\}_{s=1,\dots,p=n-k; i=k+1,\dots,n}$ is nonsingular, therefore using Cramer's rules and vectorial product we can calculate the normal vectors to the graphs

$$\phi_i : \{|x_i - c_i|\} < \delta_i, i = 1, \dots, k\} \to \mathbb{R} e_i, \quad i = k+1, \dots, n.$$

Let us observe that we have expressed the coordinates of normal vectors in terms of a_i^s, therefore we have a system of Pfaffian forms:

$$\overline{\omega_{k+1}} = B_1^{k+1} dx_1 + \cdots + B_k^{k+1} dx_k + B_{k+1}^{k+1} dx_{k+1}$$

$$\overline{\omega_n} = B_1^n dx_1 + \cdots + B_k^n dx_k + B_n^n dx_n$$

with B_i^j analytic in a neighbourhood of the origin in $\mathbb{R}^n$.

Now, one of the following two possibilities holds:

a) there exists at least one nonintegrable form in the family $\overline{\Omega} = \{\overline{\omega_{k+1}}, \dots, \overline{\omega_n}\}$,

b) each form from the family $\overline{\Omega}$ is integrable.

Consider the possibility a). Let us suppose that $\overline{\omega_i}$ is nonintegrable. In this case Γ is contained in the analytic set $W = \{\overline{\omega_i} \wedge d\overline{\omega_i} = 0\}$. It is clear that $\dim W < n$. Using the adapted stratification compatible with W (cf. 1.7) we reduce the codimension of Γ and we conclude by induction.

Consider now the possibility b). In this case $\overline{\omega_{k+1}}, \dots, \overline{\omega_n}$ are transverse to U_j and Γ is their common solution. We can assume that Γ is bounded. Let $\pi_i : \mathbb{R}^n \to \mathbb{R}^k + \mathbb{R} e_i$ be the natural projection; $i = k+1, \dots, n$. Since $\overline{\omega_{k+1}}, \dots, \overline{\omega_n}$ are integrable, and from their expression it is visible that $\pi_i^{-1}(\pi_i(\Gamma))$ is a solution of $\overline{\omega_i}$ for $i = k+1, \dots, n$. Moreover $\Gamma \subset \cap_{k+1}^n \pi_i^{-1}(\pi_i(\Gamma))$. Therefore it is enough to show that

$$(***) \qquad \dim \cap_{i=k+1}^n d(\pi_i^{-1}(\Gamma_i)) < k, \text{where } \Gamma_i = \pi_i(\Gamma).$$

Each $\Gamma_i, i = k+1, \dots, n$ is a subpfaffian set and therefore it has finitely many connected components. Using a semianalytic triangulation of $\mathbb{R}^k + \mathbb{R} e_i$ compatible with $\operatorname{Sing} \overline{\omega_i}$ (see e.g. [7]), we can assume that Γ_i is a finite union of Pfaffian hypersurfaces which are separant (cf. 1.2). By Lemma (5.1.2a) we can choose $d\Gamma_i$

such that its projection on $\mathbb{R}^k$ is finite and this is sufficient to get the condition (***) (cf. 2.6.3).

References

[1] Cano, F., Lion, J-M., Moussu, R., The frontier of a Pfaffian hypersurface. To appear.

[2] Denkowska, Z., Lojasiewicz, S., Stasica, J., Certaines propriétés élémentaires des ensembles sous-analytiques. Bull. Acad. Polon. Sci., Sér. Sci. Math. Vol XXVII (7-8) (1979), 529–536.

[3] Denkowska, Z., Lojasiewicz, S., Stasica, J., Sur le théorème du complémentaire pour les ensembles sous-analytiques. Bull. Acad. Polon. Sci., Sér. Sci. Math. Vol XXX (7-8) (1981), 333–335.

[4] Gabrielov, A. M., Projections des ensembles semi-analytiques. Funktsional. Anal. i Prilozhen. 2 (n^0 4) (1968), 18–30.

[5] Galbiati, M., Sur l'image d'un morphisme analytique réel propre. Ann. Scuola Norm. Sup. Pisa Vol. III. 1 (1976), 311-319.

[5a] Hironaka, H., Subanalytic sets. Lecture Notes of Istituto Matematico "L. Tonelli", 1973.

[6] Lion, J-M., Etude des hypersurfaces pfaffiennes. Thesis, Université de Bourgogne (1991).

[7] Lojasiewicz, S., Zurro, M.A., Introducción a la geometría semi y subanalítica. Preprint, Universidad de Valladolid.

[8] Lojasiewicz, S., Sur l'adhérence d'un ensemble partiellement semi-algébrique. Publ. Math. IHES 68 (1989), 205–210.

[9] Lojasiewicz, S., Ensembles semianalytiques. Publ. Inst. Hautes Etudes Sci., Bures-sur-Yvette, 1965.

[10] Moussu, R., Roche, C., Problème de Dulac et théorie de Hovanskii. Invent. Math. 105 (1991), 431–441.

[11] Moussu, R., Roche, C., Théorèmes de finitude uniformes pour les variétés Pfaffiennes de Rolle. Ann. Inst. Fourier 42 (1992), 393-420.

[11a] Pawłucki, W., Points de Nash des ensembles sous-analytiques. Memoirs of AMS 425, vol. 88, (1990).

[12] Roche, C., Densities for certain leaves of real analytic foliations. Prépublication de l'Université de Bourgogne, n^0 03 (1992).

[13] van den Dries, L., Tarski's problem and Pfaffian functions. Logic Colloquium 84, Paris, Willie and Wilmers (ed.), North-Holland, (1986), 59–90.

[14] Denef, J., van den Dries, L., p-adic and real subanalytic sets. Ann. of Math. 128 (1988), 79–138.

Groups of monodromy of non-singular curves of degree 6

Ilia V. Itenberg

Abstract. The paper is devoted to the calculation of the groups of monodromy of plane projective real algebraic non-singular curves of degree 6. For this calculation we use the theory of K3 surfaces. This approach was developed by V.M. Kharlamov and V.V. Nikulin to obtain classifications up to rigid isotopy of non-singular surfaces of degree 4 in $\mathbb{R}P^3$ and of non-singular curves of degree 6 in $\mathbb{R}P^2$ respectively.

1991 Mathematics Subject Classification: 14H99.

1. Introduction

Consider the space $\mathbb{R}C_6$ of real algebraic plane projective curves of degree 6. It is a real projective space of dimension 27. Let D be the subset of $\mathbb{R}C_6$ whose points correspond to curves having singular points (real or imaginary).

Let A be an arbitrary non-singular curve of degree 6. The components of the set $\mathbb{R}A$ of real points of A are called *ovals* (all these components are homeomorphic to a circle and are embedded in $\mathbb{R}P^2$ two-sidedly). Let a be the point of $\mathbb{R}C_6$ corresponding to the curve A, and U_a be the connected component of $\mathbb{R}C_6 \setminus D$ containing the point a.

Each element of the fundamental group $\pi_1(U_a, a)$ defines a permutation of ovals of the curve A. So we obtain a homomorphism,

$$\pi_1(U_a, a) \longrightarrow S_A,$$

where S_A is the group of permutations of ovals of A. We will call the image of $\pi_1(U_a, a)$ in S_A *the monodromy group* of the curve A.

In 1979 V. Nikulin [Ni1] described the set of connected components of $\mathbb{R}C_6 \setminus D$ using the theory of K3 surfaces.

V. Kharlamov [Kh] noticed that the scheme used by Nikulin gives a method for the calculation of the monodromy groups of non-singular curves of degree 6. Kharlamov [Kh] calculated the monodromy groups of non-singular M-curves of degree 6 (an *M-curve* is a curve with the maximal possible number of components of the real point set for the given degree; this number is equal to 11 for degree 6).

In the present paper the monodromy groups of all non-singular curves of degree 6 are calculated.

2. Calculation of monodromy groups

We will use the system of notations suggested by O. Viro [Vir] for the schemes of disposition of ovals of non-singular curves.

If the real point set of a non-singular curve divides the complex point set of this curve, then we say that the curve is of *type* I. In the other case we say that the curve is of *type* II.

Theorem 1. *The monodromy groups of non-singular curves of degree* 6 *coincide with the ones given in the table.*

Remark. The curves whose corresponding points lie in the same component of $\mathbb{R}C_6 \setminus D$ are called *rigidly isotopic*. Evidently, rigidly isotopic curves have isomorphic monodromy groups. So, in the table we point out only rigid isotopy types of curves. The rigid isotopy type of a non-singular curve of degree 6 is defined by the scheme of disposition of ovals and the type (I or II) of this curve (see [Ni1]).

Table

NN	Rigid isotopy type	Monodromy group
1	$<1<1> \amalg 9>$	S_3
2	$<1<5> \amalg 5>$	trivial
3	$<1<9> \amalg 1>$	$\mathbb{Z}_2$
4	$<10>$	A_5
5	$<1<1> \amalg 8>$	D_8
6	$<1<4> \amalg 5>$	trivial
7	$<1<5> \amalg 4>$	trivial
8	$<1<8> \amalg 1>$	$\mathbb{Z}_2$
9	$<1<9>>$	S_3
10	$<9>$, type I	S_9
11	$<9>$, type II	S_9
12	$<1<1> \amalg 7>$	D_7
13	$<1<2> \amalg 6>$	$\mathbb{Z}_2$
14	$<1<3> \amalg 5>$	trivial
15	$<1<4> \amalg 4>$, type I	trivial
16	$<1<4> \amalg 4>$, type II	trivial
17	$<1<5> \amalg 3>$	trivial
18	$<1<6> \amalg 2>$	trivial
19	$<1<7> \amalg 1>$	$\mathbb{Z}_2$
20	$<1<8>>$, type I	D_4
21	$<1<8>>$, type II	D_8
22	$<8>$	S_8
23	$<1<1> \amalg 6>$	D_6
24	$<1<2> \amalg 5>$	$\mathbb{Z}_2$
25	$<1<3> \amalg 4>$	trivial

26	$< 1 < 4 > \amalg 3 >$	trivial
27	$< 1 < 5 > \amalg 2 >$	trivial
28	$< 1 < 6 > \amalg 1 >$	$\mathbb{Z}_2$
29	$< 1 < 7 >>$	D_7
30	$< 7 >$	S_7
31	$< 1 < 1 > \amalg 5 >$, type I	D_5
32	$< 1 < 1 > \amalg 5 >$, type II	D_5
33	$< 1 < 2 > \amalg 4 >$	$\mathbb{Z}_2$
34	$< 1 < 3 > \amalg 3 >$, type I	$\mathbb{Z}_2$
35	$< 1 < 3 > \amalg 3 >$, type II	trivial
36	$< 1 < 4 > \amalg 2 >$	trivial
37	$< 1 < 5 > \amalg 1 >$, type I	$\mathbb{Z}_2$
38	$< 1 < 5 > \amalg 1 >$, type II	$\mathbb{Z}_2$
39	$< 1 < 6 >>$	D_6
40	$< 6 >$	S_6
41	$< 1 < 1 > \amalg 4 >$	D_4
42	$< 1 < 2 > \amalg 3 >$	$\mathbb{Z}_2 \times \mathbb{Z}_2$
43	$< 1 < 3 > \amalg 2 >$	$\mathbb{Z}_2$
44	$< 1 < 4 > \amalg 1 >$	$\mathbb{Z}_2$
45	$< 1 < 5 >>$	D_5
46	$< 5 >$	S_5
47	$< 1 < 1 > \amalg 3 >$	S_3
48	$< 1 < 2 > \amalg 2 >$, type I	$\mathbb{Z}_2 \times \mathbb{Z}_2$
49	$< 1 < 2 > \amalg 2 >$, type II	$\mathbb{Z}_2 \times \mathbb{Z}_2$
50	$< 1 < 3 > \amalg 1 >$	$\mathbb{Z}_2$
51	$< 1 < 4 >>$, type I	S_3
52	$< 1 < 4 >>$, type II	D_4
53	$< 4 >$	S_4
54	$< 1 < 1 > \amalg 2 >$	$\mathbb{Z}_2$
55	$< 1 < 2 > \amalg 1 >$	$\mathbb{Z}_2$
56	$< 1 < 3 >>$	S_3
57	$< 3 >$	S_3
58	$< 1 < 1 > \amalg 1 >$	trivial
59	$< 1 < 2 >>$	$\mathbb{Z}_2$
60	$< 1 < 1 < 1 >>>$	trivial
61	$< 2 >$	$\mathbb{Z}_2$
62	$< 1 < 1 >>$	trivial
63	$< 1 >$	trivial
64	$< 0 >$	trivial

In this table S_n is the group of permutations of n elements, A_n is the group of even permutations of n elements, D_n is the group of symmetries of a regular n-gon.

Proof. Let A be an arbitrary non-singular curve of degree 6, and $\mathbb{C}A$ be the set of complex points of this curve. The set $\mathbb{R}A$ of real points of the curve A divides the real projective plane $\mathbb{R}P^2$ into two sets with the common boundary $\mathbb{R}A$. One of them is non-orientable. Denote it by $\mathbb{R}P^2_-$ and the other by $\mathbb{R}P^2_+$.

Consider the two-sheeted branched covering space X of the complex projective plane $\mathbb{C}P^2$ with branch locus $\mathbb{C}A$. The variety X is a K3 surface.

The involution of complex conjugation acting on $\mathbb{C}P^2$ is lifted to X in two different ways. Let us denote the antiholomorphic involution on X with the set of the fixed points lying above $\mathbb{R}P^2_-$ by $conj_-$. The other antiholomorphic involution on X (its set of the fixed points lies above $\mathbb{R}P^2_+$) will be denoted by $conj_+$. Choose one of these involutions and denote it by $conj$.

We can identify the group $H_2(X\,;\mathbb{Z})$ together with the intersection form B_X and the involution $conj_*$ induced by $conj$ with a lattice L having a bilinear form B and an involution ϕ. Denote the eigenspaces of the involution ϕ corresponding to eigenvalues 1 and -1 by L_+ and L_- respectively. Let h' be the class of hyperplane section in $H_2(X\,;\,\mathbb{Z})$ and let h be the corresponding element of L.

It is known (see, for example, [Sh]) that

- the rank of L is equal to 22,
- the signature of B is equal to -16,
- the diagonal forms over $\mathbb{Q}$ of the restrictions of B on L_+ and L_- have one and two positive squares respectively,
- $B(h,h) = 2, \quad \phi(h) = -h.$

We will call an isomorphism of lattices $f : L_1 \longrightarrow L_2$ *an isomorphism of the quadruples*

$$(L_1, B_1, \phi_1, h_1)$$

and

$$(L_2, B_2, \phi_2, h_2)$$

if

$$f \circ \phi_1 = \phi_2 \circ f, \quad f(h_1) = h_2.$$

Let us fix a quadruple

$$(L, B, \phi, h) \tag{1}$$

satisfying all the above conditions. We will call a K3 surface X with fixed isomorphism of quadruples

$$\varphi : (H_2(X;\mathbb{Z}), B_X, conj_*, h') \longrightarrow (L, B, \phi, h)$$

a marked K3-surface.

Let a subspace $H(X)$ of the complex vector space $H^2(X, \mathbb{C})$ be the image under the de Rham isomorphism of the space $\mathcal{H}^{2,0}(X)$ consisting of cohomology classes of holomorphic 2-forms. The dimension of $H(X)$ is equal to 1. Denote by H the real one-dimensional subspace of $H(X)$ consisting of vectors x satisfying the following condition: the image of x under the involution of $H^2(X,\mathbb{C})$ induced by the involution of complex conjugation of coefficients coincides with $conj^*(x)$. We will denote by H' the subspace of $H_2(X,\mathbb{C})$ generated by a class dual (in the sense of duality of Poincaré) to a class generating space H.

Define (as it was done in [Ni1, Ni2]) a space

$$\Omega = \{0 \neq \omega \in L \otimes \mathbb{C} \mid B_C(\omega,\omega) = 0, \;\; B_C(\omega,\bar{\omega}) > 0,$$

$$B_C(\omega,h) = 0, \;\; \phi(\omega) = \bar{\omega}\}/\mathbb{R}^*$$

where B_C is the extension of the bilinear form B on L to the bilinear form on $L\otimes\mathbb{C}$, and $\mathbb{R}^*$ acts by multiplication. The period mapping bringing the one-dimensional subspace $\varphi(H')$ of $L \otimes \mathbb{C}$ in correspondence with a marked K3 surface makes the space Ω the space of periods of marked K3 surfaces with involution equivalent to ϕ and with linear system $\mid h' \mid$ (see [Ni1]).

We will call periods which can be transformed one into another by an automorphism of quadruple (1) *equivalent periods.*

Let us describe the image of the period mapping up to equivalence (this description was obtained by Nikulin [Ni1]).

Let an element $\omega \in L \otimes \mathbb{C}$ generate the class $[\omega] \in \Omega$ and

$$\omega = \omega_+ + i\omega_-$$

where $\omega_+ \in L_+ \otimes \mathbb{R}, \;\; \omega_- \in L_- \otimes \mathbb{R}$.

Denote the orthogonal complement in L_- of the vector h by L_{-h}. Note that

$$\omega_- \in L_{-h} \otimes \mathbb{R}, \;\; \omega_+^2 = \omega_-^2 > 0,$$

and that diagonal forms over $\mathbb{Q}$ of the restrictions of B on L_+ and L_{-h} have one positive square. Thus the space Ω consists of two connected components. Each of these components can be transformed to the other by the involution $-\phi$. We are interested in periods only up to equivalence and thus we can consider the quotient space

$$\Omega/\{1, \; -\phi\}$$

as the space of periods.

Let

$$\mathcal{L}_+ = \{x \in L_+ \otimes \mathbb{R} \mid B_C(x,x) > 0\}/\mathbb{R}^*,$$

$$\mathcal{L}_- = \{x \in L_{-h} \otimes \mathbb{R} \mid B_C(x,x) > 0\}/\mathbb{R}^*$$

be the spaces of Lobachevsky obtained by the standard procedure from the spaces $L_+ \otimes \mathbb{R}$ and $L_{-h} \otimes \mathbb{R}$, respectively. It is easy to verify that

$$\Omega/\{1, \; -\phi\} = \mathcal{L}_+ \times \mathcal{L}_-.$$

Periods $([\omega_+], [\omega_-])$ belonging to $\mathcal{L}_+ \times \mathcal{L}_-$ are the periods of non-singular marked K3-surfaces if there is no such a vector v in the lattice L that

$$B_C(v,\omega) = 0, \;\; B(v,h) = 0, \;\; B(v,v) = -2.$$

Consider a reflection with respect to a hyperplane in $\mathcal{L}_+$ orthogonal to a vector $v \in L_+$ with the square -2 (this reflection transforms x to $x + B_C(x,v)v$). It can be transposed with ϕ leaving the vector h in its place. Thus the reflection described gives equivalent periods. It also holds true for a reflection with respect to a hyperplane in $\mathcal{L}_-$ orthogonal to a vector $v \in L_{-h}$ having the square -2.

Let Ω_+ and Ω_- be fundamental domains of groups generated by reflections with respect to hyperplanes in $\mathcal{L}_+$ and $\mathcal{L}_-$ orthogonal to vectors with squares -2 in L_+ and L_{-h} respectively. We have shown that the image of the period mapping belongs to $\Omega_+ \times \Omega_-$ up to equivalence.

Nikulin [Ni1] proved that the image of the period mapping coincides up to equivalence with

$$\Omega_+ \times \Omega_- \setminus Z,$$

where Z is a subset of $\Omega_+ \times \Omega_-$ of codimension 2.

He has also proved that each class

$$[\omega] \in \Omega_+ \times \Omega_- \setminus Z$$

defines only one class of the projective equivalence of real K3 surfaces being two-sheeted covering spaces of $\mathbb{C}P^2$ branched over the complex point sets of non-singular curves of degree 6.

Assume that the scheme of disposition of ovals of the curve A does not coincide with $< 1 < 1 < 1 >>>$ (evidently, a curve with the scheme $< 1 < 1 < 1 >>>$ has trivial monodromy group).

Let a face of the polytope Ω_- be orthogonal to a vector $v \in L_{-h}$ such that

$$v \not\equiv h (\text{mod } 2L).$$

The points in the interior of this face correspond to K3 surfaces with one non-degenerate double point. These K3 surfaces can be considered as two-sheeted branched covering spaces of the complex projective plane $\mathbb{C}P^2$ with branch loci coinciding with the complex point sets of the results of some simplest degenerations of the curve A.

If the involution *conj* coincides with $conj_-$ (resp. with $conj_+$), then even empty (resp. odd) ovals of the curve A take part in these degenerations (an oval of a non-singular curve is called *even* (*odd*) if it lies inside an even (odd) number of ovals of this curve; an oval is called *empty* if it does not contain any oval of the curve inside of itself).

Let C be the Coxeter scheme (see, for example, [Vin]) of the polytope Ω_-. Remove all thick and dotted edges of the scheme C (two vertices of C are connected by a thick (or a dotted) edge if the bilinear form B has the value equal to (or more than) 2 on corresponding vectors). Let C' stand for the scheme obtained.

Consider vertices $e_1, \dots, e_l$ of the scheme C corresponding to the classes of contractions of even empty ovals (if the involution *conj* coincides with $conj_-$) and of odd ovals (if the involution *conj* coincides with $conj_+$) of the given curve.

Let S be the group of symmetries of the scheme C' generated by automorphisms of the quadruple

$$(L, B, conj, h)$$

and let S_1 be the subgroup of S consisting of the symmetries leaving the vertices $e_1, \dots, e_l$ fixed.

If M is some collection of ovals of the curve A, then we will call the image of $\pi_1(U_a, a)$ in the group of permutations of ovals belonging to M *monodromy group of* M.

It is easy to see that the following statement holds true.

Proposition 2. *The quotient group* S/S_1 *is isomorphic to the group of monodromy of even empty ovals in the case* $conj = conj_-$ *and to the group of monodromy of odd ovals in the case* $conj = conj_+$.

The group S acts on the scheme C'. Let C'' be the corresponding quotient scheme, and K be the connected component of C'' containing the images of the vertices $e_1, \dots, e_l$ (these images, always, belong to one component).

In order to calculate the group S/S_1 we first have to calculate the Coxeter scheme C of the polytope Ω_-. If the scheme C is finite, the calculation can be done using the algorithm from [Vin]. If the scheme C is infinite, it is possible to calculate a finite subscheme of C, containing all the information required. Then we have to find vertices $e_1, \dots, e_l$ in the scheme C. Let us show how this can be done.

Let A be a curve of degree 6 with a non-degenerate double point. There are two types of smoothings of singular points of this curve. Let A_t $(t \geq 0)$ be a smoothing of one of these types and A_t $(t \leq 0)$ be a smoothing of the other type. The number T_0 of ovals of the non-singular curve A_{t_0} $(t_0 < 0)$ can be more, less or equal to the number T_1 of ovals of the non-singular curve A_{t_1} $(t_1 > 0)$. If $T_0 > T_1$, then the family A_t $(t_0 < t \leq 0)$ will be called *the simplest decreasing degeneration* of curve A_{t_0}.

There are 6 cases of the simplest decreasing degenerations of a non-singular curve of degree 6 (if its scheme of disposition of ovals does not coincide with $< 1 < 1 < 1 >>>$):

1) conjunction of an even empty oval with the non-empty one;
1') conjunction of an odd oval with the non-empty one;
2) conjunction of two even empty ovals;
2') conjunction of two odd ovals;
3) contraction of an even empty oval;
3') contraction of an odd oval.

Define two graphs P and P' for an arbitrary non-singular curve of degree 6. The vertices of the graph P (resp. P') correspond to the rigid isotopy classes of the results of degenerations of types 1 - 3 (resp. 1' - 3') of the given curve. Two vertices are connected by an edge if the class corresponding to one of these vertices can be realized as a conjunction of ovals s and t, and the class corresponding to the other vertex can be realized as a contraction of the oval t.

Proposition 3. *The graph P (resp. P') of a non-singular curve of degree 6 is isomorphic to the scheme K if the involution conj coincides with $conj_-$ (resp. with $conj_+$).*

So, if a vertex of the scheme K is connected by edges with more than two other vertices, then this vertex is the image of one of the vertices $e_1, \ldots, e_l$.

Proposition 3 allows to find vertices $e_1, \ldots, e_l$ in all cases required.

It is easy to calculate the monodromy group of an arbitrary non-singular curve of degree 6 after the calculations of monodromy groups of even and of odd ovals of this curve.

The calculation of the monodromy groups of non-singular curves of degree 6 is closely connected with the question on the classification up to rigid isotopy of curves of degree 6 with one non-degenerate double point. This classification obtained in [It] makes the results of the calculations of the monodromy groups clearer.

References

[It] Itenberg, I.V., Curves of degree 6 with one non-degenerate double point and groups of monodromy of non-singular curves. In: Real Algebraic Geometry, Proceedings, Rennes 1991, 267–288, Lecture Notes in Math. 1524, Springer-Verlag 1992.

[Kh] Kharlamov, V.M., Private communication.

[Ni1] Nikulin, V.V., Integer symmetric bilinear forms and some of their geometric applications. Math. USSR-Izv. 14 (1979), no. 1, 103–167 (1980).

[Ni2] Nikulin, V.V., Involutions of integer symmetric bilinear forms and applications to real algebraic geometry (in Russian). Izv. Akad. Nauk SSSR. Ser. Mat. 47 (1983), no. 1, 109–188.

[Sh] Shafarevič, I.R. (Ed.), Algebraic Surfaces. Providence, American Mathematical Society, 1967.

[Vin] Vinberg, E.B., The groups of units of certain quadratic forms. Math. USSR–Sb. 87 (1972), 17–35.

[Vir] Viro, O.Ya., Curves of degree 7, curves of degree 8, and Ragsdale conjecture. Soviet Math. Dokl. 22 (1980), 566–569.

Blowing-up construction of maximal smoothings of real plane curve singularities

V.M. Kharlamov and J.-J. Risler

Abstract. We address the problem of construction of smoothings of real plane curve singularities with the maximum possible of connected components. The technic of construction is by using blowing-ups, and lower bounds on the maximal number of connected components are given in the general case (i.e., in the case of a non irreducible curve). The case of a curve with branches having distinct tangents is especially emphasized.

1991 Mathematics Subject Classification: 14B05, 14B07, 14H20, 14P15.

Introduction

Let $f(x,y)=0$ be the germ at $O \in \mathbb{R}^2$ of a (reduced) real analytic singularity. We identify it with its analytic representative in a suitable neighborhood of O and we look at local real deformations f_ε of it into a smooth curve (in other words, at local level sets of real morsifications of the singularity); we call such curves *smoothings of f*. The first natural question, which comes back to Arnol'd ([A 1]), is: what topological types of smoothings are possible for a given singularity?

More precisely, let $B_\eta \subset \mathbb{R}^2$ be a real part of a Milnor ball $\mathbb{C}B_\eta \subset \mathbb{C}^2$ for f of radius η (see [Mi]) and let $f_\varepsilon(x,y)=0$ be a real smooth deformation of f, i.e., a real analytic germ in (ε,x,y) such that the curve C_ε, defined in, may be, smaller Milnor ball $\mathbb{C}B_\eta$ by $f_\varepsilon(x,y)=0$, is transversal to its boudary $\partial\mathbb{C}B_\eta$ and has no singular points inside this ball for ε small enough. Then, the topological type we are looking for is the one of $C_\varepsilon \cap B_\eta$, ε taken small enough. A deformation $f_\varepsilon(x,y)$ being chosen, this topological type is constant for small valuers of ε of a same sign and does not depend on a choice of the radius of the Milnor ball. Note two facts:

1) The similar question over the complex numbers has a simple solution: there is only one topological type for the nearby smooth complex curve $\mathbb{C}C_\varepsilon \cap \mathbb{C}B_\eta$ (see, for example, [Mi]): it is homeomorphic to a sphere with g handles and r holes, r being the number of branches of the initial singularity (branch, as usually, means irreducible component) and $2g+r-1$ being the Milnor number μ , defined for

instance by $\mu = \dim_{\mathbb{C}} \frac{\mathbb{C}\{x,y\}}{(f'_x, f'_y)}$. Equivalent topological description of the Milnor number: $\mathbb{C}C_\varepsilon \cap \mathbb{C}B_\eta$ has the homotopy type of a wedge of μ circles.

2) In the real case, there is a finite number of topological types of smoothings for a given singularity f. If f has r real branches (as usually, *a branch is called real if it is invariant under complex conjugation*; in other words, if it has a one-dimensional real-zero set), $C_\varepsilon \cap B_\eta$ has r *non-closed components* - they are homeomorphic to a segment and they meet the boundary $S_\eta = \partial B_\eta$, each one in two distinct points - and some number p of *closed components* (they are called traditionally *ovals*, even if they are not convex), each one homeomorphic to a circle.

Let us for instance take $f = Y^2 - X^3$ (purely quadratic case corresponds to so-called Morse surgeries). Here $r = 1$, and two (and only two) topological types of smoothings are possible: one is given by $f_1 = Y^2 - X^3 - \varepsilon^2 X = 0$, for which $p = 0$, and the second by $f_2 = Y^2 - X^3 + \varepsilon^2 X = 0$, for which $p = 1$.

In this paper, we do not discuss the topology of an embedding $C_\varepsilon \cap B_\eta \subset B_\eta$, which is a more delicate task, but only the possible values of the integer p which determine the topological type of a smoothing, trying to get smoothings with the greatest possible number of ovals. Let us illustrate this point by Figure 1. It shows five embeddings for $r = 1$ and $p = 2$; they are different up to isotopies small at the boundary and they present 3 different types up to arbitrary isotopies.

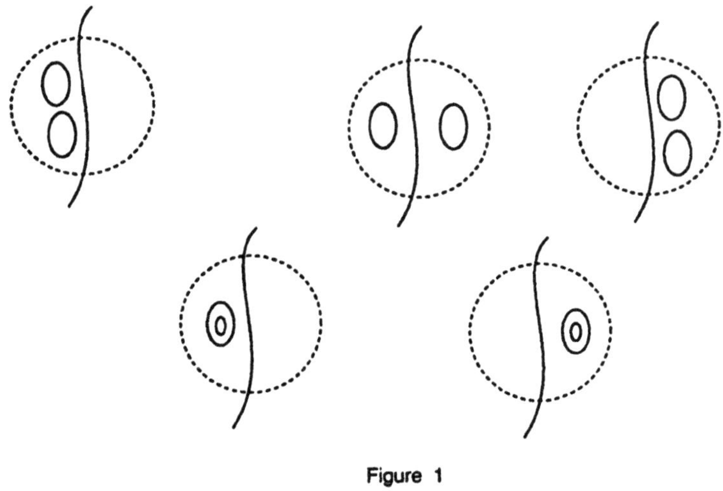

Figure 1

In dots : A Milnor ball B_η

For a complete classification of smoothings of some special singularities, see [Loo] and [Vi 2]; see also [Loe], [Va] and [Kh-Vi] for some restrictions about the topology of an embedding $C_\varepsilon \cap B_\eta \subset B_\eta$.

It is natural to split the problem of topological classification of smoothings in two parts:

a) Bounds on the topology of $C_\varepsilon \cap B_\eta$ (in fact, bounds on the number p of ovals).

b) Construction of a smoothing for each possible value of p.

The aim of this note is to discuss some such bounds and their sharpness. In this paper, *we will consider only singularities with all branches being real.* Indeed, main results can be apropriately extended to singularities having imaginary branches.

1. Bounds on the topology of f_ε

The content of two first paragraphs is presented in a little bit more general setting than that of the introduction (this will be used at the proofs in Section 3). Namely, instead of curves arising from local smoothings we consider any (non-singular) Riemann surface $\mathbb{C}C$ (in general, non-connected and with a boundary) supplied with an antiholomorphic involution τ. We call it curve C and denote by ρ the number of connected components of $\mathbb{C}C$. These components will be called *irreducible components of* C. We preserve also notations introduced at the local case: g is the number of handles, r' is the number of holes, $\mu = 2g + r' - 1$, r is the number of *non-closed* and p is the number of *closed components* (called *ovals*) of the real part $\mathbb{R}C = \operatorname{Fix}\tau$. We suppose, to be in accordance with hypothesis made in the introduction, that *each irreducible component of C is invariant under τ and it has at least one hole.* Note that $\rho = 1$ and $r' = r > 0$ if C is a smoothing of a singularity and the singularity has only real branches.

Remind that in the case of a connected curve with empty boundary ($r' = 0$), there is a well known Harnack estimate: $p \le g + 1$. The Klein proof of this estimate consists in considering real components as cuttings of the complexification and applying the geometrical definition of the genus: it is the maximal number of cutings which leave the complexification to be connected. Another proof: one obtains the Harnack bound applying the general Smith inequality to the involution of the complex conjugation. (It is R. Thom [Th] who made such application; surprisingly he did not relate the result with the Harnack bound. According to the Smith inequality, if τ is an involution of a space X then $h_\star(\operatorname{Fix}(\tau)) \le h_\star(X)$, where $h_\star(Y)$ denotes the sum $\sum \dim H_i(Y, \mathbb{Z}_2)$.) Repeating in our case any of these considerations, one immediately gets:

Proposition 1 ([A 1], see also [R]). *Under the above hypothesis and, in particular, for any smoothing*

$$(1) \qquad 2p + r \le \mu + 1, \quad \textit{or equivalently} \quad p \le \frac{\mu - r + 1}{2}.$$

By analogy with the global case ($r' = 0$), we will call this inequality *Harnack's estimate.*

We will define an *M-smoothing* as a smoothing f_ε of f such that $p = \frac{\mu-r+1}{2}$. For instance, a smoothing of $f = Y^2 - X^3$ such that $p = 1$ is an M-smoothing.

The following refinement of the Harnack estimate was used by V. Kharlamov and O. Viro [Kh-Vi] to generalize the Gudkov–Arnold–Rokhlin congruence from non-singular curves to singular ones.

Proposition 2. *Under the above hypothesis,*

$$p \leq \frac{\mu - r + 1}{2} - a \tag{2}$$

where $a + \rho$ is the number of connected components of the space $\mathbb{R}C \cup \partial\mathbb{C}C/\tau$. In particular, if C is a smoothing of a singularity and the singularity has only real branches, $a+\rho = a+1$ is the number of connected components of the space obtained from the union of the r non-closed components of $C_\varepsilon \cap B_\eta$, by gluing together each two boundary points arising from the same (real) branch of the initial singularity (see Figure 2).

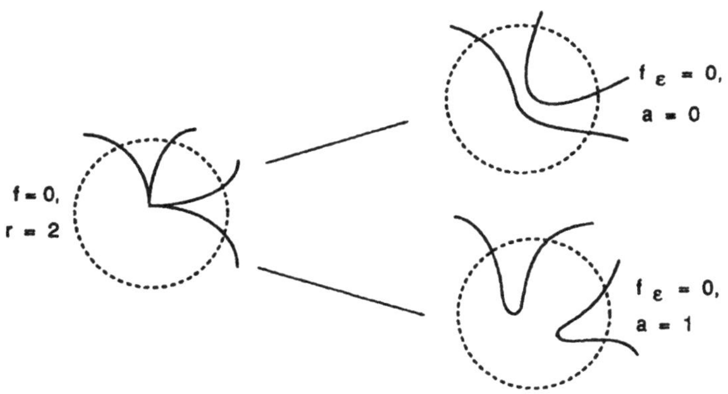

Figure 2

Similar to the original Harnack estimate, this refined estimate is proved by applying the Klein inequality or by applying the Smith one. This time they are applied not to the original Riemann surface (with boundary) $\mathbb{C}C$ but to the closed one $\mathbb{C}\Phi$ obtained from it by gluing a disk on each hole and supplied by the involution τ extending the initial one. □

Definition 1. A curve C (respectively a smoothing) satisfying equality in (2) is called an *M_a-curve* (respectively M_a-smoothing).

Note that $a \geq 0$ and that M-smoothings defined before are the same as M_0-smoothings. Therefore, for an M_a-smoothing, the integer a represents the number of ovals "loosed" in relation with an M-smoothing (for an M_a-curve, $p = M_a = M_0 - a$, $M_0 = (\mu - r + 1)/2$). The number of ovals loosed is not arbitrary, it should be equal to the precise number determined for a given smoothing by the scheme of joining of the boundary points by non-closed components. For example, at the case presented in Fig. 2: M_0-smoothing, if it exists, should have the upper right scheme; M_1-smoothing, if it exists, should have the below right scheme; as $r = 2$, there is no any M_a-smoothing with $a > 1$.

Lemma 1 ([Kh-V]). *A curve C is an M_a-curve, if and only if the real locus of each irreducible component divides the complex locus of that component in two parts, each one being homeomorphic to a disk with holes.*

Proof. Consider the same closed surface $\mathbb{C}\Phi$ as in the proof of Proposition 2. Up to holes $\mathbb{C}\Phi \setminus \mathbb{R}\Phi$ is homeomorphic to $\mathbb{C}C \setminus \mathbb{R}C$. Denote by $\mathbb{C}\Phi_i, 1 \leq i \leq \rho$ the connected components of $\mathbb{C}\Phi$, by $g(\Phi_i)$ their genus and by $p(\Phi_i)$ the number of (real) components of $\mathbb{R}\Phi_i$ (every Φ_i is invariant under τ). It is clear that

$$\Sigma p(\Phi_i) = p + a + \rho - 2a'$$

and

$$\Sigma(g(\Phi_i) + 1) = \frac{\mu - r + 1}{2} + \rho - 2a',$$

$2a'$ being the number of holes of C which have no any real point. Thus C is an M_a-curve if and only if $p(\Phi_i) = g(\Phi_i) + 1$ for each $i, 1 \leq i \leq \rho$. Klein arguments giving Proposition 2 contain an evident complementary information: $p(\Phi_i) = g(\Phi_i) + 1$ if and only if the real part of Φ_i divides Φ_i in two parts, each one being homeomorphic to a disk with holes. □

We conclude by two remarks. First, according to the definition of the parameter a given in Proposition 2, non-closed real components are decomposed in $a + \rho$ *complex groups*: two non-closed components belong to the same complex group if and only if they are contained in the same component of $\mathbb{R}C \cup \partial\mathbb{C}C$.

Second, according to Lemma 1, the real part $\mathbb{R}C$ of an M_a-curve $\mathbb{C}C$ has specific orientations, called *complex*: each one is the boundary orientation coming from the complex orientation of one half of the complex locus. One complex orientation being fixed, it defines on each complex group a cyclic order. (If the curve $\mathbb{C}C$ is connected, there are two complex orientations of the real part and they are opposite one to another. Certainly, the opposite complex orientation defines the opposite order.)

2. Control of maximality

It is known essentialy only two different systematic ways of constructing smoothings of a singularity:

a) Harnack-type variations comporting on the resolution of the singularity by blowing-ups;

b) Toric variations.

Method a) (we call it blowing-up method) was applied in [R] to construct maximal smoothings. There , it is proved that when f is irreducible and real (i.e., when $r = 1$), the inequality (1) is sharp.

Method b) (called Viro method) was invented by O. Viro ([Vi 1] and applied by him to construct maximal smoothings of some special singularities. The further results were obtained by E. Shustin, they are discussed in a forthcoming paper by Kharlamov–Risler–Shustin.

Let us describe variations appearing in the blowing-up method.

Consider some real surface P with boundary (later it will be the Milnor disc or that disc blowed up) and two real non-singular curves Γ and L of it intersecting one another transversaly and only in real points. Let their product ΓL be included in a family of real curves C_ε, $\varepsilon \geq 0$, $C_0 = \Gamma L$, such that all curves in a family are non-singular except C_0 and points of $\Gamma \cap L$ are the only singular points of a family (i.e. C_ε is a real Lefschetz family). Assume that the real parts of the curves Γ and L are oriented. We say that this variation is compatible with the given orientation of $C_0 = \Gamma L$ if there is a continous extension of the orientation to nearby elements of the family (see Figure 3, where L is a line).

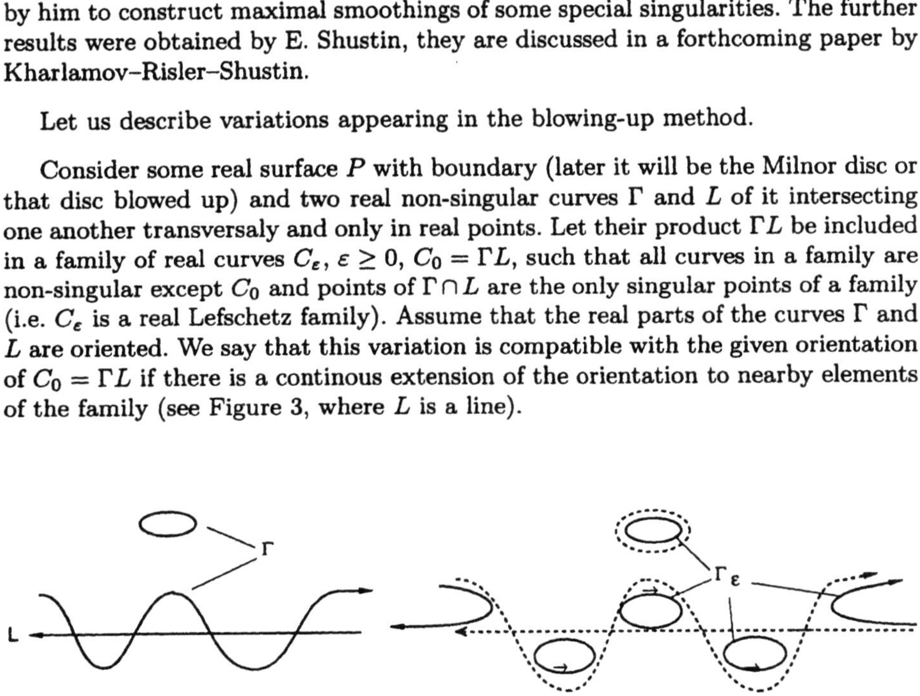

Figure 3

There is a situation where orientations of the curves appear naturally (up to the choice of one half of the complex part). That is the case when they are M_a-curves (see the last remark in Section 1).

Lemma 2. *Let Γ and L be M_a-curves intersecting transversally only in real points, and let L be irreducible. Their variation C_ε is an M_a-curve if and only if*

i) *all but one complex group of real non-closed components of L and all but one complex group of real non-closed components of each irreducible component of Γ have no points of $\Gamma \cap L$;*

ii) *the variation is compatible with some complex orientations of Γ and L;*

iii) *with respect to these orientations and the respective numbering of real non-closed components, intersection points go in the opposite order on the curves* Γ *and* L.

When one complex group of components of Γ satisfies condition iii) of Lemma 2, we say that this group *oscillates properly on* L.

This lemma follows from Lemma 1 and the elementary topological fact concerning gluings of topological discs: to obtain a topological disc with holes by gluing two discs with holes along some number of boundary arcs, the arcs should be chosen on each disc all on only one component of the boundary and moreover the cyclic orders defined on the set of these arcs by both discs should coincide. □

3. Distinct tangents

Let all branches $C_i : \{f_i = 0\}, 1 \leq i \leq r$ of a reduced real analytic singularity $C : \{f = 0\}, f = f_1 \cdots f_r$ be real. In this section, we will discuss smoothings of C which make the bound (2) exact, and, if not, which maximize the number p of ovals.

Let us call *even* (resp. *odd*) a component with even (resp. odd) multiplicity at O.

Proposition 3. *Assume that curves* $C_i, 1 \leq i \leq r$, *have pairwise distinct tangents. Then*

a) *If all components are even, for some number* a *with* $0 \leq a \leq r-1$ *there exists an* M_a*-smoothing of* C.

b) *In all cases for some number* a *with* $0 \leq a \leq r-1$ *there exists a smoothing of* C *with* $M_0 - a$ *ovals.*

Remarks.

1) We will make more precise below the values of a we can obtain, depending on the configuration of C.

2) It will be clear from the proof that if in a) or b) of Proposition 1 some a is realizable, then also any a' such that $a \leq a' \leq r-1$.

3) V. Kharlamov has proven that, if C consists in three cusps not lying in a half-plane, then there is no M_0-smoothing of C (this will be discussed in [K-R-S]): our construction will give an M_2-smoothing.

Proof of Proposition 3. For the proof, we assume $r > 1$, the case $r = 1$ being treated in [R].

Let (X, Y) be the coordinates in $\mathbb{R}^2$. Assume that C is not tangent to the Y-axis, and let $\pi : S \longrightarrow \mathbb{R}^2$ be the blowing up of O. The change of variables

$Y = XY'$ defines a chart P of S with coordinates (X, Y') (π will also denote the map $\pi|P$), and the strict transform $\tilde{C}$ of C is contained in this chart, and consists in r disjoint real germs $\tilde{C}_1, \ldots, \tilde{C}_r$ at points of $E = \pi^{-1}(O)$ corresponding to the tangents to C (Figure 4). The inverse image $\pi^{-1}(B_\eta)$ of a Milnor ball $B_\eta \subset \mathbb{R}^2$ is a Möbius strip.

The total transform of C is the union of $\tilde{C}$ and of e copies of E, e being the multiplicity of C. To construct a smoothing of C, we will construct a smoothing $\tilde{C}^o$ of $\tilde{C}$, and vary successively $\tilde{C}^o \cup E = \tilde{C}^1$, $\tilde{C}^1 \cup E = \tilde{C}^2, \ldots,$ e times, and then blow down the result, trying at each step to apply Lemma 2 above. If we can apply Lemma 2 at each step, we will end with an M_a-curve, a being the number of complex groups of components we obtain.

Fix a complex orientation of E.

The preliminary step (step 0) consists in application of the Theorem 6.1 of [R] to each irreducible germ $\tilde{C}_i$ (cf. 7.2 of [R]). According to it, there exists a smoothing $\tilde{C}_i^o$ of $\tilde{C}_i$ which is a M_0-curve, and whose the unique non closed branch oscillates properly (see end of Section 2) on E with respect to some complex orientation of $\tilde{C}_i^o$; this orientation will be used at the next step. This smoothing has $\frac{1}{2}[\mu_i - e_i(e_i - 1)]$ ovals (μ_i is the Milnor number of C_i, and e_i is its multiplicity) and it intersects E in e_i points (e_i is also the intersection multiplicity $\tilde{C}_i \cdot E$): see Figure 4, b).

For the next step (step 1), we choose points $a_i^1 \in E$ and the sign of ϵ_1 in such a way that the family

$$\tilde{C}^1 : \tilde{f}^1 = X \prod_{i=1}^{r} \tilde{f}_i^o + \epsilon_1 \prod_{i=1}^{e-1} (Y' - a_i^1)$$

satisfies Lemma 2 (with $L = E$ and $\Gamma = \tilde{C}^o = \cup \tilde{C}_i^o$). The points a_i^1 will be the intersection points of the curve $\tilde{f}^1 = 0$ with E. The curves E and $\tilde{C}_i^o$ are already oriented at step 0 and conditions i) and iii) of Lemma 2 are satisfied. To satisfy condition ii) of Lemma 2, it is enough to choose the points a_i^1 in the following manner: numerate curves $\tilde{C}_i^o, 1 \leq i \leq r$ following the orientation of E: $\tilde{C}_{i+1}^o \cap E$ be after $\tilde{C}_i^o \cap E$, $1 \leq i \leq r-1$. Let I_j be the interval of E delimited by the last point of $\tilde{C}_j^o$ and the first of $\tilde{C}_{j+1}^o$. If the oriented arc of $\tilde{C}_j^o$ coming to the begining-point of I_j and the oriented arc of $\tilde{C}_{j+1}^o$ starting from the end-point of I_j are placed on different sides of I_j, we call the interval *odd*. The number of odd intervals is less or equal to $e-1$ and of the same parity. We take $e-1$ points a_i^1 on those intervals, an odd number of points in each odd one, and an even number of points in each even one. Then, under proper choice of the sign of ϵ_1, the variation $\tilde{f}^1$ is made accordingly with the orientations of E and $\tilde{C}_j^o, 1 \leq j \leq r$ (see Figure 4, b) and c), where we have $r = 3$, and one point a_i^1 in I_1 and seven in I_2; I_1 and I_2 are odd, and I_3 is even).

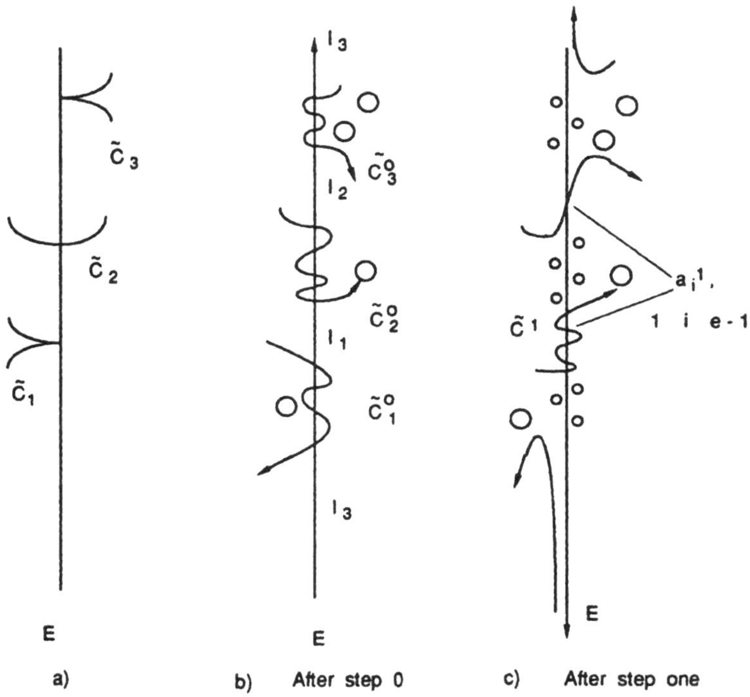

Figure 4

The curve $\tilde{C}^1$ constructed at step 1 inherites a complex orientation from $L = E$ and $\tilde{C}^o = \cup \tilde{C}_i^o$. With respect to this orientation and the reversed orientation of E, it oscillates properly on E. Futhermore, the real non-closed components of $\tilde{C}^1$ constitute one complex group. In particular, $\tilde{C}^1$ is an M_0-curve.

Now, in view of the following steps, we choose all the $e-1$ points a_i^1 in odd intervals, if the number of odd intervals is non-zero. If all intervals are even, we choose all the $e-1$ points a_i^1 in only one interval I_j.

Let us, for the next step, distinguish the two claims of Proposition 3.

a) At step 2, we have to variate $\tilde{C}^1$ with a new copy of E now oriented in the opposite way than at the preceding step. If all C_i are even curves, then e is even and to vary $\tilde{C}^1 \cup E$ in a way compatible with orientations, we can choose all the $e-2$ new points a_i^2 in only one interval of E delimited by two consecutive components of $\tilde{C}^1$. Then, for a suitable choice of the sign of ε_2, the formula

$$\tilde{f}^2 = X\tilde{f}^1 + \varepsilon_2 \prod_{i=1}^{e-2} (Y' - a_i^2)$$

gives a desired variation (see Figure 5).

As all C_i are even curves, the curve $\tilde{C}^2$ has exactly $a+1$ groups of non closed components, $a+1$ being the number of intervals containing points a_i^1 appeared at step 1 (we have therefore $0 \leq a \leq r-1$): by Lemma 2, $\tilde{C}^2$ is an M_a-curve. Furthermore, only one non-closed component of $\tilde{C}^2$ intersects E. It oscillates properly. So the proof follows now by induction exactly the same line than in [R]. Each step, including the final one, gives (by Lemma 2) an M_a-curve, with the number a of step one.

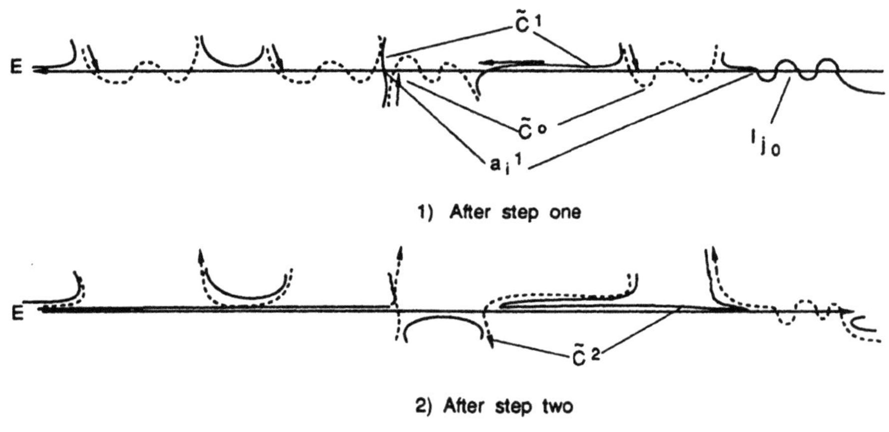

Figure 5 (Ovals are not represented)

b) Let us look now at the second part of Proposition 1, i.e., when the branches are not necesseraly even. Then at step 2, a choice of all the $e-2$ new points a_i^2 in the same interval of E delimited by two consecutive components of $\tilde{C}^1$ does not give, in principle, an M_a-curve, but it gives at least a curve with $M_0 - a$ ovals $(0 \leq a \leq r-1)$. There is one non-closed component of the obtained curve which oscillates properly on E, and applying the same procedure as in a) we obtain finally a smoothing of C with $M_0 - a$ ovals $(0 \leq a \leq r-1)$. □

Let us illustrate a difficulty arising at step 2 in case b) by a simple and typical example: assume that we have three consecutive odd components oriented in the same way (Figure 6). Then two consecutive components of $\tilde{C}^1$ we obtain at step 1 must cross E. They are oriented in the same direction (Figure 6) and at the next step, we cannot have a smoothing compatible with orientations without crossing

E (Figure 6).

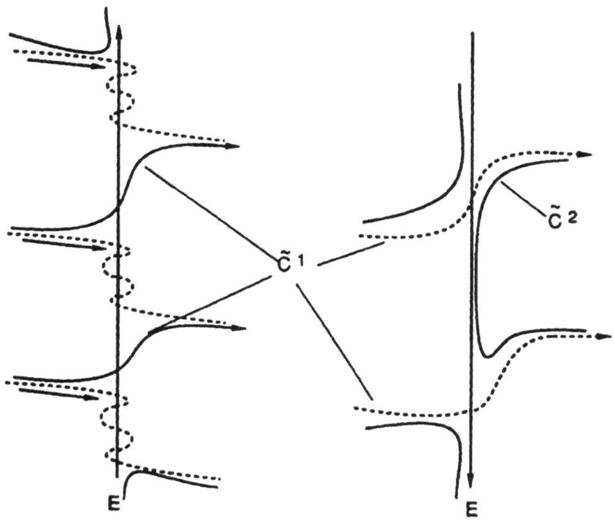

Figure 6 (ovals are not represented)

The rest of this Section is devoted to evaluation of the number a one can obtain in Proposition 1.

Let us introduce an additional object: a real line D passing by the singular point O of C under consideration. Denote by q the intersection number $C.D$ and, as before, by e the multiplicity of the singular point. It is shown in [R] that if C is irreducible, then there exists an M-smoothing C_ε of C, such that the non-closed component C'_ε of C_ε oscillates properly on D in q points, which go at opposite orders on C'_ε and D for some orientations of C'_ε and D. We ask in which case it is possible to choose freely the orientation of D, the orientation of the non-closed component of C'_ε being fixed, for example, by a choice of an order on its two boundary points.

Lemma 3. *Let C be irreducible. If D is not tangent to C (i.e., $q = e$) and e is odd, or if D is tangent to C, q is odd and e is even, we can choose freely the orientation of D.*

This lemma is illustrated in Figure 7.

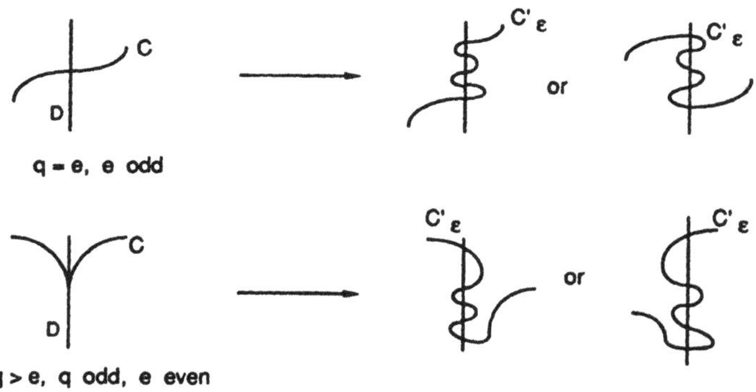

Figure 7 (ovals are not represented)

Remark. It is proved in [K-R-S] that one can choose freely the orientation of D also in the case when e and q are odd, $e > 1$, and C is non-degenerated with respect to its Newton diagram (see Figure 8, where ovals are not represented).

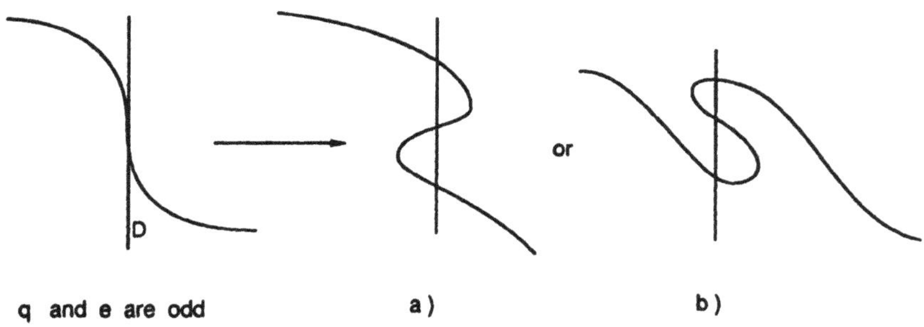

Figure 8

When $e = 1$, a deformation of type b) is clearly impossible, and when C is degenerated, the answer is not known to us.

Proof of Lemma 3. Look at the blowing-ups of C up to separation of C and D.

a) Assume that D is not tangent to C $(e = q)$ and e is odd. Then already the first blowing-up separates C and D. Let $\tilde{C}$ be the strict transform of C by the first blowing-up, $\tilde{D}$ the one of D, E the exceptional divisor, and P, Q the points where $\tilde{C}$ and $\tilde{D}$ intersect E. We have $e = \tilde{C} \cdot E$. Thus $\tilde{C} \cdot E$ is odd, and the curve $\tilde{C}$ cross E. Now at step one of the construction of a smoothing, i.e., the variation of $\tilde{C}^o \cup E$, we have to choose the points a_i^1 in one of the two intervals of E delimited by P and Q (see Figure 9). These two choices give the two opposite oscillations on $\tilde{D}$, hence also on D.

b) Assume that e is even and q is odd. Let $\tilde{e}$ be the multiplicity of $\tilde{C}$.

- If $\tilde{e} = e$, then $\tilde{C}$ is not tangent to E, and $\tilde{q} = \tilde{C} \cdot \tilde{D}$ is odd ($\tilde{q} = q - e$). We are then in the same situation than for C and D and we repeat the procedure.

- If $\tilde{e} < e$, then $\tilde{C}$ is tangent to E, the number $\tilde{q} = \tilde{C} \cdot \tilde{D}$ is again odd and is equal to $\tilde{e}$. Thus we are in the case a) and we end by the procedure corresponding to this case (see above). □

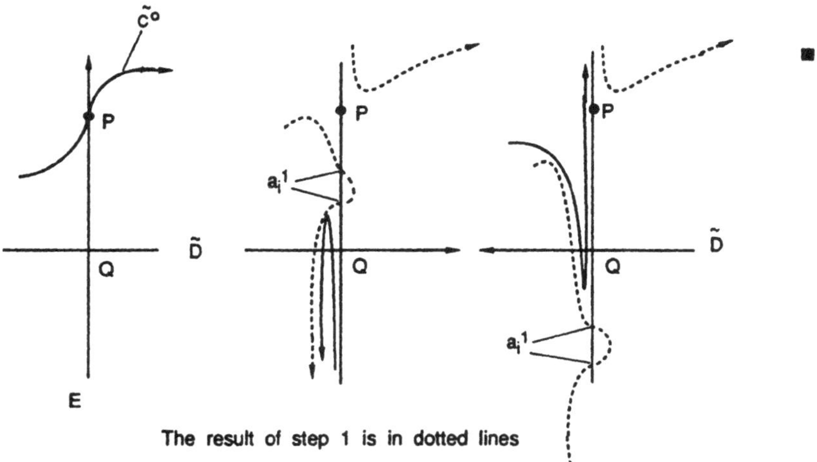

The result of step 1 is in dotted lines

Figure 9

Lemma 3 gives a justification for the following representation of the strict transform $\tilde{C}$ of C under one blowing-up (Figure 10):

For each component C_i of C, let $c_i \in E$ (E is the exceptional divisor) be the point corresponding to the tangent to C_i at O.

a) If C_i is an even component, represent it by two segments on the same side of E than $\tilde{C}_i$, linked by a little path in E containing c_i.

b) If C_i is an odd component such that $\tilde{C}_i$ is even or non tangent to E, represent it by a segment transversal to E at c_i.

c) If C_i is an odd component such that $\tilde{C}_i$ is odd and tangent to E, represent it by two segments orthogonal to E linked by a little path in E containing c_i; segments are chosen on different sides of E following the position of half branches of $\tilde{C}_i$ at c_i (see Figure 10).

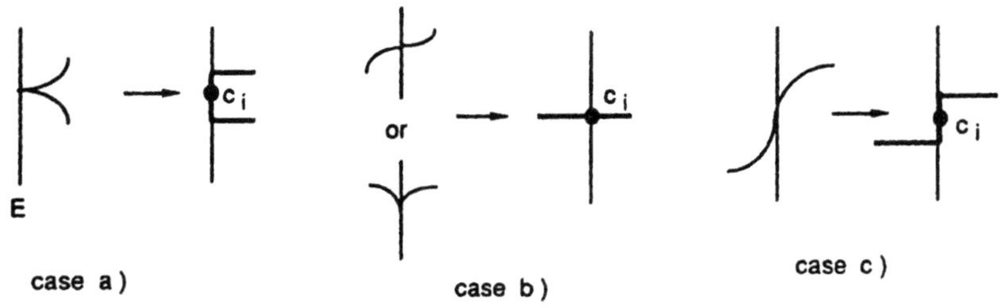

Figure 10

Now $\tilde{C}$ is represented by a (non connected) graph (the bold lines in Figure 11). Let us complete this graph in a connected one, joining neighbouring extremities (the dotted lines in Figure 11). Then the number of times we cross E doing this operation is $a+1$, where a is the number we are looking for (recall that E is not a line, but a circle S^1, and that its neighbourhood $\pi^{-1}(B)$ is a Möbius strip): see Figure 11. More precisely, application of Lemma 3 in the procedure given at the proof of the Proposition 1 shows that *the number a in Proposition 1 may be taken equal to the number a constructed by the graph above.*

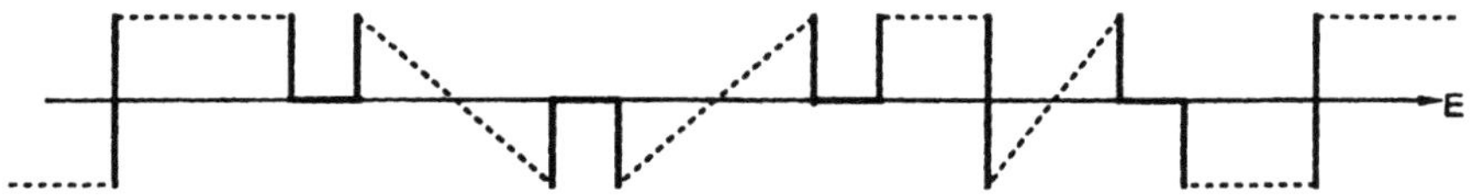

Figure 11 : a = 2

Corollary. 1) *If C have only even components and is contained in a half-plane, then there exists an M_0-smoothing.*

2) *If C has only even components in number r, there exists an M_{r-1}-smoothing; if r is even, there exists an M_{r-2}-smoothing.*

4. General case

Here we suppose again that C is reduced, but we don't demand branches to have distinct tangents at the singular point O of C.

In this case we give just a rough analysis of the construction. This time we need to control a position of smoothings with respect to two lines simultaneously.

Let D_1 and D_2 be two lines passing by O. We say that a smoothing C_ε of C *oscillates tamely on $D_1 \cup D_2$ by its (α, β) components*, if

i) C_ε intersects D_1 in $D_1 \cdot C$ (real) points and D_2 in $D_2 \cdot C$ (real) points; intersection points belong only to non-closed components of C_ε;

ii) the point O does not divide intersection points of the same branch, neither on D_1 or on D_2;

iii) components of C_ε intersecting D_1 and D_2 can be numbered and oriented in such a manner that the intersection points go in the opposite order on C_ε and on D_1 and D_2 with respect to some orientation of D_1 and D_2;

iv) there exists one (and only one) component of C_ε^i which intersects both D_1 and D_2;

v) α, $\alpha \geq 1$, is the number of components intersecting D_1 and β, $\beta \geq 1$, is the number of components intersecting D_2.

If C_ε oscillates tamely on $D_1 \cup D_2$, then the total number of (non-closed) components intersecting $D_1 \cup D_2$ is equal to $\alpha + \beta - 1$. In particular, $\alpha + \beta - 1 \leq r$, r being the number of branches of C.

Proposition 4. *Let all r branches of C be real. Then for some $a \leq 5(r-1)$, there exists a smoothing of C with $M_0 - a$ ovals.*

Proof. We do blowing-ups, until we find irreducible germs. Then, as in the proof of Proposition 3, part b), we smooth each irreducible germ and make inductively contractions inverse to the blowing-ups preceding each one by a Harnack type variation described at that proof. Proposition 4 results directly from the following (the number q involved is trivially $\leq r-1$):

Lemma 4. *Let two lines D_1 and D_2 pass by the singular point O of C and let q be the number of germs with more than one tangent appearing in the desingularization process of C. Harnack variations can be made to give a smoothing of C which oscillates tamely on $D_1 \cup D_2$ by its (α, β) components and has $M_0 - a$ ovals, where*

$$a \leq (r-1) + 4q - 2(\alpha - 1) - 2(\beta - 1).$$

If there we consider only one line D_1, Harnack variations can give a smoothing of C which oscillates propely on D_1 by its α components and has $M_o - a$ ovals, where

$$a \leq (r-1) + 4q - 2(\alpha - 1).$$

Proof. Let us prove this lemma by induction on the number of blowing-ups needed to resolve C.

We have two cases to consider:

a) C has s distinct tangents, with $1 < s \leq r$.

b) C has only one tangent.

Begin from the case a). Treat, for definiteness, the assertion concerning two lines D_1 and D_2; same arguments prove the other one.

The strict transform $\tilde{C}$ consists in s disjoint real germs $\tilde{C}_1, \ldots, \tilde{C}_s$. These germs intersect the exceptional divisor Eat the points P_i corresponding to the tangents of C. Let $\tilde{D}_1$ and $\tilde{D}_2$ be the strict transforms of the lines D_1 and D_2.

Let us assume that we have a smoothing $\tilde{C}_i^o$ of $\tilde{C}_i$ $(1 \leq i \leq s)$ with p_i ovals, and set $e_i = \tilde{C}_i \cdot E$. Then any Harnack variation provides a smoothing of C, with a number p of ovals satisfying:

$$p \leq m = \sum p_i + \sum_{i=1}^{s} (e_i - 1) + \frac{(e-1)(e-2)}{2},$$

$$e = \sum e_i.$$

If $\delta = m - p$, we say that δ ovals are lost at the variation.

Note that the construction ends with an M-smoothing of C if and only if there is no loss of ovals at any Harnack variation applied. According to the induction hypothesis, a smoothing $\tilde{C}_i^o$ of $\tilde{C}_i$ $(1 \leq i \leq s)$ can be chosen oscillating tamely on $E \cup \tilde{D}_j$ by its (α_i, β_i) components, if $\tilde{D}_j$ cuts E at P_i, and oscillating properly on E by its α_i components, if P_i is not on $\tilde{D}_1$ or $\tilde{D}_2$.

Thus there are only two reasons which can cause a loss of ovals:

1) The number of germs $\tilde{C}_i$ is greater than 1;

2) at least one of α_i is greater than 1.

Moreover, the first reason provokes a loss of $\leq s - 1$ ovals and the second one a loss of $\leq 2 \sum (\alpha_i - 1)$ ovals (cf. proof of Prop. 3(b) and see the example below).

Example. Assume $\alpha_i = 2$, $e_i = 6$ for one germ $\tilde{C}_i$ (Figure 12).

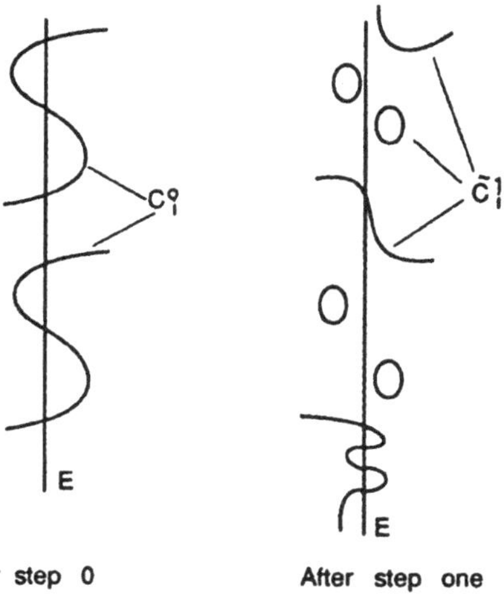

Figure 12

The total number of ovals we get after varying the curve $\tilde{C}_i^o$ with six copies of E is then $4 + 3 + 3 + 2 + 1 = 13$. We have then lost two ovals in relation to the case $\alpha_i = 1$, $e_i = 6$, where we get $5 + 4 + 3 + 2 + 1 = 15$ ovals.

Return to the proof. We have already applied the induction hypothesis and we took a smoothing of $\tilde{C}$ with $\tilde{M} - \tilde{a}$ ovals ($\tilde{M}$ being the number of ovals of an M-smoothing of $\tilde{C}$), with

$$\tilde{a} \leq 4(q-1) + \sum_{i=1}^{s}(r_i - 1) - 2\sum_{i=1}^{s}(\alpha_i - 1) - 2\sum_{\beta_i \geq 1}(\beta_i - 1),$$

the sum $\sum_{\beta_i \geq 1}(\beta_i - 1)$ having zero, one, or two terms.

Now, applying Harnack variation and contraction, we loose a' ovals more, with $a' \leq (s-1) + 2\sum(\alpha_i - 1)$ (see 1) and 2) above). We then find a smoothing of C which has $M - a$ ovals, with

$$\begin{aligned} a &\leq \tilde{a} + (s-1) + 2\sum_{i=1}^{s}(\alpha_i - 1) \\ &= 4(q-1) + \sum_{i=1}^{s}(r_i - 1) + (s-1) - 2\sum_{\beta_i \geq 1}(\beta_i - 1) \\ &\leq 4q - 4 + r - 1 - 2\sum_{\beta_i \geq 1}\beta_i + 4 \end{aligned}$$

(remember that the sum $\sum_{\beta_i \geq 1}(\beta_i - 1)$ has at most two terms).

To finish the proof of Lemma 4, part a), we have to verify that the Harnack variation may be chosen such that the two following claims are verified:

1) $\alpha + \beta \leq \sum_{\beta_i \geq 1}(\beta_i) + 2$;

2) the constructed smoothing of C oscillates tamely on $D_1 \cup D_2$.

Let us assume that $\tilde{D}_1$ cuts E at $P_i \in \tilde{C}_i$ and that $\tilde{D}_2$ cuts E at some other point $P_j \in \tilde{C}_j$ (this is the case where the sum $\sum_{\beta_i \geq 1}(\beta_i - 1)$ has two terms; the other cases are treated in the same way). At the beginning we have β_i (non-closed) components of $\tilde{C}_i^o$ which oscillate on $\tilde{D}_1$, one of them intersecting both E and $\tilde{D}_1$, and similarly for $\tilde{C}_j^o$ and $\tilde{D}_2$ (see Figure 13).

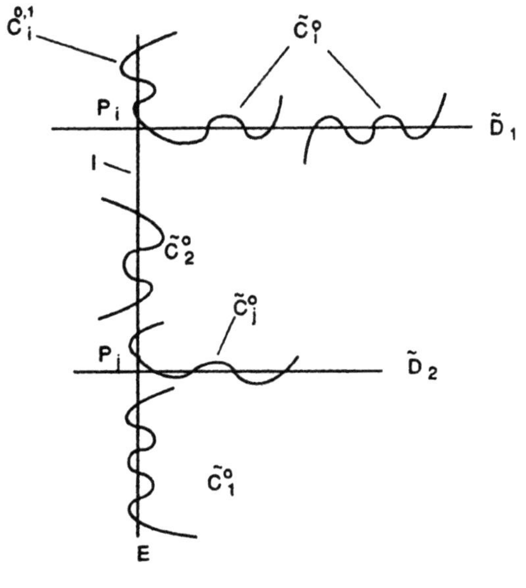

Figure 13 : after step 0

Choose among intervals I_j appearing at the first step of the Harnack variation (i.e., a smoothing of $E \cup (\cup \tilde{C}_i^o)$, see proof of Prop. 3) the interval I containing the point P_i. Let $C_i^{o,1}$ be the non-closed component of $\tilde{C}_i^o$ intersecting both $\tilde{D}_1$ and E (see Fig. 12). Choose on each odd interval (except I, if it is odd) one point (see proof of Prop. 3), take no point on even intervals (except I, if it is even) and choose all other points needed to make the Harnack variation on I on the other side of P_i with respect to $C_i^{o,1}$. Then $\tilde{C}^1$ will cut I in $e - k$ points, where k is the number of odd intervals different from I.

After step one, the situation is then the following: there are β_i branches oscillating on $\tilde{D}_1$, and one of them intersects both $\tilde{D}_1$ and E. There are β_j branches oscillating on $\tilde{D}_2$, but may be no one intersects E.

For step two, we choose points determining the Harnack variation in such a manner that the curve $\tilde{C}^2$ oscillates on E only in one interval. After this step, we may obtain a new non-closed component intersecting $\tilde{D}_2$ (in one point), but no new component intersecting $\tilde{D}_1$. Furthermore, choosing the interval properly (similarly than I), we get a component intersecting both $\tilde{D}_1$ and E.

After that, the Harnack variation proceeds exactly as in the proof of Prop. 3 (and as in the irreducible case described in [R]). This gives at the end of the process β_i branches oscillating properly on $\tilde{D}_1$ and at most $\beta_j + 2$ ones oscillating properly on $\tilde{D}_2$, with one branch intersecting both $\tilde{D}_1$ and $\tilde{D}_2$. After contraction of E, we get a smoothing oscillating tamely on $D_1 \cup D_2$ by its $(\alpha = \beta_i, \beta \leq \beta_j + 2)$ components: condition i) of the definition of tame oscillation follows from the induction hypothesis for the germs $\tilde{C}_i$, a formula $\tilde{C}_i \cdot \tilde{D}_j = C_i \cdot D_j - e_i$ (if $\tilde{D}_j$ passes by $P_i = \tilde{C}_i \cap E$) and formulae $C \cdot D_j = \sum_{i=1}^r C_i \cdot D_j$, $e = \sum_{i=1}^r e_i$; conditions ii) and iii) follow from similar conditions verified by the curves obtained at each step of the Harnack variation.

b) C has only one tangent.

Let $\tilde{C}$, $\tilde{D}_1$ and $\tilde{D}_2$ be the strict transforms of C, D_1 and D_2. We then apply the induction hypothesis to the germ $\tilde{C}$, the line E, and the line D_j ($j = 1$ or 2), if C is tangent to D_j. By the same method as above we get a smoothing with $M - a$ ovals, where

$$a \leq 4q + r - 1 + 2(\beta_1 - 1) = 4q + r - 1 - 2(\alpha - 1) - 2(\beta - 1),$$

$$\alpha = \beta_1, \ \beta = 1;$$

the smoothing obtained oscillates properly on $D_1 \cup D_2$ by its (α, β) components and again verifies the claim 1): $\alpha + \beta = 1 + \beta_1 \leq \sum_{\beta_i \geq 1}(\beta_i) + 2$. □

References

[A 1] Arnold V.I., Some open problems in the theory of singularities. Proc. Sobolev Seminar, Novosibirsk, 1976 (in Russian, English translation in Proc. Symp. Pure Math 40, 1983 , part 1).

[A 2] Arnold V.I., Index of a singular point of a vector field, the Petrovsky-Oleinik inequality, and mixed Hodge structures. Funktional. Anal. i Prilozhen. 12 (1) (1978), 1–14.

[K-R-S] Kharlamov V.M., Risler J-J., Shustin E., paper in preparation.

[Kh-Vi] Kharlamov V.M., Viro O.Y., Extension of the Gudkov–Rohlin congruence. Rohlin Seminar, Lecture Notes in Math. 1345, 357–406, Springer 1988.

[Loe] Loeser F., Un analogue local de l'inegalité de Petrovsky-Oleinik. Séminaire sur la Géom. Alg. Réelle (J-J. Risler ed.), Publ. Math. Univ. Paris VII 24 (II) (1986), 73–83.

[Loo] Loojenga E., The discriminant of a real simple singularity. Compositio Math. 37 (1978), 51–62.

[Mi] Milnor J., Singular points of complex hypersurfaces. Ann. Math. Stud. 61, Princeton, 1968.

[R] Risler J.-J., Un analogue local du théorème de Harnack. Invent. Math. 89 (1987), 119–137.

[Th] Thom R., Sur l'homologie des variétés algebriques réelles. In: Differential and Cominatorial Topology, 255–265, Princeton University Press 1965.

[Va] Varchenko A.N., On the local residue and the intersection form in vanishing cohomology. Izv. AN SSSR 49 (1) (1985), 32–54.

[Vi 1] Viro O.Y., Gluing of plane real algebraic curves and constructions of curves of degrees 6 and 7. Lecture Notes in Math. 1060, 185–200, Springer 1984.

[Vi 2] Viro O.Y., Real algebraic curves, construction with controlled topology. Leningrad. Math. J. 1 (1990), 1059–1134.

Linear bounds for resolutions of 1-boundaries in subanalytic sets

Krzysztof Kurdyka

Abstract. We show that the 1-dimensional boundary in a fiber of a subanalytic mapping bounds, in the fiber, a surface of area estimated by a linear function of the length of the boundary. In the semialgebraic case the constants of our linear estimation of area depend only on the diagram of a semialgebraic set. Those results follow from generalized L-regular stratifications.

1991 Mathematics Subject Classification: 14B05, 32B20.

Let us recall the following result of B. Teissier [T1] (see also [DK], [Ha2], [T2]):

Theorem 0. *Let X be a compact subanalytic subset of $\mathbb{R}^n$. Let $f : X \to \mathbb{R}^m$ be a continous and subanalytic mapping. Then there exists a positive constant C such that for each $y \in \mathbb{R}^m$ any two points x_0, x_1 lying in the same connected component of $f^{-1}(y)$ can be joined in $f^{-1}(y)$ by a subanalytic continous arc of the length bounded by C.*

Two points lying in the arcwise connected set can be seen as a boundary of dimension 0. Our goal is to extend this result to 1-dimensional boundaries. Let us denote by S^1 (resp. D^2) a unit circle (resp. unit closed ball) in $\mathbb{R}^2$. Let F be a subset of $\mathbb{R}^n$. We denote by $\mathrm{vol}_i(F)$ its i-dimensional volume (when it makes sense). Let A be a topological space, let $\varphi : S^1 \to A$ be a continous mapping. We say that φ is contractible if there exists a continous mapping $\Phi : D^2 \to A$ such that the restriction of Φ to S^1 equals φ. We call the mapping Φ a *resolution* of φ. Our main result is:

Theorem A. *Let X be a subanalytic compact subset of $\mathbb{R}^n$ of diameter r. Let $f : X \to \mathbb{R}^m$ be a continous and subanalytic mapping. Then there exists positive constants P and Q, such that for every $y \in \mathbb{R}^m$ the following holds.*

Let $\varphi : S^1 \to f^{-1}(y)$ be a continous mapping. Suppose that φ is contractible in $f^{-1}(y)$, and moreover that $\varphi(S^1)$ is of finite length denoted by $\mathrm{vol}_1\, \varphi(S^1)$. Then

Research partially supported by KBM grant IM PB 1219/2/91.

there exists a resolution $\Phi : D^2 \to f^{-1}(y)$ *of* φ such that

$$\mathrm{vol}_2\, \Phi(D^2) \;\le\; rP\, \mathrm{vol}_1\, \varphi(S^1) \;+\; r^2 Q.$$

Remark 1. If φ is subanalytic then Φ can also be found subanalytic.

Remark 2. We assume the compactness of X to avoid technical complications in the proof. The result holds for X bounded and subanalytic.

Remark 3. A similar formula should be true for k-boundaries (mappings from S^k), $k \ge 2$ (see the discussion in the last part of [T2] for the generalizations of Theorem 0).

A proof of Theorem A is based on a version of L-regular stratifications for families of subanalytic sets (see Theorem C below). We begin with the definition of strata of these stratifications.

Definition 1. We say that a subanalytic set A in $\mathbb{R}^n$ is an *s-cell*, if A is a point or after some orthogonal change of coordinates in $\mathbb{R}^n$ the set A is of one of the following forms

α) $$A = \{(x', x_n) \in \mathbb{R}^{n-1} \times \mathbb{R} :\ h(x') = x_n,\ x' \in A'\}$$

β) $$A = \{(x', x_n) \in \mathbb{R}^{n-1} \times \mathbb{R} :\ f(x') < x_n < g(x'),\ x' \in A'\}$$

where A' is an s-cell in $\mathbb{R}^{n-1}$. The functions f, g and h are subanalytic and continous on the closure of A', analytic on A'. We assume that $f(x') < g(x')$ for each $x' \in A'$.

Let A be an s-cell of dimension k. Then A is an analytic submanifold of $\mathbb{R}^n$. Moreover there exists a homeomorphism $\varphi : A \to B$ which extends to a homeomorhism $\bar{\varphi} : \bar{A} \to \bar{B}$, where B is an open ball in $\mathbb{R}^k$.

Definition 2. We keep the notation of Definition 1. We say that an s-cell A in $\mathbb{R}^n$ is *L-regular with constant* $M > 0$, if $| d_{x'}h | \le M$ (resp. $| d_{x'}f | \le M$ and $| d_{x'}g | \le M$) for each $x' \in A'$, where A' is an L-regular s-cell with constant M in $\mathbb{R}^{n-1}$. If A is a point we assume that A is L-regular with every constant $M > 0$.

Remark. The closure of our L-regular s-cell is a subanalytic version of the L-regular set of Parusiński (comp. Definition 3.2 of [Pa]).

In the sequel we need the following properties of L-regular s-cells:

Lemma 3. *Let* $A \subset \mathbb{R}^n$ *be an L-regular s-cell with constant* M. *Then there exists a subanalytic retraction* $r_A : \mathbb{R}^n \to \bar{A}$ *such that for all* $x, y \in \mathbb{R}^n$

$$\|r_A(x) - r_A(y)\| \le C\|x - y\|$$

where C *is a constant depending only on* n *and* M.

Corollary 4. *Let $A \subset \mathbb{R}^n$ be an L-regular s-cell with constant M. Then for every $x, y \in \bar{A}$ there exists a subanalytic arc λ joining x and y in $\bar{A}$ such that*

$$\text{length}\,\lambda \;\leq\; C\|x-y\|$$

where C is a constant depending only on n and M.

Proof. We prove Lemma 3 and Corollary 4 by induction on n.

Notice that Lemma 3 implies Corollary 4. Actually, for given points $x, y \in \bar{A}$ take the segment S joining x with y in $\mathbb{R}^n$. Then the arc $r_A(S)$ is subanalytic, joins x with y in $\bar{A}$, and its length is bounded by $C\|x-y\|$.

Suppose first that A is L-regular of the form of graph, hence

$$\bar{A} = \{(x', x_n) \in \mathbb{R}^{n-1} \times \mathbb{R} :\; h(x') = x_n,\; x' \in \bar{A}'\}$$

where A' is an L-regular s-cell with constant M in $\mathbb{R}^{n-1}$, h is a subanalytic continous function on the closure of A', analytic on A', such that $\mid d_{x'}h \mid \leq M$. By the induction we have a subanalytic retraction $r'_A : \mathbb{R}^{n-1} \to A'$ such that $\|r_A(x) - r_A(y)\| \leq C'\|x-y\|$ for all $x', y' \in \mathbb{R}^{n-1}$, where C' depends only on $n-1$ and M. Now Corollary 4 holds for A' with a constant C', hence our function h is Lipschitz with constant MC'. Let $x \in \mathbb{R}^n$ and write $x = (x', x_n) \in \mathbb{R}^{n-1} \times \mathbb{R}$ (in our coordinate system). We put

$$r_A(x) = r_A(x', x_n) = (r_{A'}(x'), h(r_{A'}(x'))).$$

Clearly r_A is Lipschitz with constant $C = C'(1+M^2)^{\frac{1}{2}}$.

Suppose now that A is bounded by the graphs of f and g (type β)), hence

$$\bar{A} = \{(x', x_n) \in \mathbb{R}^{n-1} \times \mathbb{R} :\; f(x') \leq x_n \leq g(x'),\; x' \in \bar{A}'\}$$

where $A' \subset \mathbb{R}^{n-1}$ is an L-regular s-cell with constant M, f and g are subanalytic continous functions on the closure of A', analytic on A'. Moreover $f(x') < g(x')$, $\mid d_{x'}f \mid \leq M$ and $\mid d_{x'}g \mid \leq M$ for each $x' \in A'$. By the induction there exists a retraction $r'_A : \mathbb{R}^{n-1} \to A'$ such that $\|r_A(x) - r_A(y)\| \leq C'\|x-y\|$ for all $x', y' \in \mathbb{R}^{n-1}$. By the same arguments as above we see that f and g are Lipschitz with constant MC' depending only on $n-1$ and M. Using the same notation as above we put

$$r_A(x) = r_A(x', x_n) = \begin{cases} (r_{A'}(x'), g(r_{A'}(x'))) & \text{if} \qquad\qquad\qquad x_n \geq g(r_{A'}(x')) \\ (x', x_n) & \text{if} \quad f(r_{A'}(x') \leq x_n \leq g(r_{A'}(x') \\ (r_{A'}(x'), f(r_{A'}(x'))) & \text{if} \qquad\qquad\qquad x_n \leq f(r_{A'}(x')). \end{cases}$$

Clearly in this case also r_A is Lipschitz with constant $C = C'(1+M)^{\frac{1}{2}}$. This ends the proof of Lemma 3 and Corollary 4. □

In [K] we proved the following:

Theorem B. *Let $\mathcal{A}$ be a locally finite family of subanalytic sets in $\mathbb{R}^n$. Then there exists a subanalytic stratification $\mathcal{T}$ of $\mathbb{R}^n$, compatible with the family $\mathcal{A}$, such* *that every stratum of $\mathcal{T}$ is an L-regular s-cell with a constant M, where M depends only on n.*

Remark. We can take $M = 9^n$ (see the proof of Theorem B in [K]).

Clearly if A is relatively compact then the number of strata contained in A is finite. It turns out that the number of strata can be uniformely bounded for a subanalytic family of relatively compact subanalytic sets. Precisely we have:

Theorem C. *Let $\mathcal{A}$ be a finite family of relatively compact subanalytic sets in $\mathbb{R}^n \times \mathbb{R}^p$. Let $\pi : \mathbb{R}^n \times \mathbb{R}^p \to \mathbb{R}^p$ denote the projection. Then for each $y \in \mathbb{R}^p$ there exists a subanalytic stratification $\mathcal{T}_y$ of $\mathbb{R}^n$, compatible with the family $\{\pi^{-1}(y) \cap A : A \in \mathcal{A}\}$, such that every stratum of $\mathcal{T}_y$ is an L-regular s-cell with a constant M, where M depends only on n. Moreover there exists an integer K such that every set $\pi^{-1}(y) \cap A$, where $A \in \mathcal{A}$, $y \in \mathbb{R}^p$, is a union of no more then K strata of $\mathcal{T}_y$.*

We shall explain now how to adapt the proof of Theorem B in order to get Theorem C.

Definition 5. Let $\mathcal{C}$ be a collection of analytic, connected submanifolds of M, subanalytic in M. We say that the collection $\mathcal{C}$ is *stratifying in family* if for every locally finite family $\mathcal{A}$ of subanalytic subsets of $M \times N$, where N is an analytic manifold, there exists a locally finite partition $\mathcal{T}$ of $M \times N$ such that:

- each $T \in \mathcal{T}$ is subanalytic in $M \times N$.
- for each $y \in N$ the family $\mathcal{T}_y = \{\pi^{-1}(y) \cap T : T \in \mathcal{T}\}$ is a subanalytic stratification of $M \times \{y\}$ compatible with the family $\{\pi^{-1}(y) \cap A : A \in \mathcal{A}\}$.
- for every $y \in N, T \in \mathcal{T}$ the set $\{x \in M : (x, y) \in T\}$ belongs to the collection $\mathcal{C}$.

Remark. Notice that if $\mathcal{A}$ is a finite family of relatively compact subanalytic sets, then there exists an integer K such that each set $\pi^{-1}(y) \cap A$ is the union of no more then K strata of $\mathcal{T}_y$.

Hence to prove Theorem C we need only to check that the collection of all L-regular s-cells with constant M, in $\mathbb{R}^n$, is stratifying in family.

It follows from the general method of the construction of subanalytic stratifications (see e.g. [DS], [Hal], [Ło2]) that $\mathcal{C}$ is stratifying in family if and only if:

($*_f$) *For every analytic submanifold A of $M \times N$, subanalytic in $M \times N$, there exists a closed, nowhere dense subset F of A which is subanalytic in $M \times N$ such that for each $y \in N$ every connected component of the set*

$$\{x \in M : (x, y) \in A \setminus F\}$$

belongs to the collection $\mathcal{C}$.

Notice that if N is just a point this is the condition (*) in [K]. The proof of Theorem B is based on subanalyticity of the tangent mapping ([Ło1], [DW]) and stratifications of subanalytic sets and mappings ([Ha1], [Lo2]). In fact in [K] we checked condition (*) for the collection of L-regular s-cells. It is not difficult to see that the same arguments show that the family of L-regular s-cells has property $(*_f)$, hence is stratifying in family. This ends the proof of Theorem C. □

Remark. Let us observe that Theorem 0 follows immediatly from Theorem C. Moreover, the constant C in Theorem 0 can be bounded by KrM, where r is a diameter of X, M is the constant in Theorem C and K is a bound for the number of L-regular strata of the fibers of f.

Proof of Theorem A. Observe that it is enough to prove Theorem A in the case where f is the projection $\mathbb{R}^n \times \mathbb{R}^p \to \mathbb{R}^p$ restricted to a compact set X. By Theorem C it suffices to prove the following proposition:

Proposition 6. *Suppose that the subanalytic, compact set E is a union of K strata of some stratification of $\mathbb{R}^n$ into L-regular s-cells with constant M, where M is the constant in Theorem B. Let $\varphi : S^1 \to E$ be a continous mapping, contractible in E, such that $\mathrm{vol}_1(\varphi(S^1))$ is finite. Then there exists a resolution $\Phi : D^2 \to E$ of φ such that*

$$\mathrm{vol}_2\, \Phi(D^2) \;\leq\; rP\,\mathrm{vol}_1\, \varphi(S^1) \;+\; r^2 Q$$

where r is the diameter of E. The constants P and Q depend only on n and K. Moreover, for φ subanalytic the resolution Φ is also subanalytic.

Let $E = \bigcup_{i=1}^{K} \Gamma_i$, where all Γ_i are strata of some stratification of $\mathbb{R}^n$, each Γ_i is an L-regular s-cell with constant M. We prove Propositon 6 by induction on the number of the strata Γ_i, $\dim\Gamma_i > 2$.

Clearly if $\dim E \leq 2$ then the propositon is obvious since $\mathrm{vol}_2(\Gamma_i) \leq r^2 Q_n$, for some constant Q_n depending on n and M. But M depends only on n, hence Q_n also depends only on n.

Let us denote $I = [0,1]$. The following lemma is crucial for our induction:

Lemma 7. *Let E be a subanalytic, compact subset of $\mathbb{R}^n$. Let A be an L-regular s-cell with constant M, open in E, $\dim A \geq 2$. Suppose that r is the diameter of A. Let $\varphi : S^1 \to E$ be a continous mapping, contractible in E, such that $\mathrm{vol}_1(\varphi(S^1))$ is finite. Then there exists a homotopy $H : S^1 \times I \to E$ such that:*

$$H(s,0) = \varphi(s) \text{ for } s \in S^1,$$

$$H(s,t) = \varphi(s) \text{ if } \varphi(s) \notin A \text{ for } s \in S^1, t \in I.$$

Denote $\varphi_1(s) = H(s,1)$. Then $\varphi_1(S^1) \subset E \setminus A$. Moreover

i) $$\mathrm{vol}_1(\varphi_1(S^1)) \leq a_n\, \mathrm{vol}_1(\varphi(S^1)) + b_n r,$$

ii) $$\mathrm{vol}_2\, H(S^1 \times I) \leq c_n r\, \mathrm{vol}_1(\varphi_1(S^1)) + d_n r^2$$

where the constants a_n, b_n, c_n, d_n depend only on n.

Now our induction assertion follows. Actually the set $E \setminus A$ is compact and has one stratum (of dimension greater then 2) less then the set E. Let Φ_1 be a resolution of φ_1 in $E \setminus A$, satisfying the estimates in Propositon 6. To get the resolution Φ of φ it is enough to glue Φ_1 with the homotopy H.

We prove now Lemma 7 by induction on n (dimension of ambiant space). The case $n = 2$ is easy (see step 2 below).

Step 1. Suppose that A is an L-regular s-cell with constant M, open in E, $\dim A = n$. Hence A is open in $\mathbb{R}^n$ and is of the form

$$A = \{(x', x_n) \in \mathbb{R}^{n-1} \times \mathbb{R} : \ f(x') < x_n < g(x'), \ x' \in A'\}$$

where A' is an s-cell in $\mathbb{R}^{n-1}$, f and g are subanalytic continous functions on the closure of A', analytic on A'. Moreover $f(x') < g(x')$, $| d_{x'} f | \leq M$ and $| d_{x'} g | \leq M$ for each $x' \in A'$. Let $\pi : \mathbb{R}^n \to \mathbb{R}^{n-1}$ be the projection. Recall that by Lemma 3 there exists a subanalytic retraction $r_A : \mathbb{R}^n \to \bar{A}$ which is Lipschitz with a constant C depending on n. We put

$$\Sigma = \Sigma_1 \cup (\pi^{-1}(\pi \circ r_A \circ \varphi)(S^1) \cap \bar{A})$$

where $\Sigma_1 = \pi^{-1}(L)$. By L we denote a segment joining some fixed point $x_0' \in (\pi \circ r_A \circ \varphi)(S^1) \cap A'$ with a point of the boundary of A'. Notice that if such x_0' does not exist it means that φ is contractible in $E \setminus A$ and the lemma follows with trivial homotopy. We put also

$$\Gamma_f = \{(x', f(x') : x' \in (\pi \circ r_A \circ \varphi)(S^1)\},$$

$$\Gamma_g = \{(x', g(x') : x' \in (\pi \circ r_A \circ \varphi)(S^1)\},$$

and finally $\Gamma = \Gamma_f \cup \Gamma_g \ \cup \Gamma_1$, where $\Gamma_1 = \pi^{-1}(L) \cap \partial A$. Observe that Γ is arcwise connected and $\Gamma \subset \partial A \subset (E \setminus A)$. Clearly

$$\mathrm{vol}_1 \Gamma \leq 2(1 + M^2)^{\frac{1}{2}} \mathrm{vol}_1(\pi \circ r_A \circ \varphi)(S^1) + (1 + (1 + M^2)^{\frac{1}{2}} r^2$$

$$\mathrm{vol}_2 \Sigma \leq \mathrm{vol}_1(\pi \circ r_A \circ \varphi)(S^1) r + r^2.$$

Observe also that $\mathrm{vol}_1(\pi \circ r_A \circ \varphi)(S^1) \leq (1 + C^2)^{\frac{1}{2}} \mathrm{vol}_1(\varphi)(S^1)$, where C is the Lipschitz constant of the retraction r_A.

Let us fix an orientation on S^1. The set $\varphi^{-1}(A)$ is the union of a countable family of intervals $]u_i, v_i[$ or just S^1 . Observe that φ, restricted to $[u_i, v_i]$, is homotopic in Σ with a curve joining $\varphi(u_i)$ with $\varphi(v_i)$ in Γ. If $\varphi^{-1}(A) = S^1$, then φ can be deformed continously in Σ to $\varphi_1(s) = (\pi \circ \varphi(s), f \circ \pi \circ \varphi(s)) \in \Gamma_f$ just by "pushing down". This shows how to construct the homotopy H in the case when $\dim A = n$.

Step 2. Suppose now that $\dim A = k < n$. After some orthogonal change of coordinates we have

$$A = \{(x', x'') \in \mathbb{R}^k \times \mathbb{R}^{n-k} : \ \gamma(x') = x'', \ x' \in A'\}$$

where now A' is an open L-regular s-cell with constant M in $\mathbb{R}^k$. The mapping $\gamma : \mathbb{R}^k \to \mathbb{R}^{n-k}$ is Lipschitz with the constant depending on n (this is slightly different from Definitions 1 and 2, because we want to have A' open in euclidean space, hence of the type β). Let $\pi' : \mathbb{R}^k \times \mathbb{R}^{n-k} \to \mathbb{R}^k$ be the projection. Let us apply the construction of step 1 to A', $E' = \pi'(E)$, $\varphi' = \pi \circ \varphi$, but take the homotopy only for φ restricted to $\varphi^{-1}(A)$. Now lift this homotopy to A. The mapping γ is Lipschitz with constant depending only on n, hence our estimates for volumes follow. This ends the proof of Lemma 7. □

To finish the proof Theorem A observe that if φ is subanalytic then also Σ and Γ are subanalytic. Thus using subanalytic triangulations we can construct a subanalytic homotopy H and consequently a subanalytic resolution Φ. □

Remarks on the semialgebraic case

We state below our theorems for semialgebraic sets using Yomdin's diagrams of semialgebric sets.

Definition 8 (see [Y]). Let X be a semialgebraic subset of $\mathbb{R}^n$ represented as a result of set theoretic operations over sets of the form $\{f_i = 0\}, \{f_j > 0\}$. A set theoretic formula $\mathcal{D}$ of this representation together with the dimension n and the degrees of polynomials f_i, f_j is called the *diagram* of this representation. By $\mathcal{D}(X)$ we denote the diagram of some representation of the set X.

In the sequel we need L-regular s-cells, which are semialgebraic and possibly unbounded. We assume that in Definition 1 the function f (resp. g) may be equal to $-\infty$ (resp. $+\infty$). We have

Theorem D. *Let X be a semialgebraic subset of $\mathbb{R}^n$. Then there exists a semialgebraic stratification $\mathcal{T}$ of $\mathbb{R}^n$, compatible with X, such that all strata of $\mathcal{T}$ are semialgebraic L-regular s-cells with constant M. Moreover M depends only on n, the number of strata of $\mathcal{T}$ depends only on the diagram $\mathcal{D}(X)$.*

Actually Theorem D follows immediatly from Theorem C. In the semialgebraic case the hypothesis on relative compactness in Theorem C is superfluous. We can consider a family of all semialgebraic sets having the same diagram $\mathcal{D}$ as a semialgebraic family parametrized by a finite dimensional space of coefficients of all polynomials appearing in the formula $\mathcal{D}$. Hence by the semialgebraic version of Theorem C we obtain a bound for the number of strata. □

Notice that Theorem D together with Corollary 4 easily implies the following result of Yomdin (see Theorem 3.3. in [Y]):

Theorem E. *Let X be a semialgebraic subset of $\mathbb{R}^n$, and let B_r be a ball of radius r in $\mathbb{R}^n$. Then any two points x, y belonging to the same connected component of $X \cap B_r$ can be joined in $X \cap B_r$ by a piecewise-smooth curve of length Cr, where C depends only on the diagram $\mathcal{D}(X)$.* □

From Theorem D and Proposition 6 we deduce:

Theorem F. *Let X be a compact semialgebraic subset of $\mathbb{R}^n$. Suppose that r denotes the diameter of X. Let $\varphi : S^1 \to X$ be a continous mapping. Assume that φ is contractible in X, and that $\varphi(S^1)$ is of finite length denoted by $\mathrm{vol}_1 \varphi(S^1)$. Then there exists a resolution $\Phi : D^2 \to X$ of φ such that*

$$\mathrm{vol}_2 \Phi(D^2) \leq rP \mathrm{vol}_1 \varphi(S^1) + r^2 Q$$

where the constants P and Q depend only on the diagram $\mathcal{D}(X)$. □

References

[DK] Denkowska, Z., Kurdyka, K., Une propriété métrique des fibres d'un sous-analytique. C. R. Acad. Sci. Paris, Sér. I 299 (1984), 799–801.

[DS] Denkowska, Z., Stasica, J., Sur la stratification sous-analytique. Bull. Polish Acad. Sci. Math. 30 (1982), 337–340.

[DW] Denkowska, Z., Wachta, K., Sur la sous-analycité de l'application tangente. Bull. Polish Acad. Sci. Math. 30 (1982), 329–331.

[Ha1] Hardt, R., Stratification of real analytic mappings and images. Invent. Math. 28 (1975), 193–208.

[Ha2] —, Some analytic bounds for subanalytic sets. In: Differential geometric control theory, Progr. Math. 27, Birkhäuser 1983, 259–267.

[K] Kurdyka, K., On a subanalytic stratification satisfying a Whitney property with exponent 1. In: Real algebraic geometry, Rennes 1991 (ed. by M. Coste et al.), Lecture Notes in Math. 1524, Springer (1992) 316–322.

[Ło1] Łojasiewicz, S., Sur la semi-analycité des images inverses par l'application tangente. Bull. Polish Acad. Sci. Math. 27 (1979), 525–527.

[Ło2] —, Stratifications et triangulations sous-analytiques. Seminari di Geometria, Bologna (1986), 83–97.

[Pa] Parusiński, A., Lipschitz properties of semianalytic sets. Ann. Inst. Fourier 38 (1988), 189–213.

[T1] Teissier, B., Sur trois questions de finitude en géométrie analytique réelle. Appendix to the paper: F.Trèves, On the local solvability and the local integrability of systems of vector fields. Acta Math. 151 (1983), 38–48.

[T2] —, Sur la triangulation des morphismes sous-analytiques. Inst. Hautes Études Sci. Publ. Math. 70 (1989), 169–198.

[Y] Yomdin, Y., Metric properties of semialgebraic sets and mappings and their applications in smooth analysis. In: Géométrie algébrique et applications, La Rabida 1984, Vol.3 (ed. by J.-M. Aroca et al.), Travaux en cours, Hermann (1987), 165–182.

Stewart platforms without computer?

Felice Ronga and Thierry Vust

Abstract. Let $S_0, \dots, S_5$ be 6 spheres in $\mathbb{R}^3$ and H a hexahedron with vertices $P_0, \dots, P_5$. How many ways are there to move H in such a way that P_i belongs to S_1, $i = 0, \dots, 5$? We show in this paper that generically there are at most 40 solutions. This problem is the geometric version of a control problem for Stewart robots.

1991 Mathematics Subject Classification: 70B15, 14C17, 51M20.

A (generalized) Stewart platform is a solid with 6 points $P_0, \dots, P_5$ on it attached through 6 legs to 6 fixed points $Q_0, \dots, Q_5$ in the space $\mathbb{R}^3$. Assuming that the lengths of the legs can be varied arbitrarily (within the physical limits), the problem, first considered by D. Stewart [8], is to control the position of the body.†

In mathematical terms, we can identify the space of positions of the solid with the space $SO(3) \times \mathbb{R}^3$ of rotations and translations of $\mathbb{R}^3$. We have a map

$$\Phi = \Phi_{P_i,Q_i} : SO(3) \times \mathbb{R}^3 \to \mathbb{R}^6, \quad \Phi(R,T) = \left(\|T + R(P_i) - Q_i\|^2\right)_{i=0,\dots,5}.$$

The coordinates in the target play the role of control parameters, whereas a point in the source represents a position of the solid. The problem is to study the map Φ. Note that both the source and target of Φ have dimension 6, so we might expect that for a generic choice of the points P_i and Q_i the fibers of Φ are finite. See [7] for a more detailed discussion of Stewart platforms.

The aim of this paper is to use the intersection theory in algebraic geometry to show that for a generic choice of the points P_i and Q_i the number of possible positions of the platform for 6 given lengths is at most 40.

A first version of this paper was inspired from results of D. Lazard [6], who considered the case of a planar platform, (i.e. the P_i's lying in a plane as well as the Q_i's) using formal calculus manipulations on a computer; he also found that the number of possible positions for given lengths of the legs is at most 40 and conjectured that this should hold in general, but could only prove that it is bounded by 320 in the general case. Our first version in turn inspired to D. Lazard

† In fact a similar device was designed already in 1947 by V.E. Gough; see [8].

new ideas that enabled him to simplify enormously his computations and to deal with the general case. Again, Lazard's new ideas influenced the present version.

Our starting point is the same as in [6]: write the algebraic equations defining a general fiber of Φ, complexify and compactify by adding points at infinity. [6] then uses algorithms based on the theory of Gröbner basis to find the number of solutions at finite distance using a computer. We choose a different compactification and the usual tools of intersection theory (blowing-up, Chern classes) but we don't need a computer. However, it was essential to us to be guided by the calculations of [6].

An example of configuration with 16 real solutions that goes back to R. Bricard [4] is discussed in [6], §6.

On the way we need to study the singular locus of the map Φ (§1), and we generalize a result of J.-P. Merlet [7]: the fact that given (squared) lengths $d_0, \ldots, d_5$ are a singular value of Φ depends only on the 6 lines supporting the 6 pairs (P_i, Q_i), $i = 0, \ldots, 5$.

In §4 we will consider 2 classical geometric problems related to Stewart platforms.

We are grateful to J.-J. Risler for drawing our attention to this problem. We thank D. Lazard for communicating to us his preliminary versions of [6] and B. Mourrain for helpful conversations.

Throughout the paper we use the notion of generic choice of the points P_i and Q_i and lengths $d_0, \ldots, d_5$, by which we mean that the points or lengths must be chosen in some dense (most of the time a Zariski open) subset. To say more explicitely what it means would require more work.

1. The map Φ

The results of this § hold both over the field $\mathbb{K} = \mathbb{R}$ and $\mathbb{K} = \mathbb{C}$. Let $M(n, \mathbb{K})$ denote the space of $n \times n$ matrices with coefficients in $\mathbb{K}$ and set

$$O(n, \mathbb{K}) = \{R \in M(n, \mathbb{K}) \mid RR^t = I\}$$

and

$$SO(n, \mathbb{K}) = \{R \in O(n, \mathbb{K}) \mid \det(R) = 1\}$$

where A^t denotes the transpose of the matrix A and I is the identity matrix.

1.1 Proposition. *Let $(R, T) \in SO(3, \mathbb{K}) \times \mathbb{K}^3$ be a rotation and translation of $\mathbb{K}^3$ and let P_0, P_1 and $P_2 \in \mathbb{K}^3$ not lying on a line. If*

$$T + R(P_i) = T' + R'(P_i), \quad i = 0, 1, 2, \quad (R, T), (R', T') \in SO(3, \mathbb{K}) \times \mathbb{K}^3$$

then $T = T'$ and $R = R'$.

The proof is elementary and is left to the reader. □

1.2 Corollary. *Let $P_0, \ldots, P_5$ be chosen on the solid so that they are not aligned. Then the map*

$$SO(3,\mathbb{R}) \times \mathbb{R}^3 \to (\mathbb{R}^3)^6, \quad (R,T) \mapsto (T + R(P_i))_{i=0,\ldots 5}$$

identifies $SO(3,\mathbb{K}) \times \mathbb{K}^3$ to the space of configurations of the solid equipped with the 6 points P_i. □

1.3 Proposition. *Let $R \in O(3,\mathbb{K}) \subset M(3,\mathbb{K})$. The tangent space $T_R O(3,\mathbb{K})$ to $O(3,\mathbb{K})$ at the point R is given by*

$$T_R O(3,\mathbb{K}) = \left\{X \cdot R \mid X \in M(3,\mathbb{K}),\, X + X^t = 0\right\}.$$

Proof. Consider the map

$$F : M(3,\mathbb{K}) \to \mathrm{Sym}(3,\mathbb{K}), \quad Y \mapsto YY^t - I$$

where $\mathrm{Sym}(3,\mathbb{K})$ denotes the 3×3 symmetric matrices. Then $O(3,\mathbb{K}) = F^{-1}(0)$, 0 is a regular value of F and the assertion follows by considering the kernel of the derivative of F.† □

In the next proposition we shall use the vector (or exterior) product of two vectors of $\mathbb{K}^3$. Note that we can bijectively associate to an antisymmetric matrix $X = \begin{pmatrix} 0 & a & b \\ -a & 0 & c \\ -b & -c & 0 \end{pmatrix}$ a vector $\omega = \omega_X = \begin{pmatrix} -c \\ b \\ -a \end{pmatrix} \in \mathbb{K}^3$. Then for $v \in \mathbb{K}^3$ we have

$$X \cdot v = \omega \times v,$$

that is, left multiplication by the antisymmetric matrix X equals left vector multiplication with ω_X.

1.4 Proposition. *The value of the derivative of Φ at (R,T) on the tangent vector $(\overline{\omega}, \overline{T})$ is given by*

$$d\Phi_{(R,T)}(\overline{\omega},\overline{T}) = 2\left(\langle T + R(P_i) - Q_i \,|\overline{T} - \overline{\omega} \times T\rangle - \langle (T + R(P_i)) \times Q_i \,|\overline{\omega}\rangle\right)_{i=0,\ldots 5}$$

where $\langle \ | \ \rangle : \mathbb{K}^3 \times \mathbb{K}^3 \to \mathbb{K}$ is the ordinary scalar product on $\mathbb{K}^3$:$\langle v \,|w\rangle = \sum_{i=1}^3 v_i w_i$.

Proof. Let $\varphi : SO(3,\mathbb{K}) \times \mathbb{K}^3 \to \mathbb{K}^3$, $\varphi(R,T) = T + R(P)$. We have

$$d\varphi_{(R,T)}(\overline{\omega},\overline{T}) = \overline{T} + \overline{\omega} \times R(P).$$

It follows that

$$d\Phi_{(R,T)}(\overline{\omega},\overline{T}) = \left(2\left\langle T + R(P_i) - Q_i \,|\overline{T} + \overline{\omega} \times R(P_i)\right\rangle\right)_{i=0,\ldots,5}.$$

† Note that the coefficients of $Y^tY - I$ do not belong to the ideal generated by the coefficients of $YY^t - I$, hence these do not generate the ideal of polynomials on $M(3,\mathbb{K})$ vanishing on $O(3,\mathbb{K})$!

Now

$$\begin{aligned}\langle T+R(P_i)-Q_i\,|\overline{\omega}\times R(P_i)\rangle &= \langle R(P_i)\times(T+R(P_i)-Q_i)\,|\overline{\omega}\rangle\\ = \langle (R(P_i)+T)\times(T+R(P_i)-Q_i)\,|\overline{\omega}\rangle &- \langle T\times(T+R(P_i)-Q_i)\,|\overline{\omega}\rangle\\ = -\,\langle (R(Pi)+T)\times Q_i\,|\overline{\omega}\rangle &- \langle T+R(P_i)-Q_i\,|\overline{\omega}\times T\rangle\,.\end{aligned}$$

□

There is a nice interpretation due to Merlet (see [7]) of this derivative in terms of Plücker coordinates of the 6 lines carried by the pairs of points $(T+R(P_i),Q_i)$, $i=0,\dots,5$. Recall that if P, $Q\in\mathbb{R}^3$, $P=\begin{pmatrix}x_1\\x_2\\x_3\end{pmatrix}$, $Q=\begin{pmatrix}y_1\\y_2\\y_3\end{pmatrix}$, $P\neq Q$, the Plücker coordinates of the line carried by P and Q are constituted by the 2×2 minors of the 2×4 matrix $\begin{pmatrix}1&1\\x_1&y_1\\x_2&y_2\\x_3&y_3\end{pmatrix}$ whose columns are in fact the basis of the 2-plane of $\mathbb{R}^4$ which is the cone over the line $\overline{PQ}$. Hence, up to sign, $P-Q$ and $P\times Q$ constitute the Plücker coordinates of the line $\overline{PQ}$. Note that up to the multiplication by a scalar they depend only on the line $\overline{PQ}$, not on the points P and Q chosen on it. Therefore:

1.5 Corollary (cf. Merlet [7]). *The jacobian of Φ vanishes at (R,T) if and only if the Plücker coordinates of the 6 legs belong to a hyperplane of $\mathbb{R}^6$.* □

1.6 Proposition. *For a generic choice of the points $P_0,\dots,P_5$, $Q_0,\dots,Q_5$ the map Φ_{P_i,Q_i} is dominant (i.e., the image of the complexification contains a Zariski open non empty set) and as a consequence its generic fibers are finite.*

Proof. Choose the points P_i, Q_i such that the Plücker coordinates of the 6 corresponding lines do not satisfy any non trivial linear relation. The derivative of Φ_{P_i,Q_i} at $(0,Id)$ is regular according to 1.5, and therefore Φ is dominant. □

Recall the equations defining the fibers of Φ:

$$\|T+R(P_i)-Q_i\|^2=d_i,\quad i=0,\dots,5.$$

We may assume without loss of generality that $P_0=Q_0=0$. Then our set of equations becomes

$$\|T\|^2-d_0=0 \tag{0}$$

$$\|T+R(P_i)-Q_i\|^2-d_i=0,\quad i=1,2,3. \tag{i}$$

We will denote by $\sigma_i(R,T)$, $i=0,\dots,5$ the non-homogeneous quadratic form on the left hand side of equation (i), $i=0,\dots,5$.

Now we restate the assertion that we aim at in this paper.

1.7 Theorem. *For a generic choice of the P_i's and Q_i's and d_i's, $i = 0, \ldots, 5$, the number of complex solutions of the equations*

$$\sigma_i(R,T) = 0, \quad i = 0, \ldots, 5$$

is equal to 40.

1.8 Corollary. *If the above equations have a finite number of complex solutions, then their number is at most* 40.†

Proof. Assume that for given P_i, Q_i, d_i the set

$$\{(R,T) \in SO(3,\mathbb{C}) \times \mathbb{C}^3 \mid \Phi_{P_i,Q_i}(R,T) = d_i\,,\, i = 0, \ldots 5\}$$

contains a finite number of points $t_1, \ldots t_N$ and let B_i, $i = 1, \ldots, N$ be small balls isolating the t_i's. If we perturb slightly the P_i, Q_i, d_i in order to be generic in the sense of the above theorem, the degree of $\Phi|B_i : (B_i, \partial B_i) \to (\mathbb{C}^6, \mathbb{C}^6 \setminus (d_0, \ldots, d_5))$ will not change and it is at least 1. After perturbation some new solutions might appear outside $\cup_{i=1,\ldots N} B_i$, and their total number of solutions will be 40 according to the theorem. Therefore the number of solutions of the initial equations is at most 40. □

2. Compactification of the space of configurations

Recall that the equations defining the fibers of Φ, after assuming that $P_0 = Q_0 = 0$, are equivalent to

$$\|T\|^2 - d_0 = 0 \tag{0}$$
$$\|T + R(P_i) - Q_i\|^2 - d_i = 0, \quad i = 1, \ldots, 5. \tag{i}$$

Using equation (0) we can eliminate the term $\|T\|^2$ in the expansion of equations (i), $i = 1, \ldots, 5$ and we obtain the equivalent set of equations:

$$\sigma_0(R,T) = \|T\|^2 - c_0 = 0 \tag{2–1}$$
$$\sigma_i(R,T) = \langle R(P_i)\,|T\rangle - \langle R(P_i)\,|Q_i\rangle - \langle T\,|Q_i\rangle - c_i = 0, \quad i = 1, \ldots, 5$$

where $c_0 = d_0$ and $2c_i = d_0 + \|P_i\|^2 + \|Q_i\|^2 - d_i$, $i = 1, \ldots, 5$. In order to prove the main result of our paper, Theorem 1.7, we shall use the following strategy. We construct an adequate compactification $\overline{G}$ of the group $G = SO(3,\mathbb{C}) \times \mathbb{C}^3$, that is the space of complex rotations and translations of $\mathbb{C}^3$, which is the complexification of the space of configuration of a solid (see Corollary 1.2). The functions σ_i extend as sections s_i of suitable line bundles on $\overline{G}$; if S_i denotes the divisor associated to s_i, the degree of the 0-cycle $S_0 \cdot \ldots \cdot S_5 \cdot [\overline{G}]$, where the intersection takes place

† Warning: if the number of real solutions is finite, but not the number of complex solutions, it might exceed the number of solutions of the generic case.

in the Chow ring of $\overline{G}$, will be an upper bound for the number of points in a fiber of Φ in the generic case (cf. Corollary1.8). The difficulty consists in finding a compactification $\overline{G}$ such that

$$\left(\bigcap_{i=0,\dots,5} S_i\right) \cap (\overline{G} \setminus G) = \emptyset$$

so that the degree of our 0-cycle is then exactly the number of complex solutions of equations (0) through (5).

We start with some remarks and a first compactification of G. Since all non-degenerate quadratic forms on $\mathbb{C}^3$ are equivalent, we can replace $SO(3,\mathbb{C})$ by the special orthogonal group of any non-degenerate quadratic form, rather than $\sum x_i^2$. For our purposes it will be convenient to replace $\mathbb{C}^3$ by the space $S^2\mathbb{C}^2$, that is the second symmetric power of $\mathbb{C}^2$, and to take as quadratic form the discriminant Δ:

$$\Delta(ae_1 \circ e_1 + be_1 \circ e_2 + ce_2 \circ e_2) = b^2 - 4ac$$

where e_1, e_2 is the standard basis on $\mathbb{C}^2$ and a, b, $c \in \mathbb{C}$. We shall denote by $SO(\Delta)$ the corresponding special orthogonal group. The homomorphism

$$GL(\mathbb{C}^2) \to GL(S^2\mathbb{C}^2)\,,\ A \mapsto S^2A/\det(A)$$

where $S^2A : S^2\mathbb{C}^2 \to S^2\mathbb{C}^2$ denotes the extension of A to the second symmetric power, induces a surjective homomorphism

$$GL(\mathbb{C}^2) \to SO(\Delta)$$

whose kernel is the subgroup of homoteties λI, $\lambda \in \mathbb{C}^*$ and I is the identity. Taking the quotient of $GL(\mathbb{C}^2)$ by this kernel yields an isomorphism

$$PGL(\mathbb{C}^2) \to SO(\Delta).$$

Since $PGL(\mathbb{C}^2)$ is the open subset of $\mathbb{P}(\mathrm{End}(\mathbb{C}^2))$ consisting of classes $[A]$, $A \in \mathrm{End}(\mathbb{C}^2)$ such that $\det(A) \neq 0$, and $S^2\mathbb{C}^2$ is the open subset of $\mathbb{P}(S^2\mathbb{C}^2 \oplus \mathbb{C})$ consisting of classes $[v,1]$, $v \in S^2\mathbb{C}^2$, the variety

$$X = \mathbb{P}(\mathrm{End}(\mathbb{C}^2)) \times \mathbb{P}(S^2\mathbb{C}^2 \oplus \mathbb{C}) \simeq \mathbb{P}^3 \times \mathbb{P}^3$$

is a compactification of G and the complement of G in X consists of 2 components

$$X_0 = \Sigma \times \mathbb{P}(S^2\mathbb{C}^2 \oplus \mathbb{C}) \text{ and } X_1 = \mathbb{P}^3 \times \mathbb{P}(S^2\mathbb{C}^2)$$

where $\Sigma = \{[A] \in \mathbb{P}(\mathrm{End}(\mathbb{C}^2)) \mid \det(A) = 0\}$.

Let $\mathcal{O}_{\mathbb{P}^3}(-n)$ denote the n-th tensor product of the tautological line bundle on $\mathbb{P}^3$, $n \in \mathbb{Z}$, and $\mathcal{O}_i(n)$, $i = 1,2$, the pull-back of $\mathcal{O}_{\mathbb{P}^3}(n)$ on $X = \mathbb{P}^3 \times \mathbb{P}^3$ by the projection on the i-th factor. The function $\sigma_0(R,T) = \|T\|^2 - c_0$ extends as a section s_0 of $\mathcal{O}_2(2)$ that can be written

$$s_0([A_0],[T_0,u_0])_{(T,u)} = \|T\|^2 - u^2c_0$$

where $A_0 \in \mathrm{End}(\mathbb{C}^2) \setminus \{0\}$, $(T_0,u_0) \in S^2\mathbb{C}^2 \oplus \mathbb{C} \setminus \{0\}$, and $(T,U) \in [T_0,u_0]$. Similarly, the functions $\sigma_i(R,T)$, $i = 1,\dots,5$, extend as sections s_i of $\mathcal{O}_1(2) \otimes \mathcal{O}_2(1)$

that can be written

$$s_i([A_0],[T_0,u_0])_{(A,T,u)} = \langle S^2A(P_i)\,|T\rangle - u\,\langle S^2A(P_i)\,|Q_i\rangle - \det(A)\,\langle T\,|Q_i\rangle - u\,\det(A)c_i.$$

In these formulae the norm and the scalar product are relative to the discriminant.

We denote by S_i, $i = 0,\dots,5$, the divisors on X defined as the zeroes of s_i, $i = 0,\dots,5$.

Remark. If H_i denotes the divisor of X which is the pull-back of a hyperplane in i-th factor $\mathbb{P}^3$, $1 = 1,2$, we have

$$\operatorname{degree}(S_0\cdot\ldots\cdot S_5\cdot[X]) = \operatorname{degree}\left((2H_2)(2H_1+H_2)^5\cdot[X]\right) = 160.$$

We now examine more closely $\bigcap_{i=0,\dots,5} S_i$. The variety

$$Y = \left\{([A],[T]) \in X \mid \det(A) = 0\,,\ \langle \operatorname{Im}(S^2A\,|T\rangle = 0\right\} \subset \Sigma\times\mathbb{P}(S^2\mathbb{C}^2)$$

is smooth, of dimension 3. In fact it is the projective bundle associated to the vector bundle of rank 2 over Σ:

$$\gamma = \left\{([A],T) \in \mathbb{P}(\operatorname{End}(\mathbb{C}^2))\times S^2\mathbb{C}^2 \mid \det(A) = 0\,,\ \langle \operatorname{Im} S^2A\,|T\rangle = 0\right\} \subset \Sigma\times S^2\mathbb{C}^2\,.$$

Moreover, $Y \subset X_0\cap X_1$.

2.1 Proposition.
(i) $Y \subset S_i$, $i = 1,\dots 5$;
(ii) *for a generic choice of P_i, Q_i, we have (set theoretically)*

$$\left(\bigcap_{i=0\dots,5} S_i\right)\cap(X_0\cup X_1) = S_0\cap Y.$$

Proof. Clearly, $Y \subset S_i$, $i = 1,\dots,5$. Assertion (ii) will follow from (a) and (b) below.

(a) For a generic choice of P_i, Q_i we have:

$$\bigcap_{i=0,\dots,5} S_i\cap X_1 \subset X_0.$$

Indeed, the left-hand side of this inclusion is defined by the equations:

$$\|T\|^2 = 0,\quad \langle S^2A(P_i) - \det(A)Q_i\,|T\rangle = 0,\quad i = 1,\dots,5.$$ ♣

We may assume that P_1, P_2 and P_3 are linearly independent and so we have expressions

$$P_4 = \sum_{i=1}^{3}\alpha_i^4 P_i,\quad P_5 = \sum_{i=1}^{3}\alpha_i^5 P_i$$

and we deduce from ♣ that

$$\left\langle T \,\middle|\, \det(A)\widetilde{Q}_h \right\rangle = 0, \quad h = 4, 5,$$

where $\widetilde{Q}_h = Q_h - \sum_{i=1}^{3} \alpha_i^h Q_i$. If $\det(A) \neq 0$ these equations imply that $\widetilde{Q}_h$, $h = 4, 5$ lie on a same tangent to the conic of isotropic vectors of $S^2\mathbb{C}^2$. But for a generic choice of Q_4 and Q_5 this will not happen, therefore we must have $\det(A) = 0$.

(b) For a generic choice of P_i, Q_i we have

$$\left(\bigcap_{i=1,\ldots,5} S_i \right) \cap X_0 \subset Y.$$

In order to prove this assertion we shall use the following easy lemma, whose proof is left to the reader.

Lemma. *Let $A = \alpha \otimes a$, where $\alpha \in \mathrm{Hom}(\mathbb{C}^2, \mathbb{C}) \setminus \{0\}$, $a \in \mathbb{C}^2 \setminus \{0\}$, be an endomorphism of rank 1 of $\mathbb{C}^2$. Then:*
(1) *$S^2 A$ is of rank 1 and its image is generated by $a \otimes a$;*
(2) *if $P_i = x_i \circ y_i$, $x_i, y_i \in \mathbb{C}^2$, we have $S^2 A(P_i) = \alpha(x_i)\alpha(y_i) a \circ a$, $i = 1, \ldots, 5$;*
(3) *if $P_i \neq 0$, $i = 1, \ldots, 5$ and the images of $x_1, y_1, \ldots, x_5, y_5$ in $\mathbb{P}(\mathbb{C}^2)$ are all distinct, then at most one of the $\alpha(x_i)\alpha(y_i)$ vanishes.* □

If $\det(A) = 0$, using the notations of the lemma, the left hand side of (b) is defined by:

$$\alpha(x_i)\alpha(y_i) \langle a \circ a \,|T - uQ_i \rangle = 0.$$

For a generic choice of the P_i's, after renumbering them, we may suppose that $\alpha(x_i)\alpha(y_i) \neq 0$, $i = 1, \ldots, 4$; for a generic choice we may suppose that the Q_i's satisfy some linear relation of the type $\sum_{i=1}^{4} \beta_i Q_i = 0$, where $\sum_{i=1}^{4} \beta_i \neq 0$. It follows that $\langle a \circ a \,|T \rangle = 0$, hence $u = 0$. □

Let $\pi : \widetilde{X} \to X$ denote the blowing-up of X along Y. It is also a compactification of G and our next proposition shows that it is suitable for our problem: if $\widetilde{S}_i$ denotes the strict transform of S_i, then $\left(\bigcap_{i=0,\ldots,5} \widetilde{S}_i\right) \cap \left(\widetilde{X} \setminus G\right) = \emptyset$.

2.2 Proposition. *For a generic choice of P_i, Q_i, $i = 0, \ldots, 5$, the S_i, $i = 1, \ldots, 5$, intersect cleanly at points of $S_0 \cap Y$; that is, if $z \in Y \cap S_0$, denoting by T_z the tangent space at the point z:*

$$\bigcap_{i=1,\ldots,5} T_z S_i = T_z Y.$$

Proof. Let $z = ([A], [T, u]) \in X$. For $i = 1, \ldots, 5$, setting $P_i = x_i \circ y_i$, we have:

$$\begin{aligned}d_z s_i(\overline{A},\overline{T},\overline{u}) &= \left\langle \overline{A}(x_i)\circ A(y_i) + A(x_i)\circ\overline{A}(y_i)\,|T\right\rangle \\ &+ \left\langle A(x_i)\circ A(y_i)\,|\overline{T}\right\rangle - \overline{u}\left\langle A(x_i)\circ A(y_i)\,|Q_i\right\rangle - u\left\langle \overline{A}(x_i)\circ A(y_i) + A(x_i)\circ\overline{A}(y_i)\,|Q_i\right\rangle \\ &- d_A(\det)(\overline{A})\left\langle T\,|Q_i\right\rangle - \det(A)\left\langle \overline{T}\,|Q_i\right\rangle - \overline{u}\det(A)c_i - u d_A(\det)(\overline{A})c_i.\end{aligned}$$

If $z \in Y \cap S_0$, then $u = 0$, A is of rank 1 and T is isotropic (i.e. $\langle T\,|T\rangle = 0$) and orthogonal to the image of S^2A (also of rank 1), which implies that $T \in \operatorname{Im} S^2A$. Therefore

$$d_z s_i = \left\langle A(x_i)\circ A(y_i)\,|\overline{T}\right\rangle - \overline{u}\left\langle A(x_i)\circ A(y_i)\,|Q_i\right\rangle - d_A(\det)(\overline{A})\left\langle T\,|Q_i\right\rangle.$$

With the notation of the lemma above and setting $T = ta\circ a$, $t \in \mathbb{C}$, we can write

$$d_z s_i(\overline{A},\overline{T},\overline{u}) = \alpha(x_i)\alpha(y_i)\left\langle a\circ a\,|\overline{T}\right\rangle - \alpha(x_i)\alpha(y_i)\left\langle a\circ a\,|Q_i\right\rangle\overline{u} - t\left\langle a\circ a\,|Q_i\right\rangle d_A(\det)(\overline{A}).$$

Define the 3 linear forms

$$\begin{aligned}\phi &:(\overline{A},\overline{T},\overline{u}) \mapsto \left\langle a\circ a\,|\overline{T}\right\rangle \\ \chi &:(\overline{A},\overline{T},\overline{u}) \mapsto \overline{u} \\ \psi &:(\overline{A},\overline{T},\overline{u}) \mapsto d_A(\det)(\overline{A})\end{aligned}$$

so that

$$d_z s_i = \alpha(x_i)\alpha(y_i)\phi - \alpha(x_i)\alpha(y_i)\left\langle a\circ a\,|Q_i\right\rangle\chi - t\left\langle a\circ a\,|Q_i\right\rangle\psi.$$

Since the intersection of the kernels of ϕ, χ and ψ equals T_zY, it remains to prove that for a generic choice of P_i and Q_i the 5 by 3 matrix

$$M(\alpha,a,t,(x_i,y_i,Q_i)_{i=1,\dots,5}) = \begin{pmatrix} \vdots & \vdots & \vdots \\ \alpha(x_i)\alpha(y_i) & \alpha(x_i)\alpha(y_i)\left\langle a\circ a\,|Q_i\right\rangle & \left\langle a\circ a\,|Q_i\right\rangle \\ \vdots & \vdots & \vdots \end{pmatrix}$$

is of rank 3 for all $(\alpha,a,t) \in L$, where $L = \left(\operatorname{Hom}(\mathbb{C}^2,\mathbb{C})\setminus\{0\}\right)\times(\mathbb{C}\setminus\{0\})^2\times(\mathbb{C}\setminus\{0\})$.

Consider the above defined map

$$M : L\times\left(\mathbb{C}^2\times\mathbb{C}^2\right)^5\times\left(S^2\mathbb{C}^2\right)^5 \to M(5,3,\mathbb{C})$$

where $M(5,3)$ denotes the space of 5×3 matrices with complex coefficients.

We will prove in Lemma 2.3 below that for generic (x_i,y_i,Q_i) the image of the corresponding map M consists of matrices of rank ≥ 2 and M is transversal to

$$\Sigma^2 = \{m \in M(5,3) \mid \operatorname{rank}(m) = 2\}.$$

It follows that for (x_i,y_i,Q_i) generic the restriction

$$M_0 = M|L\times\{(x_i,y_i,Q_i)\} : L\times\{(x_i,y_i,Q_i)\} \to M(5,3)$$

has the same transversality properties as M itself, and since Σ^2 is of codimension 3 in $M(5,3)$, $M_0^{-1}(\Sigma^2)$ is either empty or of dimension 2. But the group $(\mathbb{C}\setminus\{0\})^3$ acts freely on L,

$$(\tau_1,\tau_2,\tau_3)\cdot(\alpha,a,t)=(\tau_1\alpha,\tau_2 a,\tau_3 t),$$

and it leaves invariant the composite function rank $\circ M_0$. Therefore we must have

$$M_0^{-1}(\Sigma^2)=\emptyset$$

and this achieves the proof. □

2.3 Lemma. *There exists a Zariski open, nonempty subset $\mathcal{U}$ of $(\mathbb{C}^2\times\mathbb{C}^2)^5\times(S^2\mathbb{C}^2)^5$ such that the image of the map*

$$M|L\times\mathcal{U}: L\times\mathcal{U}\to M(5,3),$$

$$(a,\alpha,t,(x_i,y_i),Q_i)\mapsto\begin{pmatrix}\vdots & \vdots & \vdots\\ \alpha(x_i)\alpha(y_i) & \alpha(x_i)\alpha(y_i)\langle a\circ a\,|Q_i\rangle & \langle a\circ a\,|Q_i\rangle\\ \vdots & \vdots & \vdots\end{pmatrix}$$

consists of matrices of rank ≥ 2 and is transveral to Σ^2.

Proof.

a) We first make the genericity conditions explicit. Denote by $\mathcal{U}_{\mathcal{P}}$ the set of $(x_i,y_i)\in(\mathbb{C}^2\setminus\{0\}\times\mathbb{C}^2\setminus\{0\})^5$ such that:
(1) the images of $x_1,y_1,\ldots,x_5,y_5$ in $\mathbb{P}(\mathbb{C}^2)$ are distinct;
(2) if $x_i\circ y_i-x_j\circ y_j=x_{i,j}\circ y_{i,j}$, $x_{i,j}$, $y_{i,j}\in\mathbb{C}^2$, then the images of $x_{1,2}$, $y_{1,2}$, $\ldots,x_{4,5}$, $y_{4,5}$ are distinct.

We introduce similar conditions on the Q_i's. Denote by $\mathcal{U}_Q$ the set of $(Q_1,\ldots,Q_5)\in(S^2\mathbb{C}^2\setminus\{0\})$ such that:
(1) a tangent to the conic of $\mathbb{P}(S^2\mathbb{C}^2)$ of equation $\Delta=0$ contains at most the image in $\mathbb{P}(S^2\mathbb{C}^2)$ of one of the Q_i's, $i=1,\ldots,5$;
(2) a tangent to same conic contains at most the image of one of the Q_i-Q_j, $i,j=1,\ldots,5$, $i\neq j$.

b) Consider the maps

$$\mu:\ \mathrm{Hom}(\mathbb{C}^2,\mathbb{C})\setminus\{0\}\times\mathcal{U}_P\to\mathbb{C}^5,\quad(\alpha,(x_i,y_i))\mapsto(\alpha(x_i)\alpha(y_i))_{i=1,\ldots,5}$$
$$\nu:\ \mathbb{C}^2\setminus\{0\}\times\mathcal{U}_Q\to\mathbb{C}^5,\quad(a,Q_1,\ldots,Q_5)\mapsto(\langle a\circ a\,|Q_i\rangle)_{i=1,\ldots,5}$$

whose images $\mathcal{V}$ in both cases consists of the $(z_1,\ldots,z_5)\in\mathbb{C}^5$ such that $z_i=0$ for at most one i and $z_i-z_j=0$ for at most one pair (i,j).

Setting $\mu_i=\alpha(x_i)\alpha(y_i)$ and $\nu_i=\langle a\circ a\,|Q_i\rangle$ the map M can simply be written

$$M(\alpha,a,t,x_i,y_i,Q_i)=(\mu,\mu\star\nu,\nu)$$

where

$$\mu = \begin{pmatrix} \mu_1 \\ \vdots \\ \mu_5 \end{pmatrix} \quad \nu = \begin{pmatrix} \nu_1 \\ \vdots \\ \nu_5 \end{pmatrix} \quad \text{and } \mu \star \nu = \begin{pmatrix} \mu_1\nu_1 \\ \vdots \\ \mu_5\nu_5 \end{pmatrix}.$$

Using condition (1) for $\mathcal{U}_P$ and $\mathcal{U}_Q$ it is easy to check that the derivatives $d\mu$ and $d\nu$ are surjective at every point.

c) Because of **b)**, it suffices to prove that the map

$$N : \mathcal{V} \times \mathcal{V} \to M(5,3), \quad (\mu,\nu) \mapsto (\mu, \mu \star \nu, \nu)$$

is of rank ≥ 2 and is transversal to Σ^2. It follows from the definition of $\mathcal{V}$ that μ and $\mu \star \nu$ are always linearly independent and therefore $N(\mu,\nu)$ is always of rank at least 2.

As for the transversality to Σ^2, consider the map

$$\Lambda : (\mathbb{C}^5)^3 \simeq M(5,3) \to \Lambda^3\mathbb{C}^5 \,,\ (y_1,y_2,y_3) \mapsto y_1 \wedge y_2 \wedge y_3 .$$

Then

$$\Sigma^2 = \Lambda^{-1}(0) \text{ and } T_y\Sigma^2 = \mathrm{Ker}(d_y\Lambda).$$

Assume that $N(\mu,\nu) \in \Sigma^2$. Then there exist $r, s \in \mathbb{C}$ such that $\nu = r\mu + s\mu \star \nu$ and so

$$\begin{aligned} T_{N(\mu,\nu)}\Sigma^2 &= \{(\overline{y_1},\overline{y_2},\overline{y_3}) \mid \overline{y_1} \wedge \mu\star\nu \wedge \nu + \mu \wedge \overline{y_2} \wedge \nu + \mu \wedge \mu\star\nu \wedge \overline{y_3} = 0\} \\ &= \{(\overline{y_1},\overline{y_2},\overline{y_3}) \mid (r\overline{y_1} + s\overline{y_2} - \overline{y_3} \wedge \mu\star\nu \wedge \mu = 0\} . \end{aligned}$$

If E_k denotes the k-th vector of the canonical basis of $T_\nu\mathcal{V} \simeq \mathbb{C}^5$,

$$d_{(\mu,\nu)}N(0,E_k) = (0,\mu_kE_k,E_k) \text{ and } d_{N(\mu,\nu)}\Lambda(0,\mu_kE_k,E_k) = (s\mu_k - 1)E_k \wedge \mu\star\nu \wedge \mu .$$

In order to prove that $\mathrm{Im}(d_{(\mu,\nu)}N)$ is transversal to $Ker\left(d_{N(\mu,\nu)}\Lambda\right)$ it remains to show that $s\mu_k - 1 \neq 0$, for $k = 1,\ldots,5$, because if we choose E_{k_1}, E_{k_2}, E_{k_3} to generate a supplementary subspace to the space generated by $\mu \star \nu$ and μ, the vectors $d_{(\mu,\nu)}N(0,E_{k_i})$ $i = 1,2,3$, will generate a supplementary subspace to $T_{N(\mu,\nu)}\Sigma^2$. Observe that $r \neq 0$: otherwise $\nu_j = s\mu_j\nu_j$, $j = 1,\ldots,5$, and so $1 = s\mu_j$ for at least 4 indices , since at most 1 of the ν_j's is zero, contradicting that $\mu \in \mathcal{V}$. Finally

$$s\mu_k - 1 = 0 \Longrightarrow \mu_k = (1 - s\mu_k)\nu_k = 0 \Longrightarrow \mu_k = 0 \Longrightarrow -1 = 0\,!$$

□

3. Numeric calculations

The normal bundle of the quadric $\Sigma = \left\{[A_0] \in \mathbb{P}(\mathrm{End}(\mathbb{C}^2)) \mid \det(A_0) = 0\right\}$ in $\mathbb{P}(\mathrm{End}(\mathbb{C}^2))$ is $\mathcal{O}_1(2)$, where we recall that $\mathcal{O}_1 = \mathcal{O}_{\mathbb{P}(\mathrm{End}(\mathbb{C}^2))}$. The pull-backs by the isomorphism

$$\mathrm{Im} \times \mathrm{Ker} : \Sigma \to \mathbb{P}^1 \times \mathbb{P}^1$$

of the bundle $\mathcal{O}(n)_{\mathbb{P}^1}$ on the first, respectively the second factor will be denoted by $\mathcal{O}_1^\Sigma(n)$, respectively $\mathcal{O}_2^\Sigma(n)$, and $\mathcal{O}_1(1)|\Sigma \simeq \mathcal{O}_1^\Sigma(1) \otimes \mathcal{O}_2^\Sigma(1)$, so that the normal bundle of Σ is $\mathcal{O}_1^\Sigma(2) \otimes \mathcal{O}_2^\Sigma(2)$.

Our first task is to express all intervening bundles in terms of $\mathcal{O}_i$ and $\mathcal{O}_i^\Sigma$, $i = 1, 2$. We will sometimes use the same notation for a vector bundle and its pull-back on some space. Typically, $\mathbb{C}^2$ will denote the trivial bundle with fiber $\mathbb{C}^2$ on any space (like in (3–1) below). If $V \subset W$, V and W smooth, $\nu(V, W)$ will denote the normal bundle of V in W.

If $\det(A) = 0$, $A \neq 0$ and $a \in \mathrm{Im}(A) \setminus \{0\}$, then

$$\{T \in S^2\mathbb{C}^2 \mid \langle \mathrm{Im}(S^2A) \,|T \rangle = 0\} = a \otimes \mathbb{C}^2,$$

therefore, recalling that $\gamma = \{([A], T) \mid \langle \mathrm{Im}(S^2A) \,|T \rangle = 0\}$, we have an isomorphism

$$\gamma \simeq \mathcal{O}_1^\Sigma(-1) \otimes \mathbb{C}^2. \tag{3–1}$$

From the exact sequence

$$0 \to \mathcal{O}_1^\Sigma(-1) \otimes \mathbb{C}^2 \hookrightarrow S^2\mathbb{C}^2 \xrightarrow{\phi} \mathcal{O}_1^\Sigma(2) \to 0,$$

where $\phi(v \circ w)_{(\alpha \circ \beta)} = \alpha(v) \circ \beta(w)$, we deduce an isomorphism

$$S^2\mathbb{C}^2/\gamma \simeq \mathcal{O}_1^\Sigma(2).$$

Denote by $\mathcal{O}_\gamma(1)$ the dual of the tautological bundle over $\mathbb{P}(\gamma)$ (i.e., the bundle of linear forms on lines of $\mathbb{P}(\gamma)$). Since γ is a subbundle of the trivial bundle $\Sigma \times (S^2\mathbb{C}^2 \oplus \mathbb{C})$, it follows that

$$\mathcal{O}_2(1)|\Sigma \simeq \mathcal{O}_\gamma(1) \tag{3–2}$$

and that

$$\nu(\mathbb{P}(\gamma), \Sigma \times \mathbb{P}^3) \simeq \mathcal{O}_\gamma(1) \otimes \frac{S^2\mathbb{C}^2 \oplus \mathbb{C}}{\gamma \oplus \{0\}} \simeq \mathcal{O}_\gamma(1) \otimes (\mathcal{O}_1^\Sigma(2) \oplus \mathbb{C}).$$

From the exact sequence

$$0 \to \nu(\mathbb{P}(\gamma), \Sigma \times \mathbb{P}^3) \to \nu(\mathbb{P}(\gamma), \mathbb{P}^3 \times \mathbb{P}^3) \to p^*(\nu(\Sigma, \mathbb{P}^3)) \simeq \mathcal{O}_1^\Sigma(2) \otimes \mathcal{O}_2^\Sigma(2) \to 0$$

where $p : \mathbb{P}(\gamma) \to \Sigma$ is the natural projection, we deduce the following formula for the total Chern classes:

$$\begin{aligned} c(\nu) &= c(\nu(\mathbb{P}(\gamma), \Sigma \times \mathbb{P}^3)) \cdot c(p^*(\nu(\Sigma, \mathbb{P}^3))) = \\ &\qquad c\Big((\mathcal{O}_\gamma(1) \otimes \mathcal{O}_1^\Sigma(2)) \oplus \mathcal{O}_\gamma(1)\Big) \cdot c(\mathcal{O}_1^\Sigma(2) \otimes \mathcal{O}_2^\Sigma(2)) \end{aligned} \tag{3–3}$$

where $\nu = \nu(\mathbb{P}(\gamma), X)$ (recall that $X = \mathbb{P}^3 \times \mathbb{P}^3$).

Let $\pi : \widetilde{X} \to X$ be the blow-up of X along $\mathbb{P}(\gamma)$, with exceptional divisor $E = \mathbb{P}(\nu)$, and denote by ξ its normal bundle. In §2 we constructed sections s_i of

$\mathcal{O}_1(2) \otimes \mathcal{O}_2(1)$, $i = 1, \ldots, 5$ and s_0 of $\mathcal{O}_2(2)$. It follows from Propositions 2.1 and 2.2 that our main Theorem 1.7 will be a consequence of:

Proposition.

$$\text{degree}\Big(c_6\big(\pi^* O_2(2) \oplus \xi^* \otimes \pi^* \left[\oplus_{i=1,\ldots,5} \mathcal{O}_1(2) \otimes \mathcal{O}_2(1)\right]\big)\Big) = 40.$$

We refer to [5] for the elementary intersection theory that we shall use. The calculations that follow are conceived using the Chow ring $\mathcal{A}(Z)$, where Z is some variety, but the Chow business can be replaced by the cohomology ring. In the context of the Chow ring, the Chern classes of a vector bundle are operators on the Chow ring of the basis. We shall use the ring homomorphism $\deg = \deg_Z : \mathcal{A}(Z) \to \mathbb{Z}$.

Proof of the proposition. Let

$$\begin{aligned} c_1(\mathcal{O}_\gamma(1)) &= a \\ c_1(\mathcal{O}(1)_i) &= H_i\,,\ i = 1, 2 \\ c_1(\mathcal{O}_i^{\Sigma}(1)) &= h_i\,,\ i = 1, 2 \\ c_1(\xi) &= -t \end{aligned}$$

and note that H_1 restricts to $h_1 + h_2$ and H_2 restricts to a on $\mathbb{P}(\gamma)$. In what follows we shall use the same notation for the H_i's, $i = 1, 2$ and their pull-back $\pi^*(H_i)$ on $\widetilde{X}$. Now

$$\begin{aligned} \Big(c_6\big(\pi^* \mathcal{O}_2(2) \oplus \xi^* \otimes \pi^* \left[\oplus_{i=1,\ldots,5} \mathcal{O}_1(2) \otimes \mathcal{O}_2(1)\right]\big)\Big) &= 2H_2(t + 2H_1 + H_2)^5 \\ = \underbrace{2H_2(2H_1 + H_2)^5}_{\text{I}} &+ \underbrace{2tH_2(t^4 + 5t^3(2H_1 + H_2) + 10t^2(2H_1 + H_2)^2 + \ldots)}_{\text{II}} \end{aligned}$$

where the dots represent classes that evaluate to 0. Since

$$2H_2(2H_1 + H_2)^5 = 2H_2 \binom{5}{3} (2H_1)^3 H_2^2 = 160 H_1^3 H_2^3$$

we see that degree(I) $= 160$, as we already claimed in §2.

Let's take care of II. We can evaluate on $E = \mathbb{P}(\nu)$ the expression divided by $-t$, which is the class of the exceptional divisor E:

$$\begin{aligned} \text{II} \cap [\widetilde{X}] &= \Big(-2H_2(t^4 + 5t^3(2H_1 + H_2) + 10t^2(2H_1 + H_2)^2)\Big) \cap [\mathbb{P}(\nu)] \\ &= -\Big(2a(t^4 + 5t^3(a + 2h_1 + 2h_2) + 10t^2(6ah_1 + 4ah_2 + 8h_1h_2))\Big) \cap [\mathbb{P}(\nu)] \end{aligned} \tag{3–4}$$

where in the last equality we used that $h_1^2 = h_2^2 = 0$ and, since γ is of rank 2 and $c(\gamma) = (1 - h_1)^2 = 1 - 2h_1$,

$$0 = c_2(\gamma / \mathcal{O}_\gamma(-1)) = a(-2h_1) + a^2 \Longrightarrow a^2 = 2ah_1.$$

Consider the projection $\pi' : \mathbb{P}(\nu) \to \mathbb{P}(\gamma)$. If we denote by s_h the h-th Segre class of the ν, whose rank is 3, we have that $\pi'_*(t^k) = s_{k-2}(\nu)$. Therefore we deduce from (3–3) that

$$c(\nu) = 1 + 2a + 4h_1 + 2h_2 + 8ah_1 + 4ah_2 + 4h_1h_2 + 8ah_1h_2$$

and so

$$s(\nu) = 1 - 2a - 4h_1 - 2h_2 + 16ah_1 + 4ah_2 + 12h_1h_2 - 40ah_1h_2.$$

Replacing t^i by $s_{i-2}(\nu)$ in (3-4) we obtain

$$\mathrm{II} \cap [\widetilde{X}] = -120ah_1h_2 \cap [\mathbb{P}(\nu)]$$

and the result follows at once. □

4. Two geometric problems related to Stewart platforms

Consider a Stewart platform in which $P_0 = P_1 = 0$, $P_2 = P_3$, $P_4 = P_5$, $Q_1 = Q_2$, $Q_3 = Q_4$, $Q_5 = Q_0 = 0$. Then, for $(R,T) \in SO(3) \times \mathbb{R}^3$, setting $P'_i = T + R(P_i)$, the triangles spanned by P'_0, P'_2, P'_4 and by Q_1, Q_3, Q_5, and the 6 legs $\overline{P'_0Q_5}$, $\overline{P'_0Q_1}$, $\overline{P'_2Q_1}$, $\overline{P'_2Q_3}$, $\overline{P'_4Q_3}$, $\overline{P'_4Q_5}$ build an octahedron. Our problem in this case can be restated as follows:

How many octahedron are there with given lengths of the edges?

In [1] it is shown that there are at most 16, and an explicit example with 16 solutions is given.

We want to prove that there are at most 16 solutions with a method that is similar to the one that we used for the general case.

In fact, if we fix the lengths of the legs, the points P'_i that are at the given distance of the Q_i's (forgetting that they are tied together to be the vertices of a given triangle) describe 3 circles. Therefore the problem can be restated as follows:

Given 3 circles γ_0, γ_1 and γ_2 in the space and a triangle of vertices A_0, A_1, A_2 that we can move in the space, how many positions of the triangle such that A_i lies on γ_i are there?

Let c_i denote the center of γ_i, r_i its radius and v_i a variable point on it, that we also write $v_i = u_i + c_i$, where now u_i is of length r_i, lying in the plane of the circle γ_i translated by $-c_i$. We have

$$\begin{aligned}\|v_i - v_j\|^2 &= \|u_i - u_j - (c_i - c_j)\|^2 = \\ &= r_i^2 + r_j^2 + \|c_i - c_j\|^2 - 2\langle u_i | u_j\rangle - 2\langle u_i | c_i - c_j\rangle + 2\langle u_j | c_i - c_j\rangle .\end{aligned} \tag{4–1}$$

The maps

$$\mathbb{P}^1(\mathbb{R}) \to \mathbb{R}^3, \quad [s_i, t_i] \mapsto \frac{r_i}{s_i^2 + t_i^2}(s_i^2 - t_i^2, 2s_i t_i, 0)$$

followed by suitable rotations of $\mathbb{R}^3$ will provide parametrisations ϕ_i of the u_i's. Since the expression (4–1) is of bidegree $(1,1)$ in u_i, u_j, and the components of the ϕ_i are of degree 2, the number of solutions of our problem equals the number of points of intersection of 3 hypersurfaces of $\mathbb{P}^1 \times \mathbb{P}^1 \times \mathbb{P}^1$ of multidegree $(2,2,0)$, $(2,0,2)$ and $(0,2,2)$. If h_i denotes the pull-back on $\mathbb{P}^1 \times \mathbb{P}^1 \times \mathbb{P}^1$ of the hyperplane divisor on $\mathbb{P}^1(\mathbb{R})$ by the projection on the i-th factor, then we have

$$(2h_1 + 2h_2)(2h_1 + 2h_3)(2h_2 + 2h_3) = 16h_1h_2h_3$$

and the result follows.

The 2-dimensional analog of this problem reads as follows:

Given three circles γ_i, $i = 1, 2, 3$, in the plane and a triangle with vertices A_i, $i = 1, 2, 3$ that we can slide in the plane using translations and rotations, how many positions of the triangle such that A_i lies on γ_i are there?

This problem is solved for example in Blaschke [3], page 89, using the now forgotten methods of Kinematics: he calculates explicitely the equation of the so-called "Koppelkurve" traced by A_3 when A_1 and A_2 are tied to ly on γ_1 and γ_2 respectively and shows that it is a sextic with 2 triple points at the cyclic points (i.e., the 2 points at infinity of any circle) and therefore intersects γ_3 in at most 6 points. On the other hand there are known examples of such a sextic cutting γ_3 in exactly 6 points (see [9] page 66 and following, or [2] page 71).

References

[1] Bellido, A., Dedieu, J.-P., Yacoubson, J.-C., Combien existe-t-il d'octaèdres dont les longeurs des arêtes sont données? Annali dell'Università di Ferrara Vol. XXXVII (1991), 65–84.

[2] Brieskorn, A., Knörrer, H., Ebene algebraische Kurven. Birkhäuser, Basel, Boston, Stuttgart 1981.

[3] Blaschke, W., Ebene Kinematik. R. Oldenburg, München 1956.

[4] Bricard, R., Mémoire sur la théorie de l'octaèdre articulé. J. Math. Pures Appl. 3 (1897), 113–150.

[5] Fulton, W., Intersection Theory. Springer Verlag, Berlin, Heidelberg, New-York, Tokyo 1984.

[6] Lazard, D., Stewart platforms and Gröbner basis (preprint) (1992), 1–14.

[7] Merlet, J.-P., Singular configuration of parallel manipulators and Grassman geometry. In: Geometry and Robotics (ed. by J.-D. Boissonnat et J.-P. Laumond). Lecture Notes in Computer Science 391, Springer Verlag, Berlin, New-York, London, Paris, Tokyo, HongKong 1989.

[8] Stewart, S., A platform with 6 degrees of freedom. Proc. of the Institution of Mechanical Engineers Vol. 180 (1965-66), Part 1, number 15, 371–386.
[9] Wunderlich, W., Ebene Kinematik. Bibliographisches Institut, Mannheim 1970.

Construction of real algebraic nodal plane curves with given topology

Francisco Santos

Abstract. We study a constructive method to find an algebraic curve in the real projective plane with a (possibly singular) topological type given in advance. Our method works if the topological model T to be realized has only double singularities. In that case, it gives an algebraic curve of degree $4N + 2K$ or $4N + 2K - 1$, where N and K are the numbers of double points and connected components of T.

The construction is based on a preliminary topological manipulation of the topological model and then some perturbation techniques to obtain the polynomial defining the algebraic curve. Some algorithmic remarks are given.

1991 Mathematics Subject Classification: 14P25, 14Q05.

1. Introduction

Throughout this paper we will use the term *algebraic curve* or simply *curve* as an abbreviation for *real projective plane algebraic curve* meaning by this a real homogeneous polynomial $f \in \mathbb{R}[X, Y, Z]$ in three variables, considered up to a constant factor. Sometimes, by abuse of language, we will call curve the zero set $V(f)$ of such a polynomial in the real projective plane $\mathbb{RP}^2$. Two subsets V and W are said to have the same *topological type* if there exists a global homeomorphism of the projective plane into itself sending V to W. Note that this condition is stronger than V and W being homeomorphic.

We want to find an algebraic curve f whose zero set has the same topological type of a certain given $T \subseteq \mathbb{RP}^2$. The conditions that T must satisfy for this to be possible are contained in the following definition.

Definition 1.1. Let T be a subset of $\mathbf{RP}^2$. We say that T is a *topological model* for an algebraic curve if it has the topological type of some algebraic curve. This is equivalent to T being homeomorphic to a graph with an even (possibly zero) number of edges incident to each vertex.

Partially supported by DGICyT PB 92/0498-C02 and "POSSO", ESPRIT-BRA 6846.

We say that an algebraic curve f *realizes* T if its zero set $V(f)$ has the same topological type as T.

The equivalence between the two definitions of a topological model can be found, for example, in [BCR]. There, the second one is stated locally. The translation to our global statement is easy because of compactness of the projective plane. We want to remark that such a characterization of the topology of real algebraic sets is far from trivial in higher dimensional cases (cf. again [BCR], or [AK]) and that, even in the plane, the usual proofs of it use polynomial approximation of C^∞ functions and thus say nothing about the degree needed to *realize* a given topological model with an algebraic curve. Our method would give a new, constructive proof of the characterization.

Nevertheless, we are only able to work out the case of double singularities (see the remark just before Proposition 2.2 for details). For this reason our topological models will always be supposed to have only order 2 singular points, where we define the *order* of a point P in T to be half the number of edges incident to P if P is a vertex of the graph and 1 if P lies on an edge. A *singular point* is a point of order at least 2. Note that, here, "singular" has just a topological meaning. To state our main result (Theorem 1.3, which is a paraphrase of Theorem 3.7) let us first introduce some definitions.

Definition 1.2. Let l be an embedded circle in $\mathbb{RP}^2$. Then, l has the topological type of either a line or a circle and is called a *pseudo-line* or an *oval*, respectively.

Let T be a topological model in $\mathbb{RP}^2$ and let l be a pseudo-line transversal to T. We will say that T is *even* (resp. *odd*) if T and l have an even (resp. odd) number of intersections. The definition does not depend on the choice of l and is a topological type invariant of T.

A topological model T in $\mathbb{RP}^2$ is called *orientable* if it does not contain any pseudo-line or, equivalently, if there exists a pseudo-line not intersecting it.

Theorem 1.3 (cf. 3.7). *Let T be a topological model in $\mathbb{RP}^2$ with only singular points of order 2. Let N be the number of singular points of T and K its number of connected components. Then T can be realized with an algebraic curve of degree $4N + 2K$ if T is even, or $4N + 2K - 1$ if T is odd.* □

The first question that arises is how good the degree we obtain is. For non-singular curves our theorem gives degree $2K$ or $2K - 1$, which is trivial (for we can construct any non singular model as a product of K circles, or may be $K - 1$ circles and a line) but also optimal in the worst case (if the model consists on K nested ovals it can not be realized with an algebraic curve of degree lower than $2K$).

For singular curves, Figure 1 shows a connected topological model with 3 double points that cannot be algebraically realized with degree lower than 8. The example easily generalizes to any number of double points and to non connected models (inserting connected components one inside another), giving:

For any given sequence of numbers $n_1, \ldots, n_K$ there exists a topological model with K connected components having $n_1, \ldots, n_K$ double points on each that cannot be algebraically realized with degree lower than $d = 2\sum_1^K n_i + 2K$.

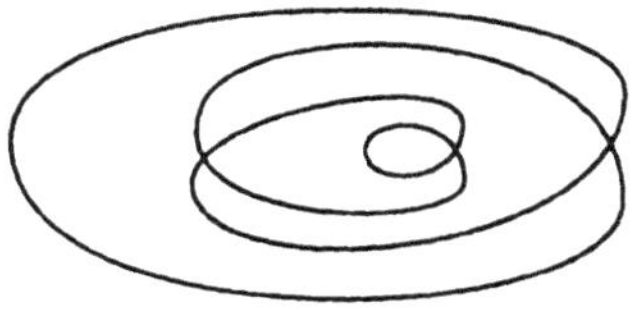

Figure 1

Thus, the degree we obtain is quite good (at most twice the optimal). We have reasons to think that the degree with which any topological model with only double points can be realized is actually this degree $2N + 2K$. In fact, in [GS] we presented (jointly with A.G. Corbalán) a draft of another method for costructing algebraic curves with given topology (in the affine plane instead of the projective one), that gave exactly that reachable bound. That construction had a conjectural step that we could not prove to work, but we are currently studying how to use some of its ideas in a different way.

Let us remark that, on the other hand, Bezout's and Harnack's theorems give lower bounds for the degree needed to algebraically realize *any* topological model. This bounds go, roughly speaking, with the square roots of N and K. Moreover, they are optimal for *some* topological models of any degree.

Our construction of an algebraic curve from a topological model consists of two parts, respectively detailed in Sections 2 and 3. We sketch here the contents of each. Section 4 is devoted to study some algorithmical aspects of the construction.

First, the problem is easily reduced to the connected case. We then make some topological desingularization of the connected topological model T to obtain what we call a *skeleton* of T. The skeleton can be put in a *standard form* that consists of a circle (or a straight line, depending on the parity of T), that we call the *core* of the skeleton, and some segments or arcs of conics –to be called *bonds*– joining different pairs of points on it. See Figure 2 for an example (bonds are the lighter, grey lines). Bonds are disjoint and represent the places where there was a singular point of T. The topological type of T can be recovered from its skeleton by substituting all bonds by double crossings.

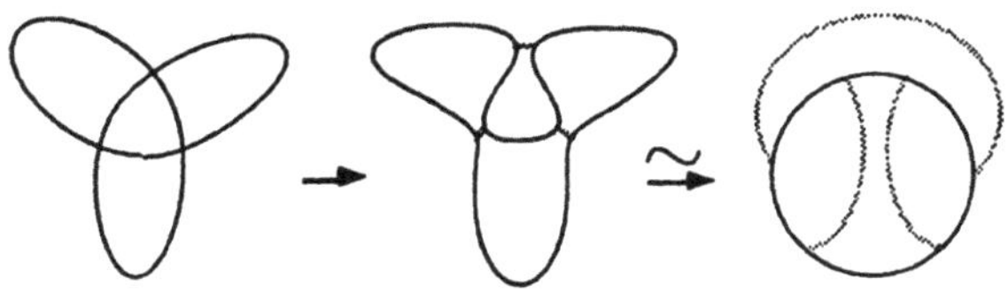

Figure 2

In the second part we introduce an eight shaped curve, or *lemniscata*, along each bond. These lemniscatae are placed so not to intersect to one another and tangent to the core at the ends of the bonds (see Figure 3.a). Each lemniscata is constructed with degree 4. Thus, the product of the core and the lemniscatae is an algebraic curve of degree $4N + 2$ (or $4N + 1$, depending on the parity of T) which has "almost" the same topological type as T. A slight perturbation of this algebraic curve will give us an algebraic curve with the same topology as T (as in Figure 3.b). The perturbation techniques used here (cf. Theorem 3.3) are similar to some classically used in the constructive part of Hilbert's 16th problem (cf. [Vi]) and are based on some results by Gudkov (cf. [Gu]).

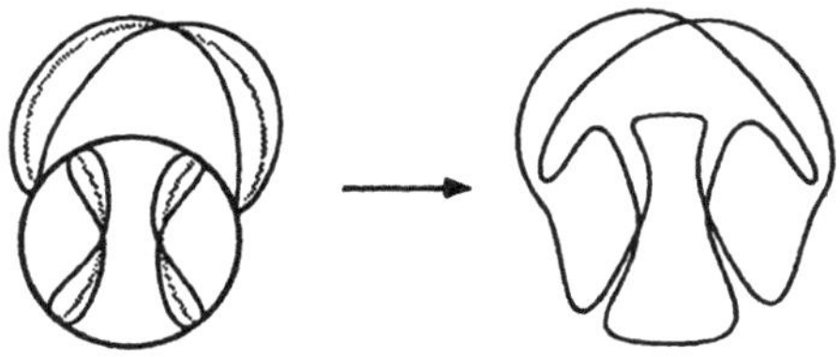

Figure 3

We would like to thank the referee for his attentive reading and useful comments on the first version of this paper.

2. Topological manipulation of the model

Throughout this section T will be a topological model for an algebraic curve in $\mathbb{RP}^2$ with only order 2 singular points. Let N denote the number of singular points of T. Next lemma will allow us to assume, moreover, that T is connected.

Lemma 2.1. *Let T be a topological model in $\mathbb{RP}^2$ and suppose that each of its connected components $T_1, \ldots, T_K$ is realizable with an algebraic curve f_i of degree d_i. Suppose, moreover, that whenever T_i is orientable the corresponding f_i does not intersect the infinity line of the projective plane.*

Then the whole model T can be realized algebraically with a curve f which is the product of a scaled translation of each f_i. Thus f has degree $\sum d_i$.

Proof. First of all, it is easy to check that any topological model has at most one non-orientable connected component. This is so because any non-orientable connected component contains a pseudo-line and two different pseudo-lines always intersect to one another.

Now, the condition that the orientable components are realized without intersecting the infinity line (i.e. in the affine part of the projective plane) implies that we can make these realizations as small as we want, just contracting the affine part of $\mathbb{RP}^2$. We can afterwards translate this contracted curves anywhere in $\mathbb{RP}^2$ by a projective translation, and none of these two operations will change the degree of the curve.

To realize the whole model T we realize first the non-orientable component (if any) and then place the realization of each orientable one, sufficiently reduced, in the appropiate place to have a curve with the topological type of T. This curve –the product of the curves realizing each component– will have degree $\sum d_i$. □

Thus, our aim is to realize T (supposed connected) with degree $4N+2$ if it is even and $4N+1$ if it is odd. This will be achieved in Theorem 3.6.

To start, let as assume that T is not a single point (which can trivially be realized with the curve $X^2+Y^2=0$). The condition that T is a connected graph with even number of edges at every vertex implies that it has an eulerian cycle, i.e., a cycle passing once through each edge. Moreover, the next lemma permits to assume that the eulerian cycle has no proper self-crossings. By this we mean that whenever the cycle passes through a vertex P, the ingoing and outgoing edges of the cycle are consecutive in the circular ordering of the edges incident to P.

Incidentally, let us remark that Proposition 2.2 is not true if the topological model has multiple points of order higher than 2. (The reader can convince himself of this by just considering a model consisting on three circles tangent to one another at the same point). In our opinion, this is the main reason why our method cannot be applied for points of higher order. Of course, the algebraic constructions to be described in Section 3 would also be more complicated in this case.

Proposition 2.2. *Let T be a topological model connected and with only double points. Then, T has an eulerian cycle that does not properly cross itself.*

Proof. Suppose that we have an eulerian cycle C on T with some proper crossings. Choose one of these crossings, at a vertex P. Call a, b, c and d the four edges incident to P, in their circular ordering (i.e. a opposite to c and b opposite to d). Without loss of generality the eulerian cycle on T is $C = ac\delta bd\gamma$, where δ and γ represent lists of edges in T. Then, the cycle $C' = ab\delta^{-1}cd\gamma$ is also eulerian and contains one proper crossing less than C. Repeating the process at every proper crossing of C we will obtain an eulerian cycle with no proper crossings. □

Once we have an eulerian cycle on T with only non-proper crossings, we can slightly perturb the cycle at each vertex in order to *desingularize* T. In this way we obtain a topological model T' which is very close to T, but non-singular, and still connected. Let us clarify this with an example. Let $T \subseteq \mathbb{RP}^2$ be the topological model in Figure 4.a and consider the eulerian cycle $C = abcfed$ which has no proper crossings. In Figure 4.b we show the nonsingular model T' obtained perturbing T according to C.

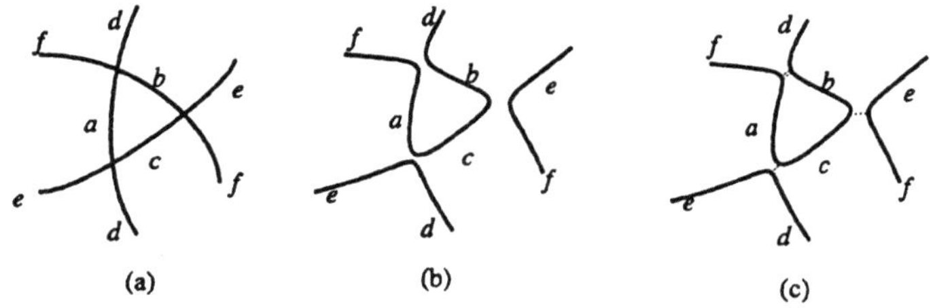

Figure 4

In order to be able to recover the topology of T from T' we introduce small lines joining each two points of T' that were identified in T. The final result T^* (as shown in Figure 4.c) will be called a *skeleton* of the topological model T. The non-singular model T' obtained separating the double crossings of T will be called the *core* of the skeleton and the lines joining separated points will be called *bonds.* In all our figures bonds will be represented as lighter, grey lines.

The definition of topological type for skeletons is the obvious one. Two skeletons T_1^* and T_2^* have the same *topological type* if there is an homeomorphism of the projective plane in itself sending the core and bonds of T_1^* respectively to the core and bonds of T_2^*.

Our next result says how to obtain from the skeleton T^* another skeleton with the same topological type than T^*, but made of a circle or a straight line for the core and line segments or arcs of conics for the bonds. This will make the following steps of our algorithm (the algebraic constructions) easier. A skeleton in the conditions of Proposition 2.3 will be said to be in *standard form.* In Figure 6 we have some examples of skeletons in standard form and the data below the figures are combinatorial descriptions of their topology. We will come back to this point in Section 4.

Proposition 2.3. *Let T be a connected topological model with only double points. Then, any skeleton of T has the same topological type than a skeleton T^* whose core is either the X-axis or the unit circle (depending on the parity of T) and such that:*

i) *if T is even, each bond of T^* is either a straight line segment joining two opposite points in the unit circle and passing through infinity, or an arc of circle joining two points of the unit circle (either through the outside or the inside) and perpendicular to it;*

ii) *if T is odd, each bound of T_0^* is either a half-circle or an arc of hyperbola joining two different points of the straight line.*

Proof. It is clear that the desingularization process that goes from T to the non-singular model (the core T') does not change parity, because it can be made outside a certain pseudo-line. In the other hand, T' is a non-singular, connected topological model, and thus it is either a pseudo-line or an oval, depending on its parity. This implies that we can find a global homeomorphism of the projective plane into itself sending the core of T^* to either the unit circle or the infinity line.

Next, it is easy to show that with another global homeomorphism (in fact, an isotopy) we can put the bonds of T^* in such a way that they cross the infinity line at most once each. In fact, suppose that the bonds of T^* are transversal to the infinity line and that they have a minimal number of crossings to the infinity line among all possible skeletons with the same topological type. In these conditions, if one of the bonds has two different crossing points P and Q with the infinity line, an isotopy in $\mathbb{RP}^2$ can make the arc PQ of the bond traverse the infinity line, thus decreasing by two the number of crossings. This contradicts the hypothesis.

Now, the topological type of the skeleton is determined by its parity, the pairs of points on the core (which is supposed to be the X-axis or the unit circle) joined by bonds and, for each pair (say (P, Q)), a letter (a), (b), (c) or (d) corresponding to the four possible ways to join P to Q with a bond (up to topological equivalence). These four ways are illustrated in Figure 5(i) for the even case and Figure 5(ii) for the odd case.

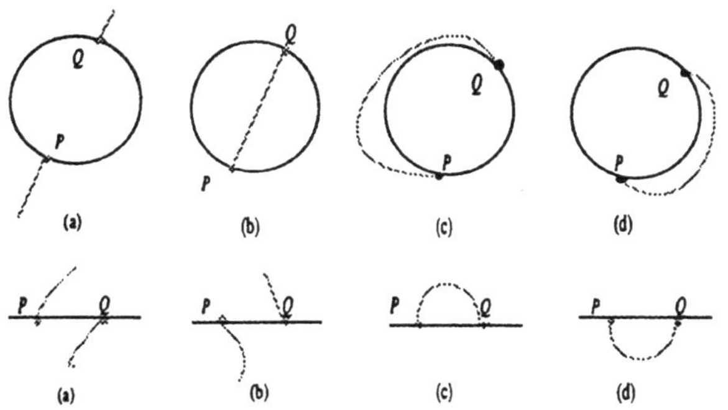

Figure 5

It just remains to move the bonds to be as in the statement of the theorem. The way to do this is explained separately for the even and the odd cases.

In the even case, we first make each pair of points joined with bonds of type (a) to be actually opposite points in the circle and its bond to be the line segment going from one to the other through infinity (moving some of the other points and bonds if necessary). Then, each of the remaining bonds (those of types (b), (c) and (d)) is substituted with the corresponding arc of circle normal to the unit circle and joining its two end-points. This process always produces disjoint bonds and does not change the topological type of the skeleton. See Figure 6.a for an example.

A small remark: if there is no (a) type bond we must be aware of the distinction between (c) and (d) bonds, and place the points in the circle in such a way that the arc of normal circle joining them is the right (c) or (d) type. Otherwise this distinction is irrelevant because only one of the two types (c) or (d) will be compatible with the fact that bonds do not intersect each other.

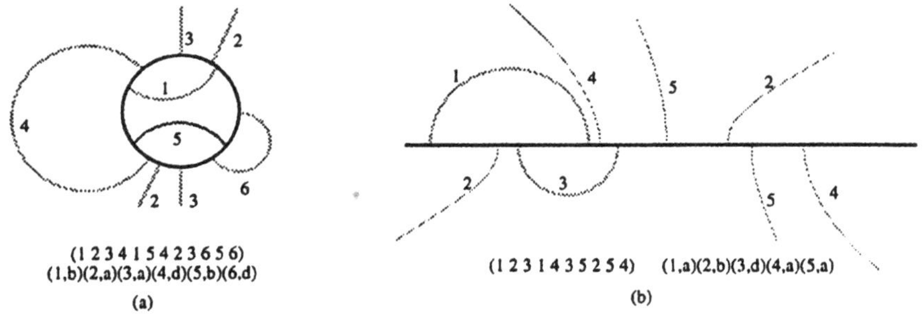

Figure 6

In the odd case, we first substitute bonds of types (c) and (d) with half-circles and then each of those of types (a) and (b) with an arc of a hyperbola whose axis are the X-axis and a vertical line passing through the midpoint of the two points to be joined. Figure 6.b shows an example. If the assymptotic lines of these hyperbolae are taken sufficiently sloped then they will not intersect the half-circles. If the exterior ones are taken more sloped than the interior ones they will not intersect each other. □

3. Construction of the algebraic curve

Throughout this section T^* is a skeleton in standard form, obtained from a topological model T as shown in Section 2. As mentioned in the introduction, the way to construct an algebraic curve which has the topological type of T will be to insert an eight shaped curve (or *lemniscata*) along each bonding line of the skeleton (tangent to the core at the end points) and then algebraically perturb the product of the core and all the lemniscatae. The insertion of lemniscatae is done in the following result.

Proposition 3.1. *For each type of bonding line described in Proposition* 2.3, *it is possible to insert an algebraic curve l of degree* 4 *with the topological type of a lemniscata in the following conditions.*

i) *l is tangent to the core at the ends of the bond.*

ii) *l is as close to the bond as we want.*

iii) *The singular point in l is a transversal crossing (i.e. it has two analytic branches which are real, non-singular and with different tangents).*

Proof. We consider each case separately:

• bonds of types 'c' and 'd', in the odd case:

The core is the X-axis and, without loss of generality, we suppose that the bond is the upper half unit circle (for case 'c', case 'd' is analogue).

Let a be any real positive constant, and consider the curve

$$f_0(X,Y,Z) = (X^2+Y^2-2aXZ-Z^2)(X^2+Y^2+2aXZ-Z^2),$$

which is the product of two circles with centers at $(a,0)$ and $(-a,0)$, both passing through the points $(0,-1)$ and $(0,1)$ (Figure 7.a). If we perturb f_0 in the form

$$f_\varepsilon(X,Y,Z) = f_0(X,Y,Z)+\varepsilon(Y-Z)^2Z^2=(X^2+Y^2-Z^2)^2-4a^2X^2Z^2+\varepsilon(Y-Z)^2Z^2,$$

with $\varepsilon > 0$, the resulting curves have the following properties, all easy to check (see Figure 7.b):

1) they have a transversal crossing at $(0,1)$,

2) they do not have any other singular point,

3) they lie in the region where $f_0(X,Y,Z)$ is negative, i.e. the two small regions between the circles that compose f_0.

4) they have the topological type of a lemniscata ("eight" shaped curve).

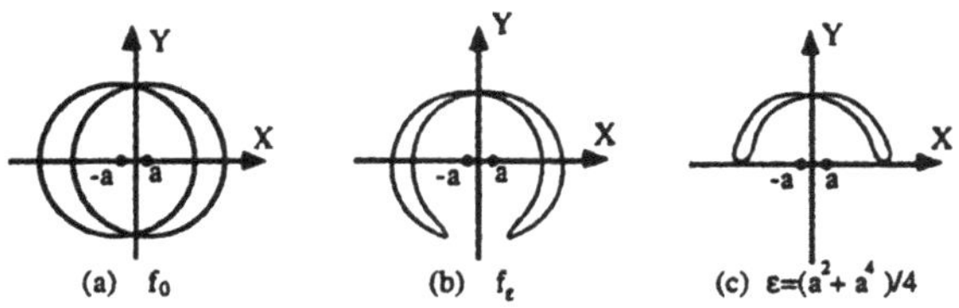

Figure 7

If the parameter a is chosen sufficiently small the curves f_ε are as close to the unit circle as we want, because of property (3). All that we need to do is to find some ε such that f_ε is tangent to the X-axis at both sides and does not intersect it elsewhere (Figure 7.c). This is obtained making $f_\varepsilon(X,0,1)$ to have two double roots, which happens for $\varepsilon = (a^2+a^4)/4$.

• bonds of types 'a' and 'b', in the odd case:

Without loss of generality we assume that the hyperbola is the standard $X^2 - Y^2 - Z^2 = 0$. By a projective change of variables $X \leftrightarrow Z$ the arc of hyperbola becomes a half unit circle and we reduce this case to the first one.

• bonds of types 'b', 'c' and 'd', in the even case:

The bond is an arc of circle normal to the unit circle. By affine transformations we can inverse the situation, and suppose that the bond is an arc of the unit circle joining two points of another circle C normal to it. We can assume moreover that this latter circle has its center on the Y-axis, and that the bond is the upper arc of the unit circle, as in Figures 8.a and 8.b. The general equation of such a circle C is $C(X,Y) = X^2+Y^2-2YZc+Z^2$, where $(0,c)$ is its center, and $\sqrt{c^2-1}$ its radius.

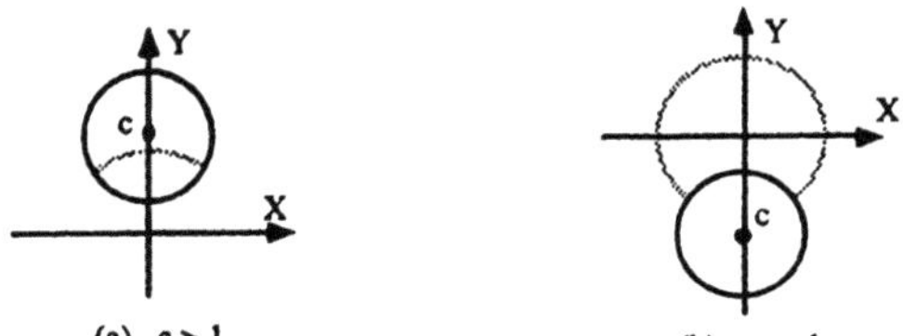

Figure 8

Our situation is now similar to the first case and we can use the f_ε lemniscatae that we defined there, except that now we want f_ε tangent to the circle C. The parameter a in the equation of f_ε is again fixed in advance and as small as we need it. It is easy to check that taking

$$\varepsilon = (a^2 + a^4)(c-1)/(c+2+a^2+1)$$

the curve f_ε is tangent to the circle C at both ends.

- <u>bonds of type 'a', in the even case:</u>

Without loss of generality we suppose that the bond is the part of the X-axis outside the core (which is the unit circle). By a projective change of coordinates $X \leftrightarrow Z$ the bond becomes the segment $[-1,1]$ of the X-axis and the core the standard hyperbola. We can take then our lemniscata to be $(X^2+Y^2/a^2)^2 - X^2Z^2 + Y^2Z^2/a^2$, for sufficiently small a. □

We now need to introduce a technical result (Theorem 3.3) on perturbation of algebraic curves, in order to change their topology in a controlled way. Let us first give some definitions.

Definition 3.2. Let f and g be two algebraic curves of the same degree. A *perturbation of f by g* is any curve of the form $f+\varepsilon g$, with $\varepsilon \in \mathbb{R}_{\neq 0}$. We say that a property is true for *small perturbations of f by g* if there exists an ε_0 such that the property is true for any perturbation of f by g with $|\varepsilon| \leq \varepsilon_0$.

We say that two curves f and g are *transversal* to one another if they intersect only at non singular points of both of them and they have different tangents at these intersection points.

We shall be concerned with curves having only double singularities. More precisely, all the singularities in our algebraic curves will consist on two real non-singular analytic branches crossing to one another at the singular point. We call such singularities *crossings*. If the two analytic branches have different tangents we will say that the crossing is *transversal* (or *non-degenerate*), and *degenerate* otherwise.

Theorem 3.3. *Let f be an algebraic curve whose singularities $Sing(f)$ are all crossings. Let D be a subset of $Sing(f)$ consisting only on transversal ones. Let g*

be a curve of the same degree which has a singular point at every point in D and intersects f transversally elsewhere.

Then, for any sufficiently small perturbation $f_\varepsilon = f + \varepsilon g$ of f by g,

(i) *f_ε has a transversal crossing at each point of D,*

(ii) *f_ε has no other singular point, and*

(iii) *the topological type of f_ε in $\mathbb{RP}^2$ can be obtained from f dissipating each of the crossings that are not in D in that of the two ways in Figure 9 compatible with the signs of f, g and ε.*

Figure 9

Proof. (i) Let $P \in D$ be one of the transversal crossings of P, and consider an affine system of coordinates around P. P being singular for f and g implies that f, g and their first order derivatives vanish at P. Thus, also $f_\varepsilon = f + \varepsilon g$ and its first derivatives vanish at P and P is a singular point of f_ε. The condition that P is a transversal crossing for f is characterized by the Hessian of f being strictly negative at P. As the Hessian of $f + \varepsilon g$ is continuous with respect to ε, P is also a transversal crossing of $f + \varepsilon g$, for sufficiently small ε.

(ii) and (iii) In the proof we will use two lemmas contained in §2 of [Gu]. Namely the *lemma on the class of a point* and the *lemma on isotopy.*

Call $x_1, \dots, x_n$ the transversal crossings in D, and $y_1, \dots, y_m$ the rest of singular points of f. Take small open balls X_i and Y_j ($i = 1, \dots, n$, $j = 1, \dots, m$) around these points. Call U the complement of the union of the Y_j's. Then, all U, X_i and Y_j are regular domains and f is regularly disposed with respect to them, according to the notation of [Gu]. This means that the boundary of each domain is a finite number of circles, and that the curve f is transversal to this boundary, has no singular point on it, and has a finite number of singular points. In these conditions we can apply the two lemmas.

The lemma on the class of a point says, simplified, that the sum of the classes of singular points in the perturbed curve is lower or equal to the sum in the old one, in any regular domain and for small perturbations. Transversal crossings have all the same class 2 (note: the class of a singular point used by Gudkov equals $m + \mu - 1$, where m is the order of the singular point and μ its Milnor number). Applying this lemma to U we conclude that $f + \varepsilon g$ has no new singular points in U, for small ε. The lemma on isotopy says that if the perturbed curve has the same singular points with the same class, in a certain regular domain where the curve is regularly disposed, then the perturbed curve is isotopic to the original one. In our case this implies that $f + \varepsilon g$ is isotopic to f in U.

It only rests to see what happens in the small circles Y_j around the other singular points y_j. First, $f + \epsilon g$ does not have any singularity there, for small ε. In fact, consider a local affine chart in one of the balls Y_i. The values ε for which

$f+\varepsilon g$ has a singular point are critical values of the rational function $-f/g$, which is defined on Y_j (if Y_j is sufficiently small) because $g(y_j) \neq 0$. By Sard's theorem, $-f/g$ has a discrete (in fact, finite) number of critical values, one of them being 0 because $f(y_i) = f_X(y_i) = f_Y(y_i) = 0$. Thus, for small $\varepsilon \neq 0$ $f+\varepsilon g$ is non-singular in Y_j.

Concerning the topology of $f+\varepsilon g$ in Y_j, if it is non singular the only possibilities are those in Figure 9, except that there could have appeared some new little ovals near the former singular point. In our case this is not possible because the curves $f+\varepsilon g$ are disjoint to one another for different ε's (g being nowhere zero in Y_j). □

In the following we will call f the product of the core of the skeleton T^* (the X-axis or the unit circle) and the lemniscatae introduced in Proposition 3.1 for each of its bonds. f has degree $4N+2$ if T was even and $4N+1$ if T was odd. Theorem 3.3 will allow us to perturb f in such a way that the perturbed curve recovers the topological type of the model T from which T^* was obtained.

Proposition 3.4. *In these conditions, there exists a curve g of the same degree than f, and such that*

(i) *g has a singular point at the transversal crossing of each lemniscata,*

(ii) *g is elsewhere transversal to f,*

(iii) *in a neighbourhood of each tangency point between the core and the lemniscatae, the sign of g coincides with the sign that f takes in the region inside the lemniscata.*

Proof. We do it separately for the even and odd cases.

- the odd case:

Let us consider first the sign conditions for g given in (iii). Let $P_1, \dots, P_{2N}$ be the $2N$ tangency points of the N lemniscatae with the X-axis, in the order they appear along the axis. Let us call *sign of P_i* the sign that f takes in the interior region of the lemniscata in aneighborhood of P_i, i.e. the sign that we want g to have at P_i. We will say that there is a *sign change* between P_i and P_{i+1} if P_i and P_{i+1} have opposite sign, according with the above.

Let $s \leq 2N-1$ be the number of sign changes along the X-axis. We can construct a curve g_1 of degree s which has the appropiate sign at each P_i as the product of s vertical lines, each crossing the X-axis in one of the sign changing intervals. These lines can be chosen transversal to f, and we can complete g_1 with $2N+1-s$ vertical lines at the right of P_{2N} (or the left of P_1) to get a curve (still notated g_1) of degree $2N+1$ with the appropiate sign at each P_i and transversal to f.

We can also find a curve g_2 of degree $2N$ with N singular points at the N transversal crossings of the lemniscatae. Namely

$$g_2 = \prod_1^n ((c_iX - a_iZ)^2 + (b_iX - a_iY)^2 + (b_iZ - c_iY))^2,$$

where (a_i, b_i, c_i) are the coordinates of the N transversal crossings. The factor g_2 is strictly positive everywhere except in those N points and, thus, the product $g = g_1g_2$ satisfies all our conditions.

- the even case:

Now f has degree $4N + 2$.

With the same notation that in the odd case, the number of sign changes along the circle will be $s \leq 2n$, and even. (The points $P_1, \ldots, P_{2N}$ are now in circular order along the unit circle and that we must consider a possible sign change between P_{2N} and P_1). Let us take g_1 to be the product of $s/2$ lines, each crossing the unit circle in two of the sign changing arcs. Again, these lines can be taken transversal to f. g_1 has in each P_i the sign that g must have and we can complete it to have degree $2N + 2$ multiplying it by $2N + 2 - s/2$ lines not intersecting the circle. Take g_2 in the same way as in the odd case, to obtain $g = g_1g_2$ of degree $4N + 2$. □

Proposition 3.5. *In the conditions of Proposition* 3.4 *the perturbed curve $f+\varepsilon g$ of f by g with ε sufficiently small and positive has the same topological type than the topological model T. Moreover, if T was orientable, then $f + \varepsilon g$ does not intersect the infinity line.*

Proof. The singular points of f are the N transversal crossings on each of the lemniscatae and the $2N$ tangency points of the lemniscatae with the core, which are degenerate crossings. It is easy to check that g satisfies all the conditions needed to apply Theorem 3.3 (with D being the N transversal crossings), so the perturbed curve for sufficiently small ε has the same topology of f but with the tangency points "desingularized". ε being positive, the desingularization of the tangency points is made towards the regions in which f and g have opposite signs, i.e. joining the lemniscatae to the core. That produces the inverse topological change that the one by which we obtained the skeleton T^* from the topological model T.

If the topological model was orientable, it is also even and its skeleton in standard form has no (a) type bonds. In this case the lemniscatae are all constructed along arcs of circle, so they do not cross the infinity line. For ε small the perturbed curve will still not touch it. □

This directly implies our main results. Theorem 3.7 follows from Theorem 3.6 using Lemma 2.1.

Theorem 3.6. *Let T be any topological model in $\mathbb{RP}^2$, connected and with only double points. Let N be its number of (topologically) singular points of T. Then T can be realized with an algebraic curve of degree $4N+2$ if it is even or $4N+1$ if it is odd. Moreover if T is orientable its realization can be made without intersecting the infinity line.* □

Theorem 3.7. *Let T be a topological model T in $\mathbb{RP}^2$ with only double points. Let K be its number of connected components, $N_1, \ldots, N_K$ the numbers of double* *points on each and $N = \sum N_i$ the total number of double points. Then, T can be realized with an algebraic curve of degree*

$$\sum_{i=1}^{K}(4N_i + 2) = 4N + 2K,$$

if it is even, or

$$\sum_{i=1}^{K}(4N_i + 2) - 1 = 4N + 2K - 1,$$

if it is odd. □

For the affine topology of plane curves we can only say the following.

Corollary 3.8. *Any compact topological model T in $\mathbb{R}^2$ with only double points can be realized with an algebraic curve of degree $d = 4N + 2K$ where N is the number of singular points in the model and K its number of connected components.*

Proof. Compactness of T in $\mathbb{R}^2$ implies that the corresponding projective model does not touch the infinity line and thus is orientable. Its projective realization with degree $4N + 2K$ can be made without touching the infinity line, and that gives in fact an affine realization of T. □

4. Algorithmic remarks

All along the construction we have not been worried about things like " *...for a sufficiently small* ε *...* " or "*...with a certain isotopy ...* " which make the construction seem not so constructive. Here we are going to analyse in a bit detail each step made to see the difficulties that arise, for example, to implement it on a machine.

a) How can we obtain the skeleton T^* in its final form (i.e. with the bond lines being straight line segments and arcs of conic) from the topological model T?

If we want an algorithm to algebraically realize a topological model with an algebraic curve in $\mathbb{RP}^2$, its input must be a combinatorial description of the topological type. To answer the question we would need to know exactly how this description is designed. A priori, one can think of many different such combinatorial descriptions, but Proposition 2.3 permits to devise one which is specially good for our purposes. The data below the two skeletons in Figure 6 shows this description, which is detailed in Corollary 4.1. Numbers from 1 to N represent the N double points (or, more precisely, the N bonds obtained from them). The letter assigned to each bond specifies its type and the list containing each twice describes the order in which the end points of the bonds appear along the X-axis, or the circular order along the unit circle. Any other natural description of the topology should be easily translatable into this one and, in fact, the first step of any algorithm using our construction with any other input would be performing this translation.

Corollary 4.1. *The topological type of a connected topological model T with N singular points of order 2 can be combinatorially described with the following data:*
- *the parity of T;*
- *a list which contains twice each number from 1 to N, and*
- *a letter 'a', 'b', 'c' or 'd' assigned to each number from 1 to N.* □

From these data we can easily construct the arcs of conic and segments which form the skeleton: for the odd case we can just take the coordinates of the touching points for the bond lines along the X-axis to be the first $2N$ integers, and then find the suitable half-circles and half-hyperbolae to join them. The even case requires a bit more care with the choosing of the points, but is also easy.

b) Can we effectively find the lemniscatae along the bonds, and the perturbing curve g?

The proof of Proposition 3.1 gives a method to find lemniscatae *"as close to the bond as we want, if a is chosen sufficiently small"*. The problem is to know how small a needs to be. Parameter a is, roughly speaking, the maximal separation of the lemniscata from the bond line, and the maximal separation permitted can be taken as half the minimum distance from one bond to another. This minimum distance is not difficult to find from the construction of the bonds.

To obtain the perturbing curve g is easy: in Proposition 3.4 g is the product of two parts g_1 and g_2. To find g_2 we just need to know the coordinates of the transversal crossing on each lemniscata, and to find g_1, a point in each of the 'sign changing' intervals of f along the X-axis or the unit circle. Both things are easily found either by construction of the lemniscatae or by standard methods.

c) How small does the ε in the perturbation $f + \varepsilon g$ need to be?

To find this ε is, without any doubt, the most difficult part of the algorithm in terms of efficiency. We can think of two possible ways to do it.

(1) Choose a "reasonably small" ε and compute the topology of $f + \varepsilon g$ to see if it coincides with the one we want. If it does not divide ε by 2 (or by a "reasonable" constant) and start again. This method seems not very efficient (in fact it is not easy to find a complexity bound for it). We must say in its defense that there exist rather fast implementations of algorithms for computing the topology of algebraic curves (cf. [Ho,][Ro]), while the method to be described below uses very hard machinery, in terms of efficiency.

(2) We know that any ε sufficiently small and positive will work. Let ε_0 be the infimum of the ε's that will not. Then, $f + \varepsilon_0 g$ has different topological type from $f + \varepsilon g$ for any $0 < \varepsilon < \varepsilon_0$ and this would not be possible (because of Theorem 3.3 on perturbations) if $f + \varepsilon_0 g$ has only the N transversal crossings of the lemniscatae. Thus, we must look for an ε_0 such that either some new singular points appear

in $f + \varepsilon_0 g$ or some of the non-degenerate 2-fold points become degenerate (or of higher order).

The second thing can be checked developping $f + \varepsilon g$ as a Taylor polynomial around each transversal crossing, and studying the degree-two terms. This is not very complicated because the points where a double crossing occurs are, up to some extent, chosen by us. In particular they can be taken rational.

The first one is more difficult, but can be made as follows. A singular point of $f + \varepsilon g$ is a common zero of it and of $f_X + \varepsilon g_X$ and $f_Y + \varepsilon g_Y$, i.e. a point where

$$f/g = f_X/g_X = f_Y/g_Y = \varepsilon.$$

Such points are exactly the common zeroes of $fg_X - gf_X$ and $fg_Y - gf_Y$ (exception made of the singular points in the lemniscata) and can be found by standard methods in real computer algebra. It suffices then to evaluate f/g in all these points and take for ε_0 the lowest positive value. In fact we require only a lower estimate of the value of ε_0 and thus numerical methods can be used with some control on the errors.

Note that this method could also be used to give a theoretical lower bound for ε_0 in terms of the number N of double points (explicitly describing a way to construct the lemniscatae and the perturbing curve g) but this bound would probably be of little practical use.

A third method could be to make a deeper theoretical study of the perturbation $f + \varepsilon g$ in the particular case of the curves f and g obtained in our construction, which seems rather difficult. Some algorithmic experimentation of our construction could be the subject of a future work.

References

[AK] S. Akbulut, H. King, *Topology of Real Algebraic Sets.* Springer-Verlag (1992).

[BCR] J. Bochnak, M. Coste, M.F. Roy, *Géometrie Algébrique Réelle.* Springer-Verlag (1987).

[Gu] D.A. Gudkov, The Topology of Real Projective Algebraic Varieties. Russian Math. Surveys 29 (1974), 1–79.

[GS] A. González-Corbalán, F. Santos, Representation of Curves in the Real Plane, and Construction of Algebraic Curves with Given Topology. Seminaire DDG sur les Structures Algébriques Ordonnées, Publ. Math. de l'Université de Paris VII (1992).

[Ho] H. Hong. Efficient Method for Analyzing Topology of Plane Real Algebraic Curves. In: IMACS SC-93, Lille 1993.

[Ro] M.F. Roy, Computation of the Topology of a Real Algebraic Curve. (In: *Proceedings of Computational Geometry and Topology*), Sevilla 1987.

[Vi] O.Ya. Viro, Real Algebraic plane curves: Constructions with controlled topology. Leningrad. Math. J. 1 (5) (1990), 1059–1134.

Purity theorems for real spectra and applications

Claus Scheiderer

Abstract. The following purity theorem is proved. Let A be an excellent regular ring and $\mathfrak{p}$ a prime ideal of A of height c such that $A/\mathfrak{p}$ is regular. If F is a locally constant sheaf on the real spectrum $\operatorname{sper} A$ then the sheaf $\underline{H}^i_{\operatorname{sper} A/\mathfrak{p}}(F)$ of cohomology with supports is for $i = c$ locally isomorphic to the restriction $F|_{\operatorname{sper} A/\mathfrak{p}}$, and is 0 for $i \neq c$. As an application we construct, on every excellent regular noetherian scheme X, a Gersten type resolution of the sheaf $\mathbb{Z}/2$ on the real spectrum of X. This resolution is used for a new construction of fundamental classes and cycle maps with values in $\mathbb{Z}/2$-cohomology of the real spectrum. As another application we give a new proof to a well-known theorem about nullhomologous divisors.

1991 Mathematics Subject Classification: 14P99, 14F20, 14C20.

Introduction

Let M be an n-dimensional topological manifold and $N \subset M$ a closed (locally flat) submanifold of dimension $n-c$. Then every $y \in N$ has an open neighborhood U in M such that $H^c_{N\cap U}(U,\mathbb{Z}) \cong \mathbb{Z}$ and $H^i_{N\cap U}(U,\mathbb{Z}) = 0$ for $i \neq c$. This can be seen by applying Poincaré duality, which shows $H^i_{N\cap U}(U,\mathbb{Z}) \cong H_{n-i}(N\cap U,\mathbb{Z}) \cong H^{i-c}(N\cap U,\mathbb{Z})$ ($i \in \mathbb{Z}$) whenever U and $N \cap U$ are oriented; here H_* denotes Borel-Moore homology. In particular, the sheaf $\underline{H}^i_N(M,\mathbb{Z})$ on M is non-zero only for $i = c$, and $\underline{H}^c_N(M,\mathbb{Z})$ is locally isomorphic to the constant sheaf $\mathbb{Z}$ on N, and can be identified with the orientation sheaf on N.

It is the purpose of the first section of this paper to prove analogues of these facts for real spectra of regular noetherian schemes. For technical reasons we have to assume that the schemes are excellent. Then everything works out as expected, see Theorem (1.7). It is a particular case (Corollary (1.11)) that, if A is an excellent regular ring and $\xi \in \operatorname{sper} A$, the pointed set of generalizations of ξ has the cohomology of a pointed c-dimensional disk, where c is the height of $\operatorname{supp}\xi$ in A. Of course, also the general problem is a local one, and an easy induction shows that one can restrict to the codimension one case. Here one first treats the case where the local ring is complete, and then tries to get control over general local rings by relating them to their completion. At one point we are using resolution of singularities, but excellence is also needed to apply Ruiz's theorem on real spectrum formal fibres.

We also consider a relative situation, proving a similar purity theorem for smooth closed pairs of schemes over some base scheme.

The remainder of the paper is devoted to applications. In §2 we produce, for every excellent and regular noetherian scheme X, a canonical resolution of the constant sheaf $\mathbb{Z}/2$ on the real spectrum X_r. The q-th component of this resolution is the direct sum, over all points $x \in X$ of codimension q, of the sheaves $(i_{x_r})_*\mathbb{Z}/2$ on X_r. Here i_{x_r} denotes the inclusion $\operatorname{sper}\kappa(x) \hookrightarrow X_r$ of the real spectrum of the residue field $\kappa(x)$ into X_r. The sheaves $(i_{x_r})_*\mathbb{Z}/2$ are acyclic on X_r, and so the resolution leads to natural complexes which calculate the cohomology (resp. cohomology with support) of X_r with coefficients $\mathbb{Z}/2$. We call this resolution "of Gersten type" since it is formally analogous to the well-known Gersten resolution in algebraic K-theory. Our resolution resp. the complexes deduced from it seem to be new even for $\mathbb{R}$-varieties, where the required purity results are of course folklore. In the case of $\mathbb{R}$-varieties our complex calculates Borel-Moore homology of $X(\mathbb{R})$ even for X singular (quasi-projective), up to a change of degrees. As we shall indicate, one can deduce this complex for smooth $\mathbb{R}$-varieties from the Bloch-Ogus resolutions for the sheaves $\mathcal{H}^n(\mu_2)$, by passing the latter to the limit $n \to \infty$ in a suitable way.

In §3 we show that the resolutions from §2 allow a very easy and natural construction of fundamental classes and cycle maps, for excellent regular noetherian schemes, with values in $\mathbb{Z}/2$-cohomology of the real spectrum. It is shown that the so defined cycle map passes through rational equivalence. In the case of $\mathbb{R}$-varieties and of compact real analytic manifolds one recovers the fundamental class as defined by Borel-Haefliger.

Finally in §4 we use our construction to add yet another proof to the following well-known theorem (which originally is due to Bröcker): On a smooth projective $\mathbb{R}$-variety X every divisor whose class in $H^1(X(\mathbb{R}), \mathbb{Z}/2)$ vanishes is linearly equivalent to a divisor which supports no real point. Our proof is simple, purely algebraic and uses a little fan theory, while the existing proofs use theorems for approximation of continuous data by algebraic data (like Stone-Weierstraß). Due to our construction of the cycle map we can prove the theorem in somewhat greater generality.

Notations. Following are some basic notations we are going to use; others will be added in the course of the paper.

If T is a topological space and $\eta, \xi \in T$, we write $\eta \succ \xi$ to indicate that $\xi \in \overline{\{\eta\}}$; then ξ is called a specialization of η, and η a generalization of ξ. One write $\operatorname{Gen}\xi$ or $\operatorname{Gen}_T \xi$ for $\{\eta \in T\colon \eta \succ \xi\}$.

All sheaves are sheaves of abelian groups, all sheaf cohomology is derived functor cohomology except when explicitly stated otherwise. If M is an abelian group then M may also denote the corresponding constant sheaf. We put $\mathbb{Z}/2 := \mathbb{Z}/2\mathbb{Z}$. The stalk of a sheaf F at $\xi \in T$ is written F_ξ. If $i\colon S \hookrightarrow T$ is the inclusion of a closed subspace then $H_S^*(T, F)$ denotes sheaf cohomology with support in S. Moreover

$i^! F$ is the sheaf $S \cap U \mapsto \ker(F(U) \to F(U - S))$ on S ($U \subset T$ open). In the literature one also finds the notations $\underline{H}^p_S(F)$ or $\underline{H}^p_S(T, F)$ instead of $i_* \mathsf{R}^p i^! F$, but we prefer the latter.

If A is a ring then spec A is its Zariski spectrum. The residue field of $\mathfrak{p} \in \operatorname{spec} A$ is denoted $\kappa(\mathfrak{p})$, and similarly on schemes. The height of $\mathfrak{p}$ is $\operatorname{ht}(\mathfrak{p}) := \dim A_{\mathfrak{p}}$. The real spectrum sper A of A is the (spectral) space of all pairs $\xi = (\mathfrak{p}, \alpha)$ with $\mathfrak{p} \in \operatorname{spec} A$ and α an ordering of $\kappa(\mathfrak{p})$. The support map supp: $\operatorname{sper} A \to \operatorname{spec} A$, $\xi \mapsto \mathfrak{p}$, is continuous (in fact, spectral). One forms the real spectrum X_r of a scheme X by glueing the real spectra of its open affine subschemes. Any scheme morphism $f: Y \to X$ induces a map $f_r: Y_r \to X_r$ of the real spectra. If $x \in X$ we frequently write x_r for $\operatorname{sper} \kappa(x)$.

If A is a local ring then $\mathfrak{m}_A$ denotes its maximal ideal. One says that A is a strictly real local ring if A is henselian and $\kappa_A = A/\mathfrak{m}_A$ is real closed. Given any local ring A and an ordering ξ of κ_A there is a unique strict real localization (or henselization) A_ξ of A with respect to ξ; and $A \to A_\xi$ induces a homeomorphism from sper A_ξ onto Gen ξ. For these facts see [AR], for example.

Acknowledgement. I am indebted to J.-L. Colliot-Thélène for helpful discussions and suggestions concerning these and related matters.

1. Purity theorems for real spectra

(1.1) Lemma. *Let k be a real closed field, and let $A := k[\![x_1, \dots, x_d]\!]$ where $d \geq 1$ is an integer. Then $\operatorname{sper} A[x_d^{-1}]$ has precisely two connected components, and $H^q(\operatorname{sper} A[x_d^{-1}], M) = 0$ for $q \geq 1$ and every abelian group M.*

Proof. From the Weierstraß preparation theorem it follows that, given an arbitrary finite family $(a_1, \dots, a_N)$ of elements of A, one can find $x'_1, \dots, x'_{d-1} \in \mathfrak{m}_A$ such that $(x'_1, \dots, x'_{d-1}, x_d)$ is again a regular sequence of parameters for A, and such that there are polynomials $p_i \in k[\![x'_1, \dots, x'_{d-1}]\!][x_d]$ and units $u_i \in A^*$ with $a_i = u_i p_i$, $i = 1, \dots, N$.

To prove the lemma we use Čech cohomology. Recall that Čech cohomology agrees with derived functor cohomology on real spectra of rings [CC, Prop. 5]. Fix $q \geq 0$ and $\alpha \in H^q(\operatorname{sper} A[x_d^{-1}], M)$. There is a finite covering $\mathcal{V} = \{V_i\}_{i \in I}$ of $\operatorname{sper} A[x_d^{-1}]$ by open constructible subsets such that α is represented by a Čech cocycle $(m_{i_0 \dots i_q}) \in Z^q(\mathcal{V}, M)$. One can assume that the section $m_{i_0 \dots i_q}$ is constant over $V_{i_0} \cap \dots \cap V_{i_q}$ for each $(i_0, \dots, i_q)$. By the Weierstraß theorem, see above, one finds $x'_1, \dots, x'_{d-1} \in \mathfrak{m}_A$ such that $(x'_1, \dots, x'_{d-1}, x_d)$ is a regular sequence of parameters, and such that the following holds: Let $B := k[\![x'_1, \dots, x'_{d-1}]\!]$, and let $h: \operatorname{sper} A = \operatorname{sper} B[\![x_d]\!] \to \operatorname{sper} B[x_d]$ be the canonical map. Then there are open constructible subsets V'_i of $\operatorname{sper} B[x_d]$ such that $V_i = h^{-1}(V'_i)$, $i \in I$.

Write $U := \operatorname{spec} B[x_d, x_d^{-1}]$, and denote the image set of h by N. Put $\tilde{V}_i := N \cap V'_i$ ($i \in I$). Then $\tilde{\mathcal{V}} = \{\tilde{V}_i\}_{i \in I}$ is an open covering of $N \cap U_r$, and so one can read $(m_{i_0 \dots i_q})$ as a cochain in $C^q(\tilde{\mathcal{V}}, M)$. It is in fact a cocycle there, and thus defines an element $\tilde{\alpha} \in H^q(N \cap U_r, M)$ which clearly pulls back to α under the restriction $\operatorname{sper} A[x_d^{-1}] \to N \cap U_r$ of h. Therefore it will be sufficient to show that $N \cap U_r$ has exactly two connected components, and that $H^q(N \cap U_r, M) = 0$ for every $q \geq 1$. It is known [AA, 3.6] that

$$N = \bigcap_{\substack{\epsilon>0 \\ \epsilon\in k}} S_\epsilon, \quad \text{where } S_\epsilon := \{\alpha \in \operatorname{sper} B[x_d]: \ |x_d(\alpha)| < \epsilon\}.$$

Therefore $H^q(N \cap U_r, M)$ is the direct limit of the groups $H^q(S_\epsilon \cap U_r, M)$, $\epsilon > 0$, and so it suffices to show for $\epsilon > 0$ that $S_\epsilon \cap U_r$ has two connected components and $H^q(S_\epsilon \cap U_r, M) = 0$ for $q \geq 1$. If B were a k-algebra of finite type it would be clear that $H^q(\operatorname{sper} B, M) \oplus H^q(\operatorname{sper} B, M) \xrightarrow{\sim} H^q(S_\epsilon \cap U_r, M)$ for every $q \geq 0$. So the actual assertion results from writing B as the union of its subalgebras of finite type and passing to the limit. □

(1.2) Proposition. *Let A be a strictly real local ring which is regular and excellent. Let $x \in \mathfrak{m}_A$ be a local parameter. Then* $\operatorname{sper} A[x^{-1}]$ *has exactly two connected components, and* $H^q(\operatorname{sper} A[x^{-1}], M) = 0$ *for $q \geq 1$ and any abelian group M.*

That x is a local parameter means that x is part of a regular system of parameters of A, or equivalently, that $x \in \mathfrak{m}_A$, $x \notin \mathfrak{m}_A^2$. (In particular, A is not a field.) Our proof follows the strategy of the proof of the corresponding result for étale cohomology [SGA4, XIX.2]. In particular we are using resolution of singularities for excellent $\mathbb{Q}$-algebras. For $N \geq 0$ let $P(N)$ denote the assertion that (1.2) holds for all A, x and M whenever $0 \leq q \leq N$. We are going to prove $P(N)$ inductively for every $N \geq 0$.

The proposition is true if A is complete (1.1). Also $P(0)$ holds. Indeed, if $\widehat{A}$ is the completion of A then $\operatorname{sper} \widehat{A}[x^{-1}]$ has two connected components. By Ruiz's theorem [R1], $\operatorname{sper} \widehat{A}[x^{-1}] \to \operatorname{sper} A[x^{-1}]$ is surjective, and so $\operatorname{sper} A[x^{-1}]$ has only two connected components, namely $\{x > 0\}$ and $\{x < 0\}$.

Recall that a (locally noetherian) scheme X is called *excellent* if X can be covered by open affine subschemes $X_\alpha = \operatorname{spec} A_\alpha$ where the A_α are excellent rings [EGA IV, 7.8.5]. The local rings of such a scheme are excellent. We use the fact that every strict real henselization of an excellent local ring is again excellent. This can be shown using resolution of singularities, see [EGA IV, 7.9.5, 18.7.5.1 and 18.9.6(iii)].

The following is a consequence of $P(0)$:

(1.3) Lemma. *Let X be an excellent regular locally noetherian scheme, and let Z be a closed regular subscheme of X. Denote by $i: Z_r \hookrightarrow X_r$ and $j: X_r - Z_r \subset X_r$ the inclusions in the real spectrum. Assume that Z is globally defined by one equation $f = 0$, $f \in H^0(X, \mathcal{O}_X)$. Let F be a locally constant sheaf on X_r. Then the choice of f determines a canonical sheaf morphism $\beta: j_*j^*F \to i_*i^*F$ on X_r for which the sequence*

$$0 \longrightarrow F \longrightarrow j_*j^*F \xrightarrow{\beta} i_*i^*F \longrightarrow 0 \tag{1}$$

is exact.

Proof. It is harmless to assume that Z does not contain any connected component of X. We define β stalkwise, so fix $\zeta \in Z_r$. Let $G := \mathrm{Gen}_{X_r}(\zeta)$ be the space of generalizations of ζ in X_r, and write $M := F_\zeta$. So $F|_G$ is the constant sheaf M on G. Let $s \in (j_* j^* F)_\zeta = H^0(G - Z_r, M)$. By $P(0)$ the subspace $G - Z_r$ of G has exactly two connected components, namely $G^+ := G \cap \{f > 0\}$ and $G^- := G \cap \{f < 0\}$. So s is constant on each of G^+, G^-, and we can define $(\beta s)_\zeta := s(\eta_+) - s(\eta_-) \in M = F_\zeta$, where $\eta_\pm \in G^\pm$ are chosen arbitrarily.

Let now W be an open quasi-compact subset of X_r, and let $s \in H^0(W - Z_r, F)$. We have to show that the element

$$((\beta s)_\zeta) \in \prod_{\zeta \in W \cap Z_r} F_\zeta$$

lies in $H^0(W \cap Z_r, F|_{W \cap Z_r})$. For this one can assume that $F|_W$ is a constant sheaf. Put $W^+ := W \cap \{f > 0\}$, $W^- := W \cap \{f < 0\}$, and write

$$W^+ = K_1 \amalg \cdots \amalg K_m, \quad W^- = L_1 \amalg \cdots \amalg L_n$$

with clopen subsets K_μ, L_ν of $W - Z_r$, such that s has constant value on each of the K_μ resp. L_ν. Then βs is constant on each of the closed subsets $\overline{K}_\mu \cap \overline{L}_\nu \cap W \cap Z_r$ of $W \cap Z_r$, and these cover $W \cap Z_r$ and are mutually disjoint. Hence $\beta s \in H^0(W \cap Z_r, F)$.

Thus a well-defined sheaf morphism $j_* j^* F \to i_* i^* F$ on X_r has been constructed. Looking at the stalks it follows from $P(0)$ that the sequence (1) is exact. □

(1.4) Lemma. *Assume that $P(N)$ holds. Let A be an excellent regular strictly real local ring, and let the sequence $(x_1, \dots, x_s)$ be part of a regular system of parameters in $\mathfrak{m}_A$. Write $U := \operatorname{spec} A[x_1^{-1}, \dots, x_s^{-1}]$. Then U_r has exactly 2^s connected components, and $H^q(U_r, M) = 0$ for $1 \le q \le N$ and any abelian group M.*

Proof by induction on s. The case $s = 0$ is trivial, and $s = 1$ is just $P(N)$. Let $s > 1$ and assume the lemma shown for smaller values of s. Let $V := \operatorname{spec} A[x_1^{-1}, \dots, x_{s-1}^{-1}]$ and $Y := \operatorname{spec} A[x_1^{-1}, \dots, x_{s-1}^{-1}]/(x_s)$, so that U is the open complement of Y in V. Let $j: U_r \subset V_r$ and $i: Y_r \hookrightarrow V_r$ denote the inclusions in the real spectrum. By (1.3) the equation x_s for Y gives rise to an exact sequence

$$0 \longrightarrow M \longrightarrow j_* M \longrightarrow i_* M \longrightarrow 0 \tag{2}$$

of sheaves on V_r. Applying $P(N)$ to the strict real henselization of $\mathcal{O}_{V,\mathrm{supp}(\eta)}$ associated with $\eta \in Y_r$ one sees that $\mathrm{R}^q j_* M = 0$ for $1 \le q \le N$. By the inductive hypothesis V_r and Y_r have exactly 2^{s-1} connected components each, and $H^p(V_r, M) = H^p(Y_r, M) = 0$ for $1 \le p \le N$. The long exact cohomology sequence associated with (2) shows therefore that U_r has exactly 2^s connected components, and that $H^p(V_r, j_* M) = 0$ for $1 \le p \le N$. Consequently, in the Leray spectral sequence

$$E_2^{pq} = H^p(V_r, \mathrm{R}^q j_* M) \implies H^{p+q}(U_r, M)$$

one has $E_2^{pq} = 0$ for $p, q \leq N$, $(p,q) \neq (0,0)$. It follows that $H^i(U_r, M) = 0$ for $1 \leq i \leq N$. □

(1.5) Lemma. *Assume that $P(N)$ holds. Let $A \to A'$ be a regular local homomorphism between excellent regular strictly real local rings. Let $U \subset \operatorname{spec} A$ be an open subscheme and put $U' := U \times_{\operatorname{spec} A} \operatorname{spec} A'$. Then $H^q(U_r, M) \to H^q(U'_r, M)$ is bijective for $0 \leq q \leq N$ and any abelian group M.*

Proof. By Hironaka's resolution of singularities ([Hi], cf. [SGA4 p. 589]) there is a proper birational morphism of schemes $f: X \to \operatorname{spec} A$ and an open immersion $h: U \to X$ over $\operatorname{spec} A$ such that X is regular and the closed reduced subscheme $Z := X - h(U)$ is locally normal crossings in X. The latter means that for every $z \in Z$ the closed subscheme $Z \times_X \operatorname{spec} \mathcal{O}_{X,z}$ of $\operatorname{spec} \mathcal{O}_{X,z}$ is defined by one equation $x_1 \cdots x_s = 0$ where $(x_1, \ldots, x_s)$ is part of a regular system of parameters in the regular local ring $\mathcal{O}_{X,z}$.

Let $f': X' \to \operatorname{spec} A'$ resp. $h': U' \to X'$ be the base extensions of f resp. h by $\operatorname{spec} A' \to \operatorname{spec} A$, and denote the morphism $X' \to X$ by g:

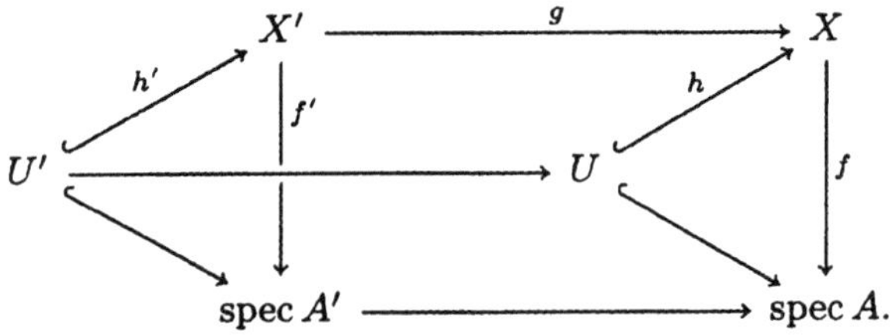

Let $Z' := Z \times_X X'$. Since $A \to A'$ is regular, also X' is a regular scheme, and Z' is locally normal crossings in X'. Now one has

$$g_r^* \mathsf{R}^q h_{r*} M \xrightarrow{\sim} \mathsf{R}^q h'_{r*} M \quad \text{for } 0 \leq q \leq N. \tag{3}$$

Indeed, this is clear for the restriction of the sheaves to $U'_r \subset X'_r$. Let therefore $\zeta' \in Z'_r$ and put $\zeta := g_r(\zeta') \in Z_r$. Then the stalk of (3) in ζ' is the map

$$H^q\Big(\big(\operatorname{spec} \tilde{\mathcal{O}}_{X,\zeta} \times_X U\big)_r, M\Big) \longrightarrow H^q\Big(\big(\operatorname{spec} \tilde{\mathcal{O}}_{X',\zeta'} \times_{X'} U'\big)_r, M\Big). \tag{4}$$

Here $\tilde{\mathcal{O}}_{X,\zeta}$ denotes the strict real henselization of $\mathcal{O}_{X,\operatorname{supp}(\zeta)}$ with respect to ζ, and similarly for $\tilde{\mathcal{O}}_{X',\zeta'}$. From (1.4) it follows that (4) is an isomorphism for $q \leq N$ (and that actually both sheaves in (3) are zero for $1 \leq q \leq N$).

Now if F is any sheaf on X_r then $H^q(X_r, F) \xrightarrow{\sim} H^q(X'_r, g_r^* F)$, by proper base change for f and f'. (See [D] resp. [S] for proper base change in the context of real spectra.) Consider the morphism of spectral sequences

$$\begin{array}{ccccc} E_2^{pq} & = & H^p(X_r, \mathsf{R}^q h_{r*} M) & \Longrightarrow & H^{p+q}(U_r, M) \\ & & \downarrow & & \downarrow \\ E_2'^{pq} & = & H^p(X'_r, \mathsf{R}^q h'_{r*} M) & \Longrightarrow & H^{p+q}(U'_r, M). \end{array} \tag{5}$$

By (3) and the remark just made, $E_2^{pq} \to E_2'^{pq}$ is an isomorphism for $0 \le q \le N$. It follows by an easy induction that the maps $E_\infty^{pq} \to E_\infty'^{pq}$ of (5) are isomorphisms as long as $q \le N$ and $p+q \le N+1$. Therefore $H^n(U_r, M) \to H^n(U_r', M)$ is bijective for $n \le N$. □

The next lemma completes the proof of Proposition (1.2):

(1.6) Lemma. *$P(N)$ implies $P(N+1)$, for $N \ge 0$.*

Proof. Let A, x be as in (1.2). Write $U := \operatorname{spec} A[x^{-1}]$ and $\widehat{U} := \operatorname{spec} \widehat{A}[x^{-1}]$ where $\widehat{A}$ is the completion of A. Let $g: \widehat{U} \to U$ be the canonical morphism. It suffices to prove

$$\mathsf{R}^q g_{r*} M = \begin{cases} M & \text{for } q = 0, \\ 0 & \text{for } 1 \le q \le N. \end{cases} \tag{6}$$

Indeed, it then follows that $H^{N+1}(U_r, M) \to H^{N+1}(\widehat{U}_r, M)$ is injective, and thus $H^{N+1}(U_r, M) = 0$ by (1.1), as desired.

To show (6) let W be a basic open constructible subset of U_r, say $W = U_r \cap \{a_1 > 0, \dots, a_m > 0\}$ with suitable elements $a_i \in A$. Then $\operatorname{spec} A\big[(a_1 \cdots a_m x)^{-1}\big]$ $=: Y$ is an open subscheme of U such that $W \subset Y_r$ and W is a clopen subset of Y_r. Let $\widehat{Y} := Y \times_U \widehat{U}$. By (1.5) one has $H^q(Y_r, M) \xrightarrow{\sim} H^q(\widehat{Y}_r, M)$ for $0 \le q \le N$. Since W is clopen in Y_r this implies $H^q(W, M) \xrightarrow{\sim} H^q(g_r^{-1}(W), M)$, $0 \le q \le N$. Letting W vary one obtains (6). □

From the above proof of (1.2) it came out that $H^q(U_r, M) \xrightarrow{\sim} H^q(\widehat{U}_r, M)$, $q \ge 0$, for every Zariski open $U \subset \operatorname{spec} A$, where A is an excellent regular strictly real local ring and $\widehat{U} := U \times_{\operatorname{spec} A} \operatorname{spec} \widehat{A}$. Equivalently, $\mathsf{R}^q h_* M = M$ resp. 0 for $q = 0$ resp. $q \ge 1$, where $h: \operatorname{sper} \widehat{A} \to \operatorname{sper} A$ is the canonical map. For $q = 0$ this was proved before by Ruiz [R2, 2.6.2]. He used Artin approximation instead of Hironaka and did not even need that A is regular. One wonders whether Artin approximation might suffice to prove this fact for all q (which would also imply (1.2), using (1.1)).

(1.7) Theorem. (Cohomological purity) *Let X be an excellent regular locally noetherian scheme, and let Z be a closed subscheme of X which is regular and purely of codimension $c > 0$. Let $i: Z_r \hookrightarrow X_r$ be the inclusion in the real spectrum. If F is any locally constant sheaf on X_r then $\mathsf{R}^q i^! F = 0$ for $q \ne c$, and $\mathsf{R}^c i^! F$ is Zariski locally isomorphic to the sheaf $i^* F$ on Z_r.*

Here and below we are saying that two sheaves F, F' on the real spectrum X_r of a scheme X are *Zariski locally isomorphic* if X can be covered by open subschemes X^λ for which $F|_{X_r^\lambda} \cong F'|_{X_r^\lambda}$.

The proof of (1.7) will show that actually $\mathsf{R}^c i^! F \cong i^* F$ holds as long as Z is definable by c global equations on X.

As usual the theorem can be reformulated in terms of the open embedding $j: X_r - Z_r \subset X_r$. The theorem says that $\mathsf{R}^q j_* j^* F = 0$ for $q \notin \{0, c-1\}$. Moreover in case $c \geq 2$ it asserts that $F \xrightarrow{\sim} j_*j^*F$, and that $\mathsf{R}^{c-1}j_*j^*F$ is Zariski locally isomorphic to i_*i^*F. For $c = 1$ the assertion is that $F \to j_*j^*F$ is injective and the cokernel is Zariski locally isomorphic to i_*i^*F.

Proof. One can assume that $X = \operatorname{spec} A$ is affine and $Z = \operatorname{spec} A/(f_1, \ldots, f_c)$ is globally defined by c equations $f_1, \ldots, f_c \in A$. Assume that $c \geq 2$. Let $Z' := \operatorname{spec} A/(f_1, \ldots, f_{c-1})$, and denote the inclusions in the real spectrum by $\mu: Z_r \hookrightarrow Z'_r$ and $\nu: Z'_r \hookrightarrow X_r$. Then Z' is regular, and $Z \hookrightarrow Z'$ resp. $Z' \hookrightarrow X$ are purely of codimension 1 resp. $c-1$. Since $i^! = \mu^! \circ \nu^!$, an inductive argument using the spectral sequence for this composite functor shows that it suffices to consider the case $c = 1$. So let $Z = \operatorname{spec} A/(f)$. Proposition (1.2), applied to the strict real henselizations of A in orderings $\zeta \in Z_r$, shows $\mathsf{R}^q j_*j^*F = 0$ for $q \geq 1$. The remaining assertion follows from Lemma (1.3). □

(1.8) Remark. Let k be some ring of coefficients, e.g. $k = \mathbb{Z}$. In the situation of the theorem one can think of $\mathsf{R}^c i^! k$ as of the "orientation sheaf" on Z_r with respect to k, an orientation corresponding to a global section of this sheaf. Indeed, if the schemes in (1.7) are $\mathbb{R}$-varieties, $\mathsf{R}^c i^! k$ corresponds to the orientation sheaf from usual topology; and in particular $Z(\mathbb{R})$ is orientable if Z is globally defined by c equations.

Assume in the general situation of (1.7) that Z is globally defined by c equations. One may ask to what extent the isomorphism $\mathsf{R}^c i^! F \cong i^*F$ can be made canonical, i.e. independent from the choice of equations. Clearly, if F is a sheaf of $\mathbb{Z}/2$-modules then $\mathsf{R}^c i^! F = i^*F$ *canonically*. For general coefficients consider the local case, i.e. let $X = \operatorname{spec} A$ with A an excellent regular local ring. Then one can show that the isomorphism $\mathsf{R}^c i^! F \cong i^*F$ is canonical up to multiplication by the sign of a unit in A; and clearly this is the best one can hope for.

(1.9) Corollary. *Under the hypotheses of* (1.7) *let F be a locally constant sheaf on X_r. Then*

$$H^q_{Z_r}(X_r, F) = \begin{cases} 0 & \text{if } q < c, \\ H^{q-c}(Z_r, \mathsf{R}^c i^! F) & \text{if } q \geq c. \end{cases}$$

In particular, if X is the spectrum of a semilocal ring then $H^q_{Z_r}(X_r, F) = 0$ for $q \neq c$ and $H^c_{Z_r}(X_r, F) \cong H^0(Z_r, F|_{Z_r})$.

Proof. The last remark uses the fact that any two locally isomorphic sheaves on Z_r are actually isomorphic, since every closed point in Z_r has arbitrarily small clopen neighborhoods. □

(1.10) Corollary. *Under the hypotheses of* (1.7) *let F be a locally constant sheaf on X_r, and write $U := X - Z$. Then one has isomorphisms*

$$H^q(X_r, F) \xrightarrow{\sim} H^q(U_r, F|_{U_r}), \quad 0 \leq q < c-1,$$

and a long exact sequence

$$0 \to H^{c-1}(X_r, F) \to H^{c-1}(U_r, F|_{U_r}) \to H^0(Z_r, \tilde{F}) \to H^c(X_r, F) \\ \to H^c(U_r, F|_{U_r}) \to H^1(Z_r, \tilde{F}) \cdots$$

with $\tilde{F} := \mathsf{R}^c i^! F$. □

(1.11) Corollary. *Let A be an excellent regular ring. Let $\xi \in \operatorname{sper} A$ and put $c := \operatorname{ht}(\operatorname{supp}\xi)$. Then $\operatorname{Gen}^*(\xi) := \operatorname{Gen}(\xi) - \{\xi\}$ has the cohomology of a punctured c-dimensional disk; i.e.*

$$H^q(\operatorname{Gen}^*(\xi), M) = \begin{cases} M & \text{for } q = 0,\ c-1, \\ 0 & \text{otherwise,} \end{cases}$$

if $c > 1$. (If $c = 0$ resp. 1 then $\operatorname{Gen}^*(\xi)$ is empty resp. a discrete two-point space.)

Proof. One may assume that A is a strictly real local ring and ξ is the closed point of $\operatorname{sper} A$. Then the assertion is a particular case of (1.10). □

(1.12) Corollary. (Semi-purity) *Let X be an excellent regular locally noetherian scheme. Let Z be any closed subscheme of X, let $i\colon Z_r \hookrightarrow X_r$ be the inclusion of the real spectra. If Z has codimension c in X then $\mathsf{R}^q i^! F = 0$ for $q < c$ and every locally constant sheaf F on X_r. Hence also $H^q_{Z_r}(X_r, F) = 0$ for $q < c$, and $H^c_{Z_r}(X_r, F) = H^0(Z_r, \mathsf{R}^c i^! F)$.*

Proof. The last remark follows from the spectral sequence $H^p(Z_r, R^q i^! F) \Rightarrow H^{p+q}_{Z_r}(X_r, F)$. Conversely, the stalk of $\mathsf{R}^q i^! F$ in $\zeta \in Z_r$ is $H^q_{Z_r \cap \operatorname{Gen}\zeta}(\operatorname{Gen}\zeta, F)$; so we can assume that X is affine, $\dim X = d < \infty$, and it suffices to prove $H^q_{Z_r}(X_r, F) = 0$ for $q < c$. Without loss of generality Z is reduced. The assertion is true if $c \geq d$ since then Z is a finite sum of spectra of fields and hence is regular. For the step from $c+1$ to c choose an open subscheme U of X such that $U \cap Z$ is regular, $U \cup Z = X$, and U contains the points of Z of codimension c in X. The long exact sequence for the closed triple $(X_r, Z_r, (X-U)_r)$, together with (1.9) and the inductive hypothesis, shows $H^q_{Z_r}(X_r, F) = 0$ for $q < c$. □

We also give a relative purity theorem. See [SGA4, XVI.3] for its analogue in étale cohomology. Recall:

Definition. Let S be a scheme. A *smooth S-pair (X, Z) of codimension c* is a closed immersion $f\colon Z \hookrightarrow X$ of smooth S-schemes such that, for every $s \in S$, the fibre Z_s has pure codimension c in X_s.

(1.13) Theorem. (Relative cohomological purity) *Let S be a scheme and (X, Z) a smooth S-pair of codimension c. Write $i := f_r\colon Z_r \hookrightarrow X_r$ for the inclusion of the real spectra. If G is any sheaf on S_r and F is its inverse image on X_r then the sheaf $\mathsf{R}^c i^! F$ on Z_r is Zariski locally isomorphic to $i^* F$, and $\mathsf{R}^q i^! F = 0$ for $q \neq c$.*

Proof. The assertion of (1.13) is obviously local on Z. For $z \in Z$ there are an open neighborhood X' of z in X and an étale S-morphism $\phi\colon X' \to \mathbb{A}^m_S$ (for some

$m \geq c$) such that $Z \cap X'$ is the preimage of $\mathbb{A}_S^{m-c} := \{t_1 = 0, \ldots, t_c = 0\}$ under ϕ [SGA1 II.4.10] (here $t_1, \ldots, t_m$ are the standard coordinates on $\mathbb{A}_S^m$):

$$\begin{array}{ccccc} Z & \hookleftarrow & Z \cap X' & \longrightarrow & \mathbb{A}_S^{m-c} \\ {\scriptstyle f}\downarrow & & \downarrow & & \downarrow{\scriptstyle \tilde{f}} \\ X & \hookleftarrow & X' & \xrightarrow{\phi} & \mathbb{A}_S^m. \end{array}$$

Since étale morphisms of schemes induce local homeomorphisms of the real spectra ([AR], [S]) it is easy to see that the restriction of $R^q i^! F = R^q f_r^! F$ to $(Z \cap X')_r$ is the restriction of $\mathrm{R}^q \tilde{f}_r^! \tilde{F}$ to $(Z \cap X')_r$, where $\tilde{F} :=$ preimage of G on $\mathbb{A}_S^m$. Thus we can replace (X, Z) by the standard pair $(\mathbb{A}_S^m, \mathbb{A}_S^{m-c})$, and even (hereby changing S) by $(\mathbb{A}_S^c, 0_S)$, where 0_S is the zero section of $\mathbb{A}_S^c \to S$. Using the filtration $0_S \subset \mathbb{A}_S^1 \subset \cdots \subset \mathbb{A}_S^c$ one reduces inductively to the case $c = 1$. Now one may as well take $(X, Z) = (\mathbb{P}_S^1, \infty_S)$, with ∞_S the section at infinity. Use the following notations for the maps in the real spectrum:

$$\begin{array}{ccccc} Z_r & & X_r & & (X - Z)_r \\ \| & & \| & & \| \\ S_r & \xhookrightarrow{i} & (\mathbb{P}_S^1)_r & \xhookleftarrow{j} & (\mathbb{A}_S^1)_r \\ & \searrow^{=} & \downarrow{\scriptstyle \varphi} & \swarrow_{\psi} & \\ & & S_r & & \end{array}$$

Note that $F = \varphi^* G$. Consider the spectral sequence

$$E_2^{pq} = \mathrm{R}^p \varphi_* \mathrm{R}^q j_* (\psi^* G) \implies \mathrm{R}^{p+q} \psi_* (\psi^* G). \tag{7}$$

The map ψ is "acyclic", i.e. $G \xrightarrow{\sim} \psi_* \psi^* G$ and $\mathrm{R}^n \psi_* (\psi^* G) = 0$, $n \geq 1$, hold for every sheaf G on S_r. (See [S, 16.7.2] for the proof.) Moreover $\mathrm{R}^p \varphi_* = 0$ for $p \geq 2$, by proper base change. Also $\mathrm{R}^q j_* (\psi^* G) = \mathrm{R}^q j_* (j^* F)$ has support on Z_r for $q \geq 1$, whence $E_2^{pq} = 0$ for $p, q \geq 1$. Summarizing one concludes that even $E_2^{pq} = 0$ for $(p, q) \neq (0, 0)$, and in particular that for $q \geq 1$

$$\mathrm{R}^{q+1} i^! F = \varphi_* \mathrm{R}^q j_* (j^* F) = E_2^{0q} = 0.$$

It remains to determine $\mathrm{R}^q i^! F$ for $q = 0, 1$. Looking at the exact sequence

$$0 \longrightarrow i_* i^! F \longrightarrow F \longrightarrow j_* j^* F \longrightarrow i_* \mathrm{R}^1 i^! F \longrightarrow 0$$

one sees that the following lemma completes the proof of (1.13):

(1.14) Lemma. *There is a natural sheaf homomorphism $\beta: j_* j^* F \to i_* i^* F$ on $(\mathbb{P}_S^1)_r$ such that the sequence*

$$0 \longrightarrow F \longrightarrow j_* j^* F \xrightarrow{\beta} i_* i^* F \longrightarrow 0$$

is exact.

Proof. The construction is similar to Lemma (1.3), therefore we only mention how to prove the following key fact: Let A be a ring. Let $\xi \in \operatorname{sper} A[t]$ with $t = 0$ in ξ, and write

$$G^+ := \mathrm{Gen}(\xi) \cap \{t > 0\}, \quad G^- := \mathrm{Gen}(\xi) \cap \{t < 0\}.$$

Then G^+, G^- are connected.

Assume $G^+ = V_1 \amalg V_2$ with proper clopen subsets V_i of G^+. Denote the projection $\mathrm{sper}\, A[t] \to \mathrm{sper}\, A$ by p, put $\xi_0 := p(\xi)$. Since p is an open map, $p|_{G^+}$ is generalizing. So $\mathrm{Gen}(\xi_0) = p(V_1) \cup p(V_2)$ where the $p(V_i)$ are pro-constructible and stable under generalization. Hence $p(V_1) \cap p(V_2) \neq \emptyset$, and so there is $\eta_0 \succ \xi_0$ such that $p^{-1}(\eta_0) \cap G^+$ is disconnected. But, identifying $p^{-1}(\eta_0)$ with the affine line $\mathrm{sper}\, k(\eta_0)[t]$ over the real closed field $k(\eta_0)$, it is easy to see that $p^{-1}(\eta_0) \cap G^+$ is a *convex* subset of $p^{-1}(\eta_0)$, and hence is connected. This proves the lemma and the theorem. □

Of course this relative purity theorem admits corollaries similar to (1.9), (1.10) and (1.12).

2. A resolution of Gersten type

Let X be a noetherian scheme. Write $X^{(c)} := \{x \in X \colon \dim \mathcal{O}_{X,x} = c\}$ for the set of points of codimension c in X. Filtering the topological space X by codimension leads in a well-known way to spectral sequences which converge against Zariski (sheaf) cohomology on X, filtered by codimension of support. See Hartshorne's exposition in [Ha, IV.1]. But one may also pull this filtration back from X to the real spectrum X_r. The purity results from §1 show that the associated spectral sequences degenerate to long exact sequences, if X is regular and excellent and locally constant sheaves are considered.

This can be regarded as a corollary to the results of §1, by the techniques of [Ha]. It may however be worthwile to present some details of the construction. In the following we do this using a technique slightly different from Hartshorne's, namely by exact couples. This approach was inspired by the paper [BO].

Given $x \in X$, recall that $x_r := \mathrm{sper}\, \kappa(x)$. Denote the inclusion $x_r \to X_r$ by i_{x_r}.

(2.1) Theorem. *Let X be an excellent regular noetherian scheme. Let W be an open constructible subset of X_r, and let F be a locally constant sheaf on W. Then there is a complex $C^\bullet(W, F)$ of abelian groups:*

$$0 \longrightarrow \bigoplus_{x \in X^{(0)}} H^0_x(W, F) \longrightarrow \bigoplus_{x \in X^{(1)}} H^1_x(W, F) \longrightarrow \bigoplus_{x \in X^{(2)}} H^2_x(W, F) \longrightarrow \cdots, \qquad (8)$$

natural in W and F, whose q-th cohomology group is canonically isomorphic to $H^q(W, F)$, $q \geq 0$. Here $H^q_x(W, F) := H^q_{x_r \cap W}(\mathrm{sper}\, \mathcal{O}_{X,x} \cap W,\ F)$. This group is isomorphic (non-canonically in general, cf. (1.8)*) to $H^0(x_r \cap W, F)$, for $x \in X^{(q)}$, $q \geq 0$.*

From the proof one can easily see that this complex is contravariantly functorial for flat morphisms of schemes.

(2.2) Corollary. *For X, W as in* (2.1) *there is a natural complex $C^\bullet(W,\mathbb{Z}/2)$:*

$$0 \longrightarrow \bigoplus_{x\in X^{(0)}} H^0(x_r\cap W,\mathbb{Z}/2) \longrightarrow \bigoplus_{x\in X^{(1)}} H^0(x_r\cap W,\mathbb{Z}/2) \longrightarrow \cdots \qquad (9)$$

whose q-th cohomology group is canonically isomorphic to $H^q(W,\mathbb{Z}/2)$, $q\geq 0$.

One may also sheafify these complexes, arriving at a canonical acyclic resolution for every locally constant sheaf on an open subset of X_r. We only mention the case of the constant sheaf $\mathbb{Z}/2$:

(2.3) Corollary. *For X as in* (2.1) *there is a canonical resolution by acyclic sheaves on X_r:*

$$0 \longrightarrow \mathbb{Z}/2 \longrightarrow \bigoplus_{x\in X^{(0)}} (i_{x_r})_*\mathbb{Z}/2 \longrightarrow \bigoplus_{x\in X^{(1)}} (i_{x_r})_*\mathbb{Z}/2 \longrightarrow \bigoplus_{x\in X^{(2)}} (i_{x_r})_*\mathbb{Z}/2 \longrightarrow \cdots \qquad (10)$$

For the proof of (2.1) we restrict to the case $W = X_r$, for notational simplicity. Let p be an integer. Write $\mathcal{Z}^p$ for the set of all closed reduced subschemes Y of X of codimension $\geq p$. Denote by $\mathcal{Z}^p/\mathcal{Z}^{p+1}$ the set of pairs (Y,Z) with $Y\in\mathcal{Z}^p$, $Z\in\mathcal{Z}^{p+1}$ and $Z\subset Y$. Both sets $\mathcal{Z}^p$ and $\mathcal{Z}^p/\mathcal{Z}^{p+1}$ are ordered (in fact, directed) by inclusion of subschemes.

Let $(Y',Z')\subset(Y,Z)$ be two pairs in $\mathcal{Z}^p/\mathcal{Z}^{p+1}$. There is a natural morphism of long exact sequences

$$\begin{array}{ccccccc}
\cdots H^n_{Z'_r}(X_r,F) & \to & H^n_{Y'_r}(X_r,F) & \to & H^n_{(Y'-Z')_r}((X-Z')_r,F) & \to & H^{n+1}_{Z'_r}(X_r,F)\cdots \\
\downarrow & & \downarrow & & \downarrow & & \downarrow \\
\cdots H^n_{Z_r}(X_r,F) & \to & H^n_{Y_r}(X_r,F) & \longrightarrow & H^n_{(Y-Z)_r}((X-Z)_r,F) & \longrightarrow & H^{n+1}_{Z_r}(X_r,F)\cdots
\end{array} \qquad (11)$$

These maps (11) form a directed system of long exact sequences, indexed by $\mathcal{Z}^p/\mathcal{Z}^{p+1}$. Defining

$$H^n_{\mathcal{Z}^p}(X_r,F) := \varinjlim_{Y\in\mathcal{Z}^p} H^n_{Y_r}(X_r,F),$$

$$H^n_{\mathcal{Z}^p/\mathcal{Z}^{p+1}}(X_r,F) := \varinjlim_{(Y,Z)\in\mathcal{Z}^p/\mathcal{Z}^{p+1}} H^n_{(Y-Z)_r}((X-Z)_r,F)$$

for $p,n\in\mathbb{Z}$ one obtains therefore a long exact sequence for every $p\in\mathbb{Z}$, passing (11) to the limit:

$$\cdots\ H^n_{\mathcal{Z}^{p+1}}(X_r,F) \xrightarrow{i} H^n_{\mathcal{Z}^p}(X_r,F) \xrightarrow{j} H^n_{\mathcal{Z}^p/\mathcal{Z}^{p+1}}(X_r,F) \xrightarrow{k} H^{n+1}_{\mathcal{Z}^{p+1}}(X_r,F)\ \cdots \qquad (12)$$

Let $\mathbb{Z} \oplus \mathbb{Z}$-graded abelian groups D, E be defined by

$$D^{pq} := H^{p+q}_{\mathcal{Z}^p}(X_r, F), \quad E^{pq} := H^{p+q}_{\mathcal{Z}^p/\mathcal{Z}^{p+1}}(X_r, F)$$

$(p, q \in \mathbb{Z})$. Then (12) is an exact couple

$$\begin{array}{ccc} D & \xrightarrow{\ i\ } & D \\ & {\scriptstyle k}\nwarrow \quad \swarrow{\scriptstyle j} & \\ & E & \end{array} \tag{13}$$

in which $\deg i = (-1, 1)$, $\deg j = (0, 0)$ and $\deg k = (1, 0)$. Note that $D^{pq} = H^{p+q}(X_r, F)$ for $p \le 0$ and $D^{pq} = 0$ for $q < 0$, the latter by semi-purity (1.12); and that $E^{pq} = 0$ if $p < 0$ or $q < 0$. From the general theory of exact couples [M] it follows that (13) gives rise to a spectral sequence

$$E_1^{pq} = E^{pq} = H^{p+q}_{\mathcal{Z}^p/\mathcal{Z}^{p+1}}(X_r, F) \implies H^{p+q}(X_r, F) \tag{14}$$

in which the d_1-differential is $j \circ k$.

The remainder of the proof of (2.1) consists in identifying the line $q = 0$ of (14) with (8) and showing $E_1^{pq} = 0$ for $q \neq 0$. Given $x \in X^{(p)}$, put $Y := \overline{\{x\}}$ and define $H_x^*(X_r, F) := \varinjlim H^*_{(Y \cap U)_r}(U_r, F)$ where the limit is taken over the Zariski open neighborhoods U of x. Clearly the restriction map $H_x^*(X_r, F) \to H^*_{x_r}(\operatorname{sper} \mathcal{O}_{X,x}, F)$ is an isomorphism. From §1 we know

$$H_x^n(X_r, F) \begin{cases} \cong H^0(x_r, F) & \text{for } n = p, \\ = 0 & \text{for } n \neq p, \end{cases}$$

since F is locally constant. Therefore (2.1) is a consequence of the following

Claim: For every $p \in \mathbb{Z}$ there is a canonical isomorphism

$$\phi\colon H^*_{\mathcal{Z}^p/\mathcal{Z}^{p+1}}(X_r, F) \xrightarrow{\sim} \bigoplus_{x \in X^{(p)}} H_x^*(X_r, F). \tag{15}$$

ϕ is constructed as follows. Given $(Y, Z) \in \mathcal{Z}^p/\mathcal{Z}^{p+1}$ and $x \in X^{(p)}$, put $V := \overline{\{x\}}$. Choose an open subscheme U of $X - Z$ containing x with $Y \cap U \subset V$. The composite maps

$$H^*_{Y-Z}\big((X - Z)_r\big) \longrightarrow H^*_{Y \cap U}(U_r) \longrightarrow H^*_{V \cap U}(U_r) \longrightarrow H_x^*(X_r)$$

(we use obvious shorthand notations) induce a map ϕ as in (15). To see that ϕ is an isomorphism one may construct its inverse: Given $x \in X^{(p)}$ and a Zariski open neighborhood U of x, put $Y := \overline{\{x\}}$ and $Z := Y - (Y \cap U)$, so that $(Y, Z) \in \mathcal{Z}^p/\mathcal{Z}^{p+1}$. By excision, $H^*_{Y \cap U}(U_r) \cong H^*_{Y-Z}\big((X - Z)_r\big)$, and these maps fit together to give a map ϕ' which goes in the opposite direction from ϕ. It is not hard to see that ϕ and ϕ' are inverses of each other. □

(2.4) Remark. Let X be an arbitrary noetherian scheme of finite dimension, and let W be an open constructible subset of X_r. Then for any sheaf F on W

there is a natural spectral sequence

$$E_1^{pq} = \bigoplus_{x \in X^{(p)}} H_x^{p+q}(W,F) \implies H^{p+q}(W,F) \tag{16}$$

in which $H_x^*(W,F)$ is defined as in (2.1). Indeed, the deduction of the spectral sequence (14) and the isomorphisms (15) in the proof of (2.1) didn't use that X is either excellent or regular, or that F is locally constant (but one needs an assumption like $\dim X < \infty$ to make sure that the spectral sequence converges).

(2.5) Remark. Let X be smooth over some base field k. Using results from [S] one may deduce Corollary (2.3) from the Bloch-Ogus resolutions. Indeed, let $\mathcal{H}^n$ be the Zariski sheaf on X associated with the presheaf $U \mapsto H_{et}^n(U, \mathbb{Z}/2)$, $n \geq 0$. By [BO] one has for each n an exact sequence of Zariski sheaves

$$0 \longrightarrow \mathcal{H}^n \longrightarrow \bigoplus_{x \in X^{(0)}} i_{x*} H_{et}^n(\kappa(x)) \longrightarrow \bigoplus_{x \in X^{(1)}} i_{x*} H_{et}^{n-1}(\kappa(x)) \longrightarrow \cdots \tag{17}$$

where $H_{et}^*(\kappa(x)) := H_{et}^*(\kappa(x), \mathbb{Z}/2)$ and i_x denotes the inclusion $\{x\} \subset X$. (We may assume $\operatorname{char} k \neq 2$.) The sequences (17) form a direct system for $n \to \infty$, where one passes from n to $n+1$ by cup-product with $(-1) \in H_{et}^1(k)$. The direct limit $\lim_{n\to\infty} \mathcal{H}^n$ is the Zariski sheaf $\operatorname{supp}_* \mathbb{Z}/2$ where supp: $X_r \to X$ is the support map [S, (19.5)]. On the other hand, $\lim_{n\to\infty} H_{et}^n(\kappa(x)) = H^0(x_r, \mathbb{Z}/2)$ was shown already in [AEJ]. The limit of (17) for $n \to \infty$ is supp_* of the sequence (10). (Under suitable hypotheses on k this limit is attained for finite n already.) Now the additive functor supp_* from sheaves on X_r to sheaves on X is exact [S, (19.2)] and faithful. So one deduces (2.3).

We now make explicit the differentials in the complex of Corollary (2.3):

(2.6) Proposition. *Let X be an excellent regular noetherian scheme. Let $y \in X^{(q)}$, $z \in X^{(q+1)}$, and denote by*

$$\partial_{y,z}\colon H^0(y_r, \mathbb{Z}/2) \longrightarrow H^0(z_r, \mathbb{Z}/2)$$

the map induced by the differential $C^q(X_r, \mathbb{Z}/2) \to C^{q+1}(X_r, \mathbb{Z}/2)$ of (9). Let $Y := \overline{\{y\}}$ (reduced closure). If $z \notin Y$ then $\partial_{y,z} = 0$. Otherwise put $\mathcal{O}_{y,z} := \mathcal{O}_{Y,z}$, a 1-dimensional local domain. For each $\zeta \in z_r$ there is a finite even number of orderings $\eta \in y_r$ which specialize into ζ (in $\operatorname{sper} \mathcal{O}_{y,z}$*). Given $s \in H^0(y_r, \mathbb{Z}/2)$ one has*

$$(\partial_{y,z} s)(\zeta) = \sum_{\substack{\eta \succ \zeta \\ \eta \in y_r}} s(\eta) \in \mathbb{Z}/2, \tag{18}$$

for $\zeta \in z_r$.

Proof. Let $(X', Y', Z') := (\operatorname{sper} \mathcal{O}_{X,z}, \operatorname{sper} \mathcal{O}_{y,z}, z_r)$, a closed triple of spaces. Going into the construction of the complex (9) one sees that

$$\begin{array}{ccc} H^q_{Y'-Z'}(X'-Z',\mathbb{Z}/2) & \xrightarrow{\ \partial\ } & H^{q+1}_{Z'}(X',\mathbb{Z}/2) \\ \downarrow\wr & & \downarrow\wr \\ H^0(Y'-Z',\mathbb{Z}/2) & \xrightarrow{\ \partial_{y,z}\ } & H^0(Z',\mathbb{Z}/2) \end{array}$$

commutes. Here ∂ is the connecting homomorphism in the long exact sequence for (X',Y',Z'), and the vertical maps are the (canonical!) isomorphisms deduced from (1.7). Note $Y'-Z'=y_r$ and $Z'=z_r$.

First assume that $\mathcal{O}_{y,z}$ is normal. Then every $\zeta \in z_r$ has exactly two generalizations in y_r, and formula (18) follows from (1.3) and its proof, for example.

In the general case there is a morphism $h:\tilde{X} \to \operatorname{spec}\mathcal{O}_{X,z}$ which is a composition of finitely many blowings-up of closed points, such that the restriction $\tilde{h}:\tilde{Y} \to \operatorname{spec}\mathcal{O}_{y,z}$ of h to the strict transform $\tilde{Y}$ of $\operatorname{spec}\mathcal{O}_{y,z}$ is the normalization of $\operatorname{spec}\mathcal{O}_{y,z}$. Then $\tilde{X}$ is again excellent and regular [EGA IV, 19.4]. Let $\tilde{z} := \tilde{h}^{-1}(z)$. Let $\zeta \in z_r$, and let $\zeta_1,\dots,\zeta_t$ be its preimages in $\tilde{z}_r$. Let η_i, η_i' be the two generalizations of ζ_i in $\tilde{Y}_r$. Since $\tilde{h}$ is finite it follows that $\eta_1,\eta_1',\dots,\eta_t,\eta_t'$ are precisely the different generalizations of ζ in y_r. Formula (18) follows easily by considering the map $(\tilde{X}_r,\tilde{Y}_r,\tilde{z}_r) \to (X',Y',Z')$ induced by h_r. □

3. Fundamental classes

Let again X be an excellent regular noetherian scheme.

(3.1) Proposition. *Let $Y \subset X$ be a closed subscheme, and let W be an open constructible subset of X_r. Consider the subcomplex $C^\bullet_Y(W,\mathbb{Z}/2)$:*

$$0 \longrightarrow \bigoplus_{x\in X^{(0)}\cap Y} H^0(x_r\cap W,\,\mathbb{Z}/2) \longrightarrow \bigoplus_{x\in X^{(1)}\cap Y} H^0(x_r\cap W,\,\mathbb{Z}/2) \longrightarrow \cdots \qquad (19)$$

of $C^\bullet(W,\mathbb{Z}/2)$. For every $q\ge 0$ its q-th cohomology group is canonically isomorphic to $H^q_{Y_r\cap W}(W,\mathbb{Z}/2)$.

Proof. This follows directly from (2.3): The sheaves in the resolution (10) are acyclic for $H^0_{Y_r\cap W}(W,-)$, since for $x\in X^{(q)}$ one has

$$H^0_{Y_r\cap W}\Big(W,\,(i_{x_r})_*\mathbb{Z}/2|_W\Big) = \begin{cases} H^0(x_r\cap W,\,\mathbb{Z}/2) & \text{if } x\in Y,\\ 0 & \text{if } x\notin Y\end{cases}$$

and $(i_{x_r})_*$ is exact. □

From (3.1) one concludes that if Y is an irreducible closed subscheme of X of codimension q in X, then $H^n_{Y_r}(X_r,\mathbb{Z}/2)$ is the $(n-q)$-th cohomology group of the complex

$$0 \longrightarrow \bigoplus_{y\in Y^{(0)}} H^0(y_r,\mathbb{Z}/2) \longrightarrow \bigoplus_{y\in Y^{(1)}} H^0(y_r,\mathbb{Z}/2) \longrightarrow \bigoplus_{y\in Y^{(2)}} H^0(y_r,\mathbb{Z}/2) \longrightarrow \cdots \qquad (20)$$

This complex depends only on Y. This suggests to think of $H_i(Y_r, \mathbb{Z}/2) :=$ $(e-i)$th cohomology group of (20) ($e := \dim Y$) as a homology theory, at least if one makes some biequidimensionality assumption (which hold for varieties over a field, for example). In any case we have:

(3.2) Corollary. *Let Y be a quasi-projective scheme over* spec $\mathbb{R}$. *Then for every $n \geq 0$ the n-th Borel-Moore homology $H_n(Y(\mathbb{R}), \mathbb{Z}/2)$ is canonically isomorphic to the n-th homology group of the complex*

$$\cdots \longrightarrow \bigoplus_{y \in Y_{(2)}} H^0(y_r, \mathbb{Z}/2) \longrightarrow \bigoplus_{y \in Y_{(1)}} H^0(y_r, \mathbb{Z}/2) \longrightarrow \bigoplus_{y \in Y_{(0)}} H^0(y_r, \mathbb{Z}/2) \longrightarrow 0, \tag{21}$$

where $Y_{(n)} := \{y \in Y\colon \dim \overline{\{y\}} = n\}$.

Proof. The corollary is true whenever Y admits a closed embedding into a non-singular $\mathbb{R}$-variety X, by (3.1) and by Poincaré duality $H^n_{Y(\mathbb{R})}(X(\mathbb{R}), \mathbb{Z}/2) \xrightarrow{\sim} H_{d-n}(Y(\mathbb{R}), \mathbb{Z}/2)$, $d := \dim X$ [Iv, p. 381]. □

Of course the corollary holds as well over any real closed field if one uses semi-algebraic Borel-Moore homology.

Returning to the more general situation considered before, recall some notations. One writes $Z^q(X)$ for the group of codimension q cycles on X, i.e. for the free abelian group on $X^{(q)}$. The support $|z|$ of $z = \sum_{x \in X^{(q)}} n_x \cdot x \in Z^q(X)$ is the closed reduced subscheme $\bigcup\{\overline{\{x\}}\colon n_x \neq 0\}$ of X. Given $x \in X^{(q-1)}$ and $a \in \kappa(x)^*$ one forms the codimension q cycle $\operatorname{div}(a) \in Z^q(X)$ [F]. The quotient of $Z^q(X)$ by the subgroup generated by these cycles is written $CH^q(X)$, the Chow group of codimension q cycles modulo rational equivalence. We sometimes denote rational equivalence by $\sim$.

Let Y be a closed irreducible subscheme of X of codimension q, with generic point $y \in X^{(q)}$. Denote by η_Y or η_y the constant section $1 \in H^0(y_r, \mathbb{Z}/2)$, considered as an element of $C^q_Y(X_r, \mathbb{Z}/2)$, cf. (3.1). It follows immediately from (2.6) that $\eta_Y \in \ker\left(C^q_Y(X_r, \mathbb{Z}/2) \to C^{q+1}_Y(X_r, \mathbb{Z}/2)\right)$. So η_Y determines a class in $H^q_{Y_r}(X_r, \mathbb{Z}/2)$. More generally, if $z = n_1 Y_1 + \cdots + n_s Y_s \in Z^q(X)$ is a cycle ($n_i \in \mathbb{Z}$) we may regard $\eta_z := n_1 \eta_{Y_1} + \cdots + n_s \eta_{Y_s}$ as an element of $C^q_Z(X_r, \mathbb{Z}/2)$, $Z := |z|$. This element η_z determines a class in $H^q_{Z_r}(X_r, \mathbb{Z}/2)$.

Definition.
a) Let $z \in Z^q(X)$ be a cycle, $Z := |z|$. We write $[z]$ or $[Z]$ for the class of η_z in $H^q_{Z_r}(X_r, \mathbb{Z}/2)$, and call this element the *fundamental class* of z.
b) The image of $[z]$ under the natural map $H^q_{Z_r}(X_r, \mathbb{Z}/2) \to H^q(X_r, \mathbb{Z}/2)$ is denoted by $cl^q(z)$ or $cl(z)$. The map $cl^q\colon Z^q(X) \to H^q(X_r, \mathbb{Z}/2)$, $z \mapsto cl^q(z)$ is called the *cycle map*.

(3.3) Proposition. *Let X be an excellent regular noetherian scheme, let $z \in Z^q(X)$ and $Z := |z|$.*

a) *For every* $\zeta \in Z_r$ *for which the support of* ζ *is a regular point of* Z, *the image of* $[Z]$ *in* $H^q_{Z_r \cap \mathrm{Gen}\,\zeta}(\mathrm{Gen}\,\zeta, \mathbb{Z}/2)$ *is the non-zero element.*
b) *The fundamental class* $[Z]$ *is characterized in* $H^q_{Z_r}(X_r, \mathbb{Z}/2)$ *by property a).*

Proof. Let $\tilde{X} := \mathrm{spec}\,\tilde{\mathcal{O}}_{X,\zeta}$ where $\tilde{\mathcal{O}}_{X,\zeta}$ is the strict real henselization of $\mathcal{O}_{X,\mathrm{supp}\,\zeta}$ in ζ, and put $\tilde{Z} := Z \times_X \tilde{X}$. Then $\mathrm{Gen}\,\zeta = \tilde{X}_r$ and $Z_r \cap \mathrm{Gen}\,\zeta = \tilde{Z}_r$. Since $C^{q-1}_{\tilde{Z}}(\tilde{X}_r, \mathbb{Z}/2) = 0$ (cf. (3.1)) it is clear that $[\tilde{Z}]$ is non-zero in $H^q_{\tilde{Z}_r}(\tilde{X}_r, \mathbb{Z}/2)$; this group is $\mathbb{Z}/2$ by (1.9). b) follows from semi-purity (1.12): The singular locus S of Z is a closed subscheme of X of codimension $> q$, so $H^q_{S_r}(X_r, \mathbb{Z}/2) = 0$, and b) follows from the exact sequence $H^q_{S_r}(X_r, \mathbb{Z}/2) \longrightarrow H^q_{Z_r}(X_r, \mathbb{Z}/2) \longrightarrow H^q_{(Z-S)_r}((X-S)_r, \mathbb{Z}/2) = H^0((Z-S)_r, \mathbb{Z}/2)$. □

For example, if X is a non-singular algebraic variety over $\mathbb{R}$ then the cycle map defined above coincides with the cycle map defined by Borel and Haefliger [BH]. For another example let M be a compact real analytic manifold and $\mathcal{O}(M)$ its ring of global (real) analytic functions. Then $\mathcal{O}(M)$ is an excellent regular ring and $\mathrm{sper}\,\mathcal{O}(M)$ has the same cohomology as M. If $Z \subset M$ is a closed analytic subset of dimension p then the vanishing ideal $I := I_Z$ of Z has height $n-p$ in $\mathcal{O}(M)$ ($n := \dim M$), and so the above construction associates to I a class in $H^{n-p}_{\mathrm{sper}\,\mathcal{O}(M)/I}(\mathrm{sper}\,\mathcal{O}(M), \mathbb{Z}/2) = H^{n-p}_Z(M, \mathbb{Z}/2) \cong H_p(Z, \mathbb{Z}/2)$. This is the fundamental class of Z as constructed in [BH].

It is obvious that $cl(Z)$ vanishes for every prime cycle Z whose function field is non-real.

We do not intend to study the cycle map in greater detail in this paper. Let us only mention here that the construction passes through rational equivalence, i.e. induces cycle maps $cl^q\colon CH^q(X) \to H^q(X_r, \mathbb{Z}/2)$:

(3.4) Proposition. $cl(Z) = 0$ *for every cycle* Z *which is rationally equivalent to zero. More precisely, if* $y \in X^{q-1}$ *and* $a \in \kappa(y)^*$ *then*

$$\eta_{\mathrm{div}(a)} = \partial(\mathrm{sgn}\,a) \quad \text{in } C^q(X_r, \mathbb{Z}/2),$$

where $\mathrm{sgn}\,a \in H^0(y_r, \mathbb{Z}/2)$ *is the characteristic function of* $\{a < 0\} \subset y_r$ *and* $\partial\colon C^{q-1}(X_r, \mathbb{Z}/2) \to C^q(X_r, \mathbb{Z}/2)$ *is the differential.*

Proof. Put $Y := \overline{\{y\}}$, and let $h\colon Y' \to Y$ be the normalization of Y. Then $\mathrm{div}\,a = \sum_{z \in Y^{(1)}} n_z z$, where

$$n_z = \sum_{z' \in h^{-1}(z)} [\kappa(z') : \kappa(z)] \,.\, v_{z'}(a) \tag{22}$$

and $v_{z'}$ is the discrete valuation of $\kappa(y)$ associated to z' [F]. On the other hand, if $z \in Y^{(1)}$ and $\zeta \in z_r$ then the generalizations of ζ in y_r correspond $2:1$ to the preimages ζ' of ζ in $h^{-1}(z)_r$: Each ζ' induces two generalizations of ζ, and all these are different. Since for $z' \in h^{-1}(z)$ the number of extensions of ζ to z'_r is $\equiv [\kappa(z') : \kappa(z)] \pmod 2$, one reads off the desired equality $\eta_{\mathrm{div}(a)} = \partial(\mathrm{sgn}\,a)$ from (22), using (2.6). □

(3.5) Remark. Let k be a field of characteristic not 2 and X a smooth k-scheme. Using the notation of Remark (2.5), there are natural maps $CH^q(X) \to H^q(X, \mathcal{H}^q)$, $q \geq 0$, which induce isomorphisms $CH^q(X)/2 \xrightarrow{\sim} H^q(X, \mathcal{H}^q)$. This is a well-known consequence of the work of Bloch and Ogus [BO]. As already mentioned, the limit of the direct system $\mathcal{H}^q \xrightarrow{(-1)} \mathcal{H}^{q+1} \xrightarrow{(-1)} \mathcal{H}^{q+2} \cdots$ of Zariski sheaves is the sheaf $\mathrm{supp}_* \mathbb{Z}/2$. By exactness of supp_* one obtains a natural composite map

$$CH^q(X) \longrightarrow H^q(X, \mathcal{H}^q) \longrightarrow H^q(X, \mathrm{supp}_*\mathbb{Z}/2) = H^q(X_r, \mathbb{Z}/2), \qquad (23)$$

for every $q \geq 0$. This map (23) is nothing but the cycle map cl^q defined above. This follows readily from Remark (2.5) and our construction of cl^q. Note in particular the case when k is real closed, in which $H^q(X_r, \mathbb{Z}/2)$ is semi-algebraic cohomology $H^q_{sa}(X(R), \mathbb{Z}/2)$ and cl^q is the (semi-algebraic) Borel-Haefliger map.

4. A new proof of E.P.T. (1)

As a final application we now explain how our construction of the cycle map leads to a new proof of the following

(4.1) Theorem. *Let X be a smooth projective R-variety, where R is an* archimedean *real closed field. Then every Weil divisor D on X with $cl^1(D) = 0$ in $H^1_{\mathrm{sa}}(X(R), \mathbb{Z}/2)$ is linearly equivalent to a divisor whose support contains no real point.*

(4.2) Some remarks are in order. Of course H^*_{sa} denotes semi-algebraic cohomology; in case $R = \mathbb{R}$ this is singular cohomology. The above theorem has seen many proofs. Probably the first was given by Bröcker in 1980 [Br]. Others are due to Bochnak-Kucharz-Shiota [BKS] and Ischebeck-Schülting [IS]. The latter show that the theorem is a consequence of a result by Evans from 1971. Very likely there are other places in the literature where this or a related result is proved.

As pointed out in [IS], all these proofs (formulated for $R = \mathbb{R}$) use approximation theorems like the Stone-Weierstraß theorem or approximation of continuous by algebraic sections in algebraic vector bundles. These theorems, although they make sense over any real closed field R, are known to be false if R is not archimedean. In fact it seems not to be known whether (4.1) holds over such R.

The proof presented below doesn't use any of these tools and is "purely algebraic". However it is interesting to observe how the hypothesis that R is archimedean enters again in a crucial way: If R were not archimedean it would admit non-trivial real valuations, and our key argument which uses fan theory would break down.

The essential step for the proof is to show that any divisor D with $cl^1(D) = 0$ is linearly equivalent to a divisor $D' + 2D''$, where $D' = \sum_i P_i$ with non-real prime

(1) E.P.T. = Everybody proved this theorem. (According to L. Bröcker.)

divisors P_i. (A prime divisor P is called non-real if its function field is non-real.) The proof that D' resp. $2D''$ are in turn linearly equivalent to divisors supporting no real point can be carried out separately and by elementary means. (Although the arguments for this latter reduction step are certainly known to the experts, I found myself unable to locate them in the literature. Therefore I will include them in an appendix to this section.)

The more general situation in which we constructed the cycle map allows us to prove the above theorem in greater generality. Let X be a noetherian scheme and $z = \sum n_i x_i \in Z^p(X)$ a cycle ($x_i \in X^{(p)}$ pairwise different, $0 \neq n_i \in \mathbb{Z}$). If the field $\kappa(x_i)$ is non-real for every i then z will be called *non-real*. If the real spectrum of $|z| = \bigcup \overline{\{x_i\}}$ is empty we say that z is *completely non-real*. Below we refer to the elements of $Z^1(X)$ resp. $X^{(1)}$ as *divisors* resp. *prime divisors* of X. We will prove:

(4.3) Theorem. *Let Λ be a domain such that Λ is contained in $H(K)$, the real holomorphy ring of the quotient field K of Λ. Put $S := \operatorname{spec} \Lambda$. Assume that X is a connected excellent regular noetherian scheme and $\pi: X \to S$ is a proper surjective scheme morphism.*

Then every divisor z on X with $cl^1(z) = 0$ in $H^1(X_r, \mathbb{Z}/2)$ is rationally equivalent to a divisor $z' + 2z''$ with z' nonreal.

Applying the proposition proved in the appendix below one can, under suitable hypotheses (e.g., π projective), actually conclude that z is rationally equivalent to a completely non-real divisor.

Proof of (4.3). Recall that the real holomorphy ring $H(K)$ of K is the intersection of all valuation rings of K with a formally real residue field [Be]. It will only be used that π is surjective and universally closed. Let x be the generic point of X, and let z be a divisor with $cl^1(z) = 0$. By our construction of the cycle map there is $a \in H^0(x_r, \mathbb{Z}/2)$ such that $\partial a = \eta_z$ in $C^1(X_r, \mathbb{Z}/2)$. (∂ refers to the complex $C^\bullet(X_r, \mathbb{Z}/2)$.) It will suffice to find $f \in \kappa(x)^*$ with $a = \operatorname{sgn} f$, cf. (3.4). Indeed, we then have

$$\eta_{\operatorname{div} f} \;=\; \partial(\operatorname{sgn} f) \;=\; \eta_z$$

by (3.4), which shows that modulo 2 and modulo non-real divisors, z is equal to $\operatorname{div} f \sim 0$.

The existence of such f can be decided by looking at fans, which are special subsets of $x_r = \operatorname{sper} \kappa(x)$, cf. [L] for example. In fact it suffices to consider an arbitrary 4-element fan $F = \{\xi_1, \xi_2, \xi_3, \xi_4\}$ in x_r and to show

$$\sum_{i=1}^{4} a(\xi_i) \;=\; 0 \quad \text{in } \mathbb{Z}/2. \tag{24}$$

By fan trivialization [L] there are a valuation ring B of $\kappa(x)$ and orderings $\bar{\eta}$, $\bar{\eta}' \in \operatorname{sper} \kappa_B$ ($\kappa_B :=$ residue field of B) such that, after relabelling, $\xi_1, \xi_2 \succ \bar{\eta}$ and $\xi_3, \xi_4 \succ \bar{\eta}'$ in $\operatorname{sper} B$.

Since π is surjective it induces an embedding $K \hookrightarrow \kappa(x)$. The hypothesis on Λ implies $\Lambda \subset B \cap K$. So from the valuative criterion for properness we get $y \in X$

such that B dominates $\mathcal{O}_{X,y}$. Hence there are $\eta, \eta' \in y_r$ (namely the restrictions of $\tilde{\eta}, \tilde{\eta}'$ to y_r) with $\xi_1, \xi_2 \succ \eta$ and $\xi_3, \xi_4 \succ \eta'$ in X_r.

The divisor z is principal in a Zariski neighborhood U of y. So there is $h \in \kappa(x)^*$ such that the restriction of $\operatorname{div}(h) - z \in Z^1(X)$ to U vanishes. Thus $\partial(\operatorname{sgn} h) = \eta_{\operatorname{div} h} = \eta_z = \partial a$ after restriction to U, and so there is $\epsilon \in H^0(U_r, \mathbb{Z}/2)$ with

$$a(\xi) \;=\; \epsilon(\xi) + \operatorname{sgn} h(\xi) \quad \text{for every } \xi \in x_r.$$

Since $\epsilon(\xi_1) = \epsilon(\xi_2) = \epsilon(\eta)$ and $\epsilon(\xi_3) = \epsilon(\xi_4) = \epsilon(\eta')$, formula (24) follows. □

Appendix

In the following we indicate how one can reduce the proof of Theorem (4.1) to the situation considered in (4.3).

Proposition. *Let $X = \operatorname{spec} A$ with A an excellent regular ring. For every divisor z on X, $2z$ is rationally equivalent to a completely non-real divisor. Also, every non-real divisor is rationally equivalent to a completely non-real divisor.*

Of course the proposition applies to many non-affine (excellent regular) schemes X as well, namely if the real spectrum X_r is contained in an open affine subscheme. This is the case e.g. if X is quasi-projective over some ring.

Lemma 1. *Let $y \in X^{(1)}$ be a non-real prime divisor, put $Y := \overline{\{y\}}$, and assume $Y_r \neq \emptyset$. Then $y \sim \sum_i n_i y_i$ where the prime divisors y_i are such that $(Y_i)_r$ is a proper subset of Y_r ($Y_i := \overline{\{y_i\}}$).*

Lemma 2. *Let y be any prime divisor. Then $2y$ is rationally equivalent to a non-real divisor.*

The proposition follows from these two lemmas, by noetherian induction. This induction, using Lemma 1, is applied to the Zariski closure of $\operatorname{supp} Y_r := \{\operatorname{supp}(\eta)\colon \eta \in Y_r\}$ in X. (Note that the y_i in Lemma 1 are automatically non-real.) We assume in the following that X is connected, and denote by K the quotient field of A. If $y \in X^{(1)}$ we write v_y for the associated discrete valuation of K.

The proof of Lemma 2 is immediate: Choose $a \in A$ with $v_y(a) = 1$. Let $y_1, \dots, y_s$ be the real prime divisors $\neq y$ for which $v_{y_i}(a) > 0$. By the approximation lemma find $b \in A$ with $v_y(b) = 1$ and $v_{y_i}(b) = 0$, $i = 1, \dots, s$. Then $\operatorname{div}(a^2 + b^2) \equiv 2y$ modulo non-real divisors.

For the proof of Lemma 1 denote by Z the reduced Zariski closure of $\operatorname{supp} Y_r$. So Z is a non-empty proper closed subscheme of Y. Let $\mathfrak{p}$ be the prime ideal of A corresponding to y. Consider the semilocalization $B := \mathcal{O}_{X,Z}$ of A in the irreducible components of Z. We need

Lemma 3. *Let B be an excellent regular semilocal domain, with quotient field K. Let $\mathfrak{p} \in \operatorname{spec} B$, $\operatorname{ht}(\mathfrak{p}) = 1$, such that $\kappa(\mathfrak{p})$ is not real. Then $\mathfrak{p} = \pi B$ with some $\pi \in B \cap \Sigma K^2$.*

Postponing the proof of this lemma, apply it to find $\pi \in B \cap \Sigma K^2$ with $\pi B = \mathfrak{p}B$. Writing $\pi = a/b$ with $a, b \in A$ and $b \in \mathcal{O}^*_{X,Z}$ one sees that, after replacing π by ab, one can assume $\pi \in A$. Note that $\pi \geq 0$ on sper A, by regularity of A.

We have to shrink the zero set of π in sper A. Let V be its Zariski closure, i.e. the Zariski closure of $\{\mathfrak{q} \in X : \pi \in \mathfrak{q}$ and $\kappa(\mathfrak{q})$ is real$\}$. Every irreducible component of Z is also one of V, but V may have others. Let W be such an irreducible component of V which is not contained in Z. Then it is easy to see that there is $a \in A$, not vanishing identically on W, such that π divides a in $B = \mathcal{O}_{X,Z}$. (Look at the prime factorization of π in $\mathcal{O}_{X,Z\cup W}$.) Replacing a with na for some integer $n \neq 0$, if necessary, one can arrange that $1 + \pi^{-1}a^2$ is a unit in B. Thus $\tilde{\pi} = \pi + a^2 \in A$ is a generator for $\mathfrak{p}B$, is a sum of squares in K, and the Zariski closure of its real zero set is strictly smaller than that of π since it doesn't contain W. Iteration of this argument leads to $\pi' \in A \cap \Sigma K^2$ with $\pi' B = \mathfrak{p}B$ and sper $A/\pi' A =$ sper $A/\mathfrak{p} = Z_r$. Hence div $\pi' = y + z$ where z is an effective divisor on X such that the Zariski closure of its real zero set (i.e. of $\operatorname{supp}(|z|_r)$) is strictly smaller than Z.

It remains to prove Lemma 3. Since B is factorial (it is locally factorial and the class group is zero) there is $b \in B$ with $\mathfrak{p} = bB$. We claim that b is locally semidefinite on sper B; that is, writing $U^+ := \{b > 0\}$, $U^- := \{b < 0\} \subset$ sper B, that $S := \overline{U^+} \cap \overline{U^-}$ is empty. Assume this is false, and choose an element $\xi \in S$ which is minimal in S under $\succ$. Write $G :=$ Gen(ξ) and $G^\pm := G \cap U^\pm$. Then $G^* := G - \{\xi\}$ is the disjoint union of its intersections with $\overline{G^+}$ and $\overline{G^-}$, which shows that G^* is disconnected. Thus ht(supp $\xi) = 1$ by (1.11), and so necessarily supp $\xi = \mathfrak{p}$ since clearly $b(\xi) = 0$. This contradicts that $\kappa(\mathfrak{p})$ is not real.

So sper $B = \overline{U^+} \amalg \overline{U^-}$. We want to know that there is a unit u of B with $bu \geq 0$ on sper B. By fan theory one has to take any 4-element fan $F \subset (\text{sper } B)^{\max}$ and to show $\#(F \cap \overline{U^\pm}) = 2$. But since B is regular, F can be lifted to a 4-element fan $\tilde{F}$ in sper K (i.e. such that $F \subset \overline{\tilde{F}}$), and the assertion follows. □

References

[EGA IV] A. Grothendieck, Éléments de Géométrie Algébrique, chap. IV. Inst. Hautes Études Sci. Publ. Math. **20, 24, 28, 32**, 1964–1967.

[SGA1] A. Grothendieck, Revêtements Étales et Groupe Fondamental (SGA1). Lecture Notes in Math. **224**, Springer, Berlin etc., 1971.

[SGA4] M. Artin, A. Grothendieck, J. L. Verdier, Théorie des Topos et Cohomologie Étale des Schemas (SGA4), tome 3. Lecture Notes in Math. **305**, Springer, Berlin etc., 1973.

[AA] M. E. Alonso, C. Andradas, Real spectra of complete local rings. Manuscripta Math. **58**, 155–177 (1987).

[AR] M. E. Alonso, M. F. Roy, Real strict localizations. Math. Z. **194**, 429–441 (1987).

[AEJ] J. Kr. Arason, R. Elman, B. Jacob, The graded Witt ring and Galois cohomology, I. In: Quadratic and Hermitian Forms, Conf. Hamilton, Ontario, 1983, Canad. Math. Soc. Conf. Proc. **4** (1984), pp. 17–50.

[Be] E. Becker, Valuations and real places in the theory of formally real fields. In: Géométrie Algébrique Réelle et Formes Quadratiques, Proc. Rennes 1981, ed. J.-L. Colliot-Thélène, M. Coste, L. Mahé, M.-F. Roy, Lecture Notes in Math. **959**, Springer, Berlin etc., 1982, pp. 1–40.

[BO] S. Bloch, A. Ogus, Gersten's conjecture and the homology of schemes. Ann. Sci. École Norm. Sup. (4) **7**, 181–202 (1974).

[BKS] J. Bochnak, W. Kucharz, M. Shiota, The divisor class groups of some rings of global real analytic, Nash or rational regular functions. In: Géométrie Algébrique Réelle et Formes Quadratiques, Proc. Rennes 1981, ed. J.-L. Colliot-Thélène, M. Coste, L. Mahé, M.-F. Roy, Lecture Notes in Math. **959**, Springer, Berlin etc., 1982, pp. 218–248.

[BH] A. Borel, A. Haefliger, La classe d'homologie fondamentale d'un espace analytique. Bull. Soc. Math. France **89**, 461–513 (1961).

[Br] L. Bröcker, Reelle Divisoren. Arch. Math. **35**, 140–143 (1980).

[CC] M. Carral, M. Coste, Normal spectral spaces and their dimensions. J. Pure Appl. Algebra **30**, 227–235 (1983).

[D] H. Delfs, Homology of Locally Semialgebraic Spaces. Lecture Notes in Math. **1484**, Springer, Berlin etc., 1991.

[F] W. Fulton, Intersection Theory. Ergeb. Math. Grenzgeb. (3), Bd. **2**, Springer, Berlin etc., 1984.

[Ha] R. Hartshorne, Residues and Duality. Lecture Notes in Math. **20**, Springer, Berlin etc., 1966.

[Hi] H. Hironaka, Resolution of singularities of an algebraic variety over a field of characteristic zero. Ann. of Math. **79**, 109–326 (1964).

[IS] F. Ischebeck, H.-W. Schülting, Rational and homological equivalence for real cycles. Invent. Math. **94**, 307–316 (1988).

[Iv] B. Iversen, Cohomology of Sheaves. Springer, Berlin etc., 1986.

[L] T. Y. Lam, Orderings, Valuations and Quadratic Forms. Conf. Board Math. Sciences **52**, Amer. Math. Soc., Providence, R. I., 1983.

[M] S. Mac Lane, Homology. Grundlehren Math. Wiss. **114**. Springer, Berlin etc., 1967.

[R1] J. M. Ruiz, On Hilbert's 17th problem and real Nullstellensatz for global analytic functions. Math. Z. **190**, 447–454 (1985).

[R2] J. M. Ruiz, On the connected components of a global semianalytic set. J. Reine Angew. Math. **392**, 137–144 (1988).

[S] C. Scheiderer, Real and Étale Cohomology. Lecture Notes in Math. **1588**, Springer, Berlin etc., 1994.

Geometry of subanalytic and semialgebraic sets: abstract

Masahiro Shiota

Abstract. The present paper explains the main results of the book [S3], which the author is preparing, with sketches of proofs. The book intends to clarify where good properties of subanalytic and semialgebraic sets come from, by showing that any family of subsets of Euclidean spaces, which satisfies certain axioms, retains the properties.

1991 Mathematics Subject Classification: 14P10, 32S60, 58A35, 03C40.

0. Introduction

We call a family of subsets of Euclidean spaces $\mathfrak{X}$ if it satisfies the axioms (i), ... ,(iv) in §2. We introduce the axioms so that the families of all subanalytic sets and of all semialgebraic sets are typical examples of $\mathfrak{X}$-families. Knight, Pillay, Steinhorn, van den Dries and others considered the concept of an 0-minimal Tarski system (= an axiomatically defined family of subsets of Euclidean spaces). The axioms of $\mathfrak{X}$-family are similar to the axioms of an 0-minimal Tarski system, and an 0-minimal Tarski system containing all semialgebraic sets is an $\mathfrak{X}$-family. But the family of subanalytic sets is not an 0-minimal Tarski system.

We can prove some properties of $\mathfrak{X}$-sets (= elements of an $\mathfrak{X}$-family) easily or in the same way as in the subanalytic or semialgebraic case, and the proofs are not very long. But this is not the case for the $\mathfrak{X}$-Hauptvermutung theorem (2.9) which implies uniqueness of $\mathfrak{X}$-triangulations of $\mathfrak{X}$-sets and the triviality theorem (2.18) which states that a proper C^r $\mathfrak{X}$-submersion (r = a positive integer) from a bounded C^r $\mathfrak{X}$-manifold to an open disk is C^r $\mathfrak{X}$-trivial. The original proof of the first theorem in the subanalytic case proceeds mainly in the analytic category [S-Y]. But in the $\mathfrak{X}$ case we can not expect analyticity of $\mathfrak{X}$-sets and $\mathfrak{X}$-maps (see the example (b) below). Hence we need to reconstruct the proof in the C^r category. For that we introduce certain sufficient conditions for two compact polyhedra to be PL homeomorphic in term of usual differential topology. We see easily that the conditions are satisfied by any $\mathfrak{X}$-homeomorphism between compact polyhedra. But for the proof that the conditions are sufficient we need a long detailed research of Whitney stratifications and then a result on C^r triangulations of some compact sets (= a generalization of Cairns–Whitehead Theorem). The second theorem was proved in [C-S$_1$] and [C-S$_2$] in the Nash case (= C^ω + semialgebraic) by the theory of real spectrum. In the subanalytic and $\mathfrak{X}$ case we can not apply the theory and, hence, the proofs become much longer (but easier for a layman of real spectrum). As a corollary of (2.18) we obtain the compactification theorem (2.20) and the algebraic model theorem (2.21), which were proved in the Nash case in [S_1]. Moreover, using (2.18) we can prove the $\mathfrak{X}$-version (2.19) of the Thom's first isotopy lemma [Th] in the same way as [C-S_2].

Some results are new also in the subanalytic and semialgebraic case ((2.15) and (2.22)). (2.15) states subanalytic triangulations of complex analytic functions and it is a globalization of a local result of [S_2]. It follows from a more general investigation of $\mathfrak{X}$-triangulations of $\mathfrak{X}$-maps (2.14). About $\mathfrak{X}$-triangulations of $\mathfrak{X}$-maps we have Conjecture 3 which is the $\mathfrak{X}$-version of the Thom's conjecture on triangulations of stratified maps sans éclatement. (2.22) means that the $\mathfrak{X}$-category lies between C^0 and C^1 categories in some sense.

The simplest axiomatically defined non-trivial family of subsets of Euclidean spaces may be the family of semilinear sets, which are defined by linear functions in the same way as semialgebraic sets. (Here "linear" means "affine linear".) This family is not $\mathfrak{X}$. In order to include this family we weaken the axioms of an $\mathfrak{X}$-family. Then we define a $\mathfrak{Y}$-family. $\mathfrak{Y}$-sets retain some of the properties of $\mathfrak{X}$-sets. But (2.20) and (2.21) do not always hold for $\mathfrak{Y}$-sets (3.20) and (3.21). By strengthening the axioms of an $\mathfrak{X}$-family and a $\mathfrak{Y}$-family, we define also an $\mathfrak{X}^0$-family and a $\mathfrak{Y}^0$-family, which have, moreover, some finite properties (§3) and the first of which coincides with an 0-minimal Tarski system which contains all algebraic sets.

Examples of $\mathfrak{X}$-families other than the families of subanalytic sets and of semialgebraic sets are the following.

(a) The family of all locally semialgebraic sets.

(b) An 0-minimal Tarski system generated by the relations $x+y=z$, $x\cdot y=z$ and $\exp(x)=y$ (its existence was shown by Wilkie [$W_{1,2}$]). Note that the closure of the graph of the function $y=\exp(-1/x^2)$ is an element of this system and C^∞ smooth but not C^ω smooth.

(c) Let $X\subset\mathbb{R}^n$ be a closed semialgebraic set, and let f_i, $i=1,\dots,k$, be Pfaffian functions on an open neighborhood U of X, i.e., for each i

$$\frac{\partial f_i}{\partial x_j}(x)=P_j(x,f_i(x))\quad\text{on } U,\ j=1,\dots,n,$$

for some polynomial functions P_j. Then by [W_1] each element of the smallest family $\mathfrak{X}$ containing graph $f_i|_X$ is described by f_i and a finite number of polynomial functions.

1. Subanalytic, semialgebraic and semilinear sets

We define axiomatically a subanalytic set as follows. The family of all *subanalytic sets* in Euclidean spaces is by definition the smallest family $\mathfrak{S}_1$ of subsets of all Euclidean spaces which satisfies the following axioms:

Axiom (i)$_1$ Every analytic set in a Euclidean space is an element of $\mathfrak{S}_1$.

Axiom (ii)$_1$ For elements $X_1 \subset \mathbb{R}^n$ and $X_2 \subset \mathbb{R}^n$ of $\mathfrak{S}_1$, $X_1 \cap X_2$ and $X_1 - X_2$ are elements of $\mathfrak{S}_1$.

Axiom (iii)$_1$ Let $X \subset \mathbb{R}^n$ be an element of $\mathfrak{S}_1$, and let $p : \mathbb{R}^n \to \mathbb{R}^m$ be a linear map such that the restriction of p to the closure of X in $\mathbb{R}^n$ is proper. Then $p(X)$ is an element of $\mathfrak{S}_1$.

Since $\mathfrak{X}$-sets retain almost all properties of subanalytic sets and we explain important properties of $\mathfrak{X}$-sets in §2, we point out here the properties which hold only for subanalytic sets. Let $r = 1, \dots, \infty$ or ω and let X be a subanalytic set in $\mathbb{R}^n$. Let $\Sigma_r X$ denote the C^r *singular point set* of X (we call a point x of X C^r *singular* if x has no neighborhood U in $\mathbb{R}^n$ such that $X \cap U$ is a C^r submanifold of dimension $= \dim X$). Then

(1.1) *$\Sigma_\infty X$ and $\Sigma_\omega X$ are identical, subanalytic and of dimension* $< \dim X$. *If X is bounded then we have*

$$\Sigma_r X = \Sigma_{r+1} X = \cdots = \Sigma_\omega X$$

for some integer r [Ta].

(1.2) Lojasiewicz inequality. *For compact subanalytic sets X and Y in $\mathbb{R}^n$ there exist positive numbers c and d such that*

$$\operatorname{dis}(x, X \cap Y) \le c(\operatorname{dis}(x, Y))^d \quad \textit{for } x \in X.$$

Note that by the above example (b), (1.1) and (1.2) do not always hold for general $\mathfrak{X}$.

We define similarly the family of all semialgebraic sets in Euclidean spaces to be the smallest family $\mathfrak{S}_2$ of subsets of Euclidean spaces which satisfies the following axioms:

Axiom (i)$_2$ Every algebraic set in a Euclidean space is an element of $\mathfrak{S}_2$.

Axiom (ii)$_2$ The same as (ii)$_1$.

Axiom (iii)$_2$ The same as (iii)$_1$.

For a semialgebraic set X in $\mathbb{R}^n$ we have

(1.3) *$\Sigma_\infty X$ and $\Sigma_\omega X$ are semialgebraic, and for some r*

$$\Sigma_r X = \Sigma_{r+1} X = \cdots = \Sigma_\omega X.$$

The family of all *semilinear sets* in Euclidean spaces is by definition the smallest family $\mathfrak{S}_3$ of subsets of Euclidean spaces which satisfies the following axioms:

Axiom (i)$_3$ Every linear space in a Euclidean space is an element of $\mathfrak{S}_3$.

Axiom (ii)$_3$ For elements $X_1 \subset \mathbb{R}^n$ and $X_2 \subset \mathbb{R}^n$ of $\mathfrak{S}_3$, $X_1 \cap X_2$, $X_1 - X_2$ and $X_1 \times X_2$ are elements of $\mathfrak{S}_3$.

Axiom (iii)$_3$ The same as (iii)$_1$.

Axiom (iv)$_3$ Every interval in $\mathbb{R}$ is an element of $\mathfrak{S}_3$.

(1.4) *A subset of $\mathbb{R}^n$ is semilinear if and only if it is semialgebraic and locally semilinear at each point of $\mathbb{R}^n$.*

Proof. "Only if" is clear. Hence we prove "if". Let X be a semialgebraic set in $\mathbb{R}^n$ and locally semilinear. We can assume that X is closed in $\mathbb{R}^n$ for the following reason. Consider the finite sequence of sets

$$X_0 = X, \quad X_i = \overline{X_{i-1}} - X_{i-1},\ i = 1, \dots .$$

Then X_i and $\overline{X_i}$, $i = 1, 2, \dots$, also are semialgebraic and locally finite unions of open simplexes and of simplexes, respectively. Moreover, we have

$$X = \overline{X_0} - X_1 = \overline{X_0} - (\overline{X_1} - X_2) = (\overline{X_0} - \overline{X_1}) \cup X_2 = \\ \cdots = (\overline{X_0} - \overline{X_1}) \cup (\overline{X_2} - \overline{X_3}) \cup \cdots .$$

Hence it suffices to show that each $\overline{X_i}$ is semilinear. Therefore, we can assume that X is closed in $\mathbb{R}^n$.

We prove the fact by induction on dim X. Clearly $\Sigma_\omega X$ is closed, semialgebraic and of dimension < dim X. Moreover, by elementary methods of PL topology we can show that $\Sigma_\omega X$ is a polyhedron. Hence by induction hypothesis, $\Sigma_\omega X$ is semilinear. Let $W_1, \dots, W_k$ denote the connected components of $X - \Sigma_\omega X$. Then each W_i is an open subset of a linear subspace Π_i of $\mathbb{R}^n$ and a connected component of $\Pi_i - \Sigma_\omega X$. Hence $\overline{W_i}$ and, hence, X are semilinear. □

2. $\mathfrak{X}$-sets

Let r always denote a positive integer. Let $\mathfrak{X}$ be a family of subsets of Euclidean spaces, which satisfies the axioms:

Axiom (i) Every algebraic set in a Euclidean space is an element of $\mathfrak{X}$.

Axiom (ii) For elements $X_1 \subset \mathbb{R}^n$ and $X_2 \subset \mathbb{R}^n$ of $\mathfrak{X}$, $X_1 \cap X_2$, $X_1 - X_2$ and $X_1 \times X_2$ are elements of $\mathfrak{X}$.

Axiom (iii) Let $X \subset \mathbb{R}^n$ be an element of $\mathfrak{X}$ and let $p : \mathbb{R}^n \to \mathbb{R}^m$ be a linear map such that the restriction of p to $\overline{X}$ is proper. Then $p(X)$ is an element of $\mathfrak{X}$.

Axiom (iv) For an element $X \subset \mathbb{R}$ of $\mathfrak{X}$, each point of X has a neighborhood in X which is a finite union of points and intervals.

An $\mathfrak{X}$-*set* is an element of $\mathfrak{X}$, and an $\mathfrak{X}$-*map* is a continuous map between $\mathfrak{X}$-sets with $\mathfrak{X}$-graph. A C^r $\mathfrak{X}$-*submanifold* M of (or $\mathfrak{X}$-*manifold* in) $\mathbb{R}^n$ is an $\mathfrak{X}$-subset of $\mathbb{R}^n$, which is locally C^r $\mathfrak{X}$-diffeomorphic to an open simplex. We define also a C^0 $\mathfrak{X}$-submanifold of $\mathbb{R}^n$ in the same way. Note that a C^r submanifold M of $\mathbb{R}^n$ is a C^r $\mathfrak{X}$-submanifold if M is an $\mathfrak{X}$-subset of $\mathbb{R}^n$. But this note does not hold in the case of C^0 $\mathfrak{X}$-submanifolds. An C^r $\mathfrak{X}$-*stratification* of an $\mathfrak{X}$-set is a stratification whose strata are C^r $\mathfrak{X}$-submanifolds. An $\mathfrak{X}$-*triangulation* of an $\mathfrak{X}$-set X is a pair of a simplicial complex K in some $\mathbb{R}^{n'}$ and an $\mathfrak{X}$-homeomorphism $\pi : |K| \to X$ such that $|K|$ the underlying polyhedron of K is an $\mathfrak{X}$-set in $\mathbb{R}^{n'}$. A C^r $\mathfrak{X}$-*stratification* of an $\mathfrak{X}$-map $f : X \to Y$ is a pair of C^r $\mathfrak{X}$-stratifications $\{X_i\}_i$ of X and $\{Y_j\}_j$ of Y such that for each i, $f|_{X_i}$ is a surjective C^r submersion to some Y_j. An $\mathfrak{X}$-*triangulation* of an $\mathfrak{X}$-function $f : X \to \mathbb{R}$ is an $\mathfrak{X}$-triangulation (K, π) of X such that $f \circ \pi$ is linear on each simplex of K. An $\mathfrak{X}$-*triangulation* of an $\mathfrak{X}$-map $f : X \to Y$ is a pair of $\mathfrak{X}$-triangulations (K_1, π_1) of X and (K_2, π_2) of Y such that $\pi_2^{-1} \circ f \circ \pi_1 : K_1 \to K_2$ is a simplicial map. An R-$\mathfrak{X}$-*triangulation* of an $\mathfrak{X}$-map $f : X \to Y$ is an $\mathfrak{X}$-triangulation (K, π) of X such that Y is a polyhedron and for each simplex $\sigma \in K$, $f \circ \pi|_\sigma$ is linear.

From now on we state main facts on $\mathfrak{X}$-sets and $\mathfrak{X}$-maps. If we can prove a fact easily or in the same way as in the subanalytic or semialgebraic case, then we omit its proof.

(2.1) *For a nonempty $\mathfrak{X}$-set $X \subset \mathbb{R}^n$, $\overline{X}$ is an $\mathfrak{X}$-set, and $\overline{X} - X$ is of dimension* $< \dim X$. *(Here we define* $\dim \varnothing = -1$.*) Consequently, an $\mathfrak{X}$-set $\subset \mathbb{R}$ is locally a finite union of points and intervals at each point of $\mathbb{R}$.*

Proof. Although the proof is easy, we prove the fact that $\overline{X}$ is an $\mathfrak{X}$-set because the proof is interesting. Let g denote the distance function from X in $\mathbb{R}^n$, and set

$$Y = \{(y,t) \in \mathbb{R}^n \times \mathbb{R} : \ g(y) < t\}.$$

By (iii), Y is an $\mathfrak{X}$-set, because Y is the image of the $\mathfrak{X}$-set

$$\{(x,y,t) \in X \times \mathbb{R}^n \times \mathbb{R} : \ |x - y| < t\}$$

under the projection $X \times \mathbb{R}^n \times \mathbb{R} \to \mathbb{R}^n \times \mathbb{R}$. Hence by (ii), the set

$$Y^c \cap (\mathbb{R}^n \times (]0, \infty[)) = \{(y,t) \in \mathbb{R}^n \times \mathbb{R} : \ 0 < t \leq g(y)\}$$

is an $\mathfrak{X}$-set. The image of this set under the projection $\mathbb{R}^n \times \mathbb{R} \to \mathbb{R}^n$ equals $\overline{X}^c$. Hence it follows from (ii) and (iii) that $\overline{X}$ is an $\mathfrak{X}$-set. □

(2.2) *For a bounded $\mathfrak{X}$-set $X \subset \mathbb{R}^n$, the number of connected components of X is finite, and each connected component is an $\mathfrak{X}$-set.*

(2.3) *The curve selection lemma holds, i.e., for an $\mathfrak{X}$-set $X \subset \mathbb{R}^n$ with $\overline{X} - X \ni 0$ there exists a C^r $\mathfrak{X}$-curve $\varphi : [0,1] \to \overline{X}$ such that*

$$\varphi(0) = 0 \quad \textit{and} \quad \varphi(]0,1]) \subset X.$$

For an $\mathfrak{X}$-set $X \subset \mathbb{R}^n$, let $\Sigma_0 X$ denote the C^0 $\mathfrak{X}$-singular point set of X, where a point x of X is C^0 *$\mathfrak{X}$-singular* if x has no neighborhood in X which is $\mathfrak{X}$-homeomorphic to an open simplex of dimension $= \dim X$.

(2.4) *For a nonempty $\mathfrak{X}$-set $X \subset \mathbb{R}^n$, $\Sigma_0 X$ and $\Sigma_r X$ are $\mathfrak{X}$-sets of dimension $<$ $\dim X$.*

Conjecture 1. *For an $\mathfrak{X}$-set $X \subset \mathbb{R}^n$, $\Sigma_\infty X$ and $\Sigma_\omega X$ are $\mathfrak{X}$-sets, and if X is bounded then*

$$\Sigma_r X = \Sigma_{r+1} X = \cdots = \Sigma_\infty X$$

for some integer r.

(2.5) *Let X and Y be nonempty C^r $\mathfrak{X}$-submanifolds of $\mathbb{R}^n$ such that*

$$X \cap Y = \varnothing \quad \textit{and} \quad \overline{X} \cap Y \neq \varnothing.$$

Let Y' denote the set of points of Y where the pair of X and Y does not satisfy the Whitney condition. Then Y' is an $\mathfrak{X}$-set and of dimension $<$ $\dim Y$.

The following fact follows from (2.4) and (2.5).

(2.6) Whitney stratification theorem of $\mathfrak{X}$-sets. *An $\mathfrak{X}$-set admits the canonical C^r Whitney $\mathfrak{X}$-stratification. Here we call a C^r Whitney $\mathfrak{X}$-stratification $\{X_i\}_i$ canonical if for any other C^r Whitney $\mathfrak{X}$-stratification $\{X'_i\}_i$,*

$$\{X_i : \dim X_i \geq k\} = \{X'_i : \dim X'_i \geq k\} \quad \textit{for some } k$$

and each stratum X'_i of dimension $k-1$ is included in some stratum $X_{i'}$ of dimension $k-1$.

(2.7) Whitney stratification theorem of $\mathfrak{X}$-maps. *Let $X \subset \mathbb{R}^n$ and $Y \subset \mathbb{R}^m$ be $\mathfrak{X}$-sets, and let $f : X \to Y$ be an $\mathfrak{X}$-map such that for any bounded sets $B \subset \mathbb{R}^n$ and $C \subset \mathbb{R}^m$, $f(X \cap B)$ and $f^{-1}(C)$ are bounded. Then f admits the canonical C^r Whitney stratification. (See the book for the definition of "canonical".)*

(2.8) Triangulation theorem of $\mathfrak{X}$-sets. *A compact $\mathfrak{X}$-set admits an $\mathfrak{X}$-triangulation.*

(2.9) $\mathfrak{X}$-Hauptvermutung theorem. *Two compact $\mathfrak{X}$-homeomorphic polyhedra are PL homeomorphic. As a corollary, the $\mathfrak{X}$-triangulation in* (2.8) *is unique up to PL homeomorphism.*

Note that there are two compact homeomorphic polyhedra which are not PL homeomorphic (Milnor).

To prove (2.9) we introduce some concepts related to C^r Whitney stratifications. For simple explanation assume $r \geq 3$. Let M be a C^r submanifold of $\mathbb{R}^n$. A *tube* at M is a triple $T = (|T|, \pi, \rho)$, where $|T|$ is an open neighborhood of M in $\mathbb{R}^n$, $\pi : |T| \to M$ is a submersive C^r retraction, and ρ is a nonnegative C^r function on $|T|$ such that $\rho^{-1}(0) = M$ and each point x of M is the unique and nondegenerate critical point of the restriction of ρ to $\pi^{-1}(x)$. Let $\{X_i\}_i$ be a C^r Whitney stratification in $\mathbb{R}^n$. A *controlled tube system* for $\{X_i\}_i$ consists of one tube $T_i = (|T_i|, \pi_i, \rho_i)$ at each X_i such that for each pair i, j with $(\overline{X_j} - X_j) \cap X_i = \varnothing$,

$$\left.\begin{aligned} \pi_i \circ \pi_j(x) &= \pi_i(x), \\ \rho_i \circ \pi_j(x) &= \rho_i(x) \end{aligned}\right\} \quad \text{for } x \in |T_i| \cap |T_j|.$$

Let $\{X_i\}_i$ and $\{Y_i\}_i$ be C^r Whitney stratifications of subsets X and Y, respectively, of $\mathbb{R}^n$ such that for each i the dimensions of X_i and Y_i are equal to i. Let $k = \dim X$, let p_1 and p_2 denote the projections of $\mathbb{R}^n \times \mathbb{R}^n$ on the first and latter factors, respectively, and let $\{T_i^X = (|T_i^X|, \pi_i^X, \rho_i^X)\}_i$ be a controlled tube system for $\{X_i\}_i$. Note that for each $0 \leq i < j \leq k$ and any sequence ε of arbitrarily small numbers

$$\infty \gg \varepsilon_0 \gg \cdots \gg \varepsilon_{k-1} > 0$$

the set

$$X_{i,j,\varepsilon} = X_j \cap \rho_i^{X-1}([0, \varepsilon_i]) - \bigcup_{\ell<i} \rho_\ell^{X-1}([0, \varepsilon_\ell[)$$

is a C^r submanifold possibly with boundary and corners of $\mathbb{R}^n$. We define also $Y_{i,j,\varepsilon}$ for a controlled tube system $\{T_i^Y = (|T_i^Y|, \pi_i^Y, \rho_i^Y)\}_i$ for $\{Y_i\}_i$. A *strong isomorphism* $f : \{X_i\} \to \{Y_i\}$ is a homeomorphism from X to Y such that the following three conditions are satisfied. For each i, $f|_{X_i}$ is a C^r diffeomorphism onto Y_i. $\{Z_i\}_i = \{\text{graph } f|_{X_i}\}_i$ is a C^r Whitney stratification of Z =graph f in $\mathbb{R}^n \times \mathbb{R}^n$. For some controlled tube system for $\{X_i\}_i$, $\{Y_i\}_i$ and $\{Z_i\}_i$, each $0 \leq i < j \leq k$ and any ε, the gradients of the restrictions of ρ_i^X and $\rho_i^Z \circ p_1^{-1}$ to each face of $X_{i,j,\varepsilon}$ are nonzero and do not point to opposite directions at each point, and so are the gradients of the restrictions of ρ_i^Y and $\rho_i^Z \circ p_2^{-1}$ to each face of $Y_{i,j,\varepsilon}$, where ρ_i^Z is the function which appears in the tube system for $\{Z_i\}_i$.

Conjecture 2. *Let X and Y be compact polyhedra in $\mathbb{R}^n$. Assume $r \geq 3$. If there exists C^r Whitney stratifications $\{X_i\}_i$ of X and $\{Y_i\}_i$ of Y and a strong isomorphism $f : \{X_i\} \to \{Y_i\}$ then X and Y are PL homeomorphic.*

We can prove this under a condition.

(2.10) Triangulation theorem of stratified maps. *In Conjecture 2 assume that for each connected component C of X_i, $i = 1, 2, \dots$, $\{X_i \cap \overline{C}\}_i$ is strongly isomorphic to a C^r Whitney stratification of a solid sphere $\{x \in \mathbb{R}^{\dim C} : |x| \le 1\}$. Then X and Y are PL homeomorphic.*

Proof of (2.9). Let $r \ge 3$. Let $f : X \to Y$ be an $\mathfrak{X}$-homeomorphism between compact polyhedra. By (2.6) we have C^r Whitney $\mathfrak{X}$-stratifications $\{X_i\}_i$ of X and $\{Y_i\}_i$ of Y such that for each i, X_i and Y_i are of dimension i, $f|_{X_i}$ is a C^r diffeomorphism onto Y_i, $\{Z_i\}_i = \{\text{graph } f|_{X_i}\}_i$ is a C^r Whitney stratification of $Z = \text{graph } f$ in $\mathbb{R}^n \times \mathbb{R}^n$. Moreover, by (2.8) we can choose $\{X_i\}$ so that for each connected component C of X_i, $i = 1, 2, \dots$, there exists an $\mathfrak{X}$-homeomorphism $g_C : \overline{C} \to \{x \in \mathbb{R}^{\dim C} : |x| \le 1\}$ such that for each j, $g_C|_{X_j \cap \overline{C}}$ is a C^r imbedding and $\{g_C(X_j \cap \overline{C})\}_j$ and $\{\text{graph } g_C|_{X_j \cap \overline{C}}\}_j$ are C^r Whitney stratifications.

By using a partition of unity of class C^r and $\mathfrak{X}$, we obtain a controlled $\mathfrak{X}$-tube system for any C^r Whitney $\mathfrak{X}$-stratification. On the other hand, we see the following fact. Let f_1 and f_2 be nonnegative $\mathfrak{X}$-functions on a C^r $\mathfrak{X}$-submanifold M of $\mathbb{R}^n$ with the common zero set D. Then for some $\mathfrak{X}$-neighborhood U of 0 in M the gradients of f_1 and f_2 are nonzero and do not point to opposite directions at each point of $U - 0$. These imply that $f : \{X_i\} \to \{Y_i\}$ and each $g_C : \{X_j \cap \overline{C}\}_j \to \{g_C(X_j \cap \overline{C})\}_j$ are strong isomorphisms. Therefore, (2.9) follows from (2.10). □

To prove (2.10) we need to generalize the Cairns-Whitehead theorem on unique C^∞ triangulations of C^∞ manifolds. Let K be a simplicial complex in a Euclidean space. A C^r *map* $f : K \to \mathbb{R}^n$ is a continuous map $f : |K| \to \mathbb{R}^n$ such that $f|_\sigma$, $\sigma \in K$, are of class C^r. Let $b \in |K|$. We define the map $df_b : |\text{st}(b, K)| \to \mathbb{R}^n$ by

$$df_b(x) = d(f|_\sigma)_b(x - b) \quad \text{for } \sigma \in \text{st}(b, K),\ x \in \sigma,$$

where $\text{st}(b, K)$ denote the complex of star of b in K. We call a subset X of $\mathbb{R}^n$ C^r *triangulable* if there exists a C^r map $f : K \to \mathbb{R}^n$ such that $f(|K|) = X$ and f and df_b, $b \in |K|$, are homeomorphisms onto the images. We call (K, f) or simply $|K|$ a C^r *triangulation* of X. In the same way as [Mu] we can prove the following statement.

Statement. Let X be a subset of $\mathbb{R}^n$. Assume that for each point x of X there exists a C^r diffeomorphism germ φ of $\mathbb{R}^n$ at x such that $\varphi(X)$ is a polyhedron germ. Then X is C^r triangulable uniquely up to PL homeomorphism.

Proof of (2.10). Let $\{T_i^X = (|T_i^X|, \pi_i^X, \rho_i^X)\}_i$ and $\{T_i^Y = (|T_i^Y|, \pi_i^Y, \rho_i^Y)\}_i$ be given as above. Let ε be a sequence of sufficiently small numbers

$$\infty \gg \varepsilon_0 \gg \cdots \gg \varepsilon_{k-1} > 0.$$

For each i set

$$U_{i,\varepsilon} = X \cap \rho_i^{X-1}([0,\varepsilon_i]) - \bigcup_{\ell<i} \rho_\ell^{X-1}([0,\varepsilon_\ell[),$$

$$V_{i,\varepsilon} = Y \cap \rho_i^{Y-1}([0,\varepsilon_i]) - \bigcup_{\ell<i} \rho_\ell^{Y-1}([0,\varepsilon_\ell[),$$

$$X_{i,\varepsilon} = X_i \cap U_{i,\varepsilon} \quad \text{and} \quad Y_{i,\varepsilon} = Y_i \cap V_{i,\varepsilon}.$$

It is easy to show that these sets satisfy the conditions of the above statement. Hence they are C^r triangulable. For each i let $\tilde{X}_{i,\varepsilon}$, $\tilde{Y}_{i,\varepsilon}$, $\tilde{U}_{i,\varepsilon}$ and $\tilde{V}_{i,\varepsilon}$ denote C^r triangulations of $X_{i,\varepsilon}$, $Y_{i,\varepsilon}$, $U_{i,\varepsilon}$ and $V_{i,\varepsilon}$, respectively. By uniqueness of C^r triangulation, for the proof of (2.10) we can replace $X_{i,\varepsilon}$ with $\tilde{X}_{i,\varepsilon}$, ... , and $V_{i,\varepsilon}$ with $\tilde{V}_{i,\varepsilon}$, and it suffices to construct a PL homeomorphism from $\cup_{i=0}^k \tilde{U}_{i,\varepsilon}$ to $\cup_{i=0}^k \tilde{V}_{i,\varepsilon}$. By downward induction on i we want to define a PL homeomorphism φ_i from $\cup_{\ell=i}^k \tilde{U}_{\ell,\varepsilon}$ to $\cup_{\ell=i}^k \tilde{V}_{\ell,\varepsilon}$. By the assumption on gradients in the definition of a strong isomorphism, for each $j = 0,\dots,k$, $X_{j,\varepsilon}$ and $Y_{j,\varepsilon}$ are C^r diffeomorphic. Hence by the Cairs-Whitehead theorem $\tilde{X}_{j,\varepsilon}$ and $\tilde{Y}_{j,\varepsilon}$ are PL homeomorphic. In particular, we have a PL homeomorphism $\varphi_k : \tilde{X}_{k,\varepsilon} = \tilde{U}_{k,\varepsilon} \to \tilde{Y}_{k,\varepsilon} = \tilde{V}_{k,\varepsilon}$. Assume that for a nonnegative integer $i < k$ we have defined φ_{i+1}. By the assumption in (2.10) each connected component of $\tilde{X}_{i,\varepsilon}$ is a PL disk. Hence each connected component of $\tilde{Y}_{i,\varepsilon}$, $\tilde{U}_{i,\varepsilon}$ and $\tilde{V}_{i,\varepsilon}$ is a PL disk, a PL cone and a PL cone, respectively. Therefore, by the Alexander trick we can extend φ_{i+1} to $\tilde{U}_{i,\varepsilon} \to \tilde{V}_{i,\varepsilon}$ and then we define φ_i. □

If $\mathfrak{X} = \mathfrak{S}_1$, (2.8) and (2.9) hold for any closed subanalytic subsets of Euclidean spaces. For this generalization in the $\mathfrak{X}$ case, we need the following axiom:

Axiom (v) Let X be a subset of $\mathbb{R}^n$. Assume that X is locally $\mathfrak{X}$-set at each point of $\mathbb{R}^n$ (i.e. each point of $\mathbb{R}^n$ has a neighborhood U such that $X \cap U$ is an $\mathfrak{X}$-set). Then X is an $\mathfrak{X}$-set.

Note that $\mathfrak{S}_1$ satisfies (v) but $\mathfrak{S}_2$ or $\mathfrak{S}_3$ do not.

(2.8)′ Triangulation theorem of $\mathfrak{X}$-sets. *If the Axiom* (v) *is satisfied, then a locally closed $\mathfrak{X}$-subset of $\mathbb{R}^n$ admits an $\mathfrak{X}$-triangulation.*

(2.9)′ $\mathfrak{X}$-Hauptvermutung theorem. *If* (v) *is satisfied, then two $\mathfrak{X}$-homeomorphic polyhedra in $\mathbb{R}^n$ are PL homeomorphic. As a corollary, the $\mathfrak{X}$-triangulation in* (2.8)′ *is unique up to PL homeomorphism.*

(2.11) Triangulation theorem of $\mathfrak{X}$-functions. *An $\mathfrak{X}$-function on a compact $\mathfrak{X}$-set admits a unique $\mathfrak{X}$-triangulation. Here uniqueness means that for two $\mathfrak{X}$-triangulations (K_1,π_1) and (K_2,π_2) of an $\mathfrak{X}$-function $f : X \to \mathbb{R}$ there exists a PL homeomorphism $\tau : |K_1| \to |K_2|$ such that*

$$f \circ \pi_2 \circ \tau = f \circ \pi_1.$$

(2.11)′ Triangulation theorem of $\mathfrak{X}$-functions. *If* (v) *is satisfied, then an $\mathfrak{X}$-function on a locally closed $\mathfrak{X}$-subset of $\mathbb{R}^n$ admits a unique $\mathfrak{X}$-triangulation.*

As a corollary of (2.11) and (2.11)′ we obtain:

(2.12) Equivalence theorem of $\mathfrak{X}$-functions. *Let f_1 and f_2 be $\mathfrak{X}$-functions on an $\mathfrak{X}$-set $X \subset \mathbb{R}^n$ such that the zero point sets coincide and the positive point sets also coincide. (X is not necessarily locally closed in $\mathbb{R}^n$.) Assume that X is bounded unless $\mathfrak{X}$ satisfies* (v). *Then there exists an $\mathfrak{X}$-homeomorphism π of X such that $f_1 \circ \pi$ equals f_2 on an $\mathfrak{X}$-neighborhood of the zero set.*

(2.13) Equivalence class theorem of $\mathfrak{X}$-functions. *Consider an equivalence relation on all bounded $\mathfrak{X}$-functions on all bounded $\mathfrak{X}$-sets. Bounded $\mathfrak{X}$-functions f_1 and f_2 on bounded $\mathfrak{X}$-sets X_1 and X_2, respectively, are equivalent if there exist $\mathfrak{X}$-homeomorphisms $\pi_1 : X_2 \to X_1$ and π_2 of $\mathbb{R}$ such that*

$$f_1 \circ \pi_1 = \pi_2 \circ f_2 \quad \text{on } X_2.$$

Then the cardinal number of equivalence classes is countable.

We introduce sufficient conditions for $\mathfrak{X}$-maps to be (R-)$\mathfrak{X}$-triangulable. Note that a blowing-up map is not C^0 triangulable (which is naturally defined), e.g.

$$f(x,y) = (x, xy) \quad \text{for } (x,y) \in \mathbb{R}^2.$$

Let $X \subset \mathbb{R}^n$ and $Y \subset \mathbb{R}^m$ be locally closed $\mathfrak{X}$-sets, let $f : X \to Y$ be an $\mathfrak{X}$-map, and let $\mathbb{R}^n \times \mathbb{R}^m = \mathbb{R}^{n_0} \times \mathbb{R}^m \xrightarrow{p_1} \mathbb{R}^{n_1} \times \mathbb{R}^m \xrightarrow{p_2} \cdots \xrightarrow{p_\ell} \mathbb{R}^{n_\ell} \times \mathbb{R}^m = 0 \times \mathbb{R}^m$ be a sequence of projections which forget the respective first factors. For each $0 \le i \le \ell$ and $x \in \mathbb{R}^{n_i} \times \mathbb{R}^m$ let $\mathcal{A}_{i,x}$ be a nonempty countable family of germs at x of C^0 $\mathfrak{X}$-stratifications of neighborhoods of x in $\mathbb{R}^{n_i} \times \mathbb{R}^m$ with the frontier condition. We call $\{\mathcal{A}_{i,x}\}_{i=0,\dots\ell,\, x\in\mathbb{R}^{n_i}\times\mathbb{R}^m}$ a *local resolution* of f with respect to $p_1, \dots, p_\ell$ if the following four conditions are satisfied.

(1) For each $A_i \in \mathcal{A}_{i,x}$, $i < \ell$, there exist $A_{i+1} \in \mathcal{A}_{i+1,p_{i+1}(x)}$ and C^0 $\mathfrak{X}$-stratifications $\{X_{i,j}\}_j$ of a neighborhood U_i of x in $\mathbb{R}^{n_i} \times \mathbb{R}^m$ and $\{X_{i+1,j}\}_j$ of a neighborhood of $p_{i+1}(x)$ in $\mathbb{R}^{n_{i+1}} \times \mathbb{R}^m$ such that the germs of $\{X_{i,j}\}_j$ at x and of $\{X_{i+1,j}\}_j$ at $p_{i+1}(x)$ coincide with A_i and A_{i+1}, respectively, for each j

$$p_{i+1}(X_{i,j}) = X_{i+1,j'} \quad \text{for some } j',$$

if $p_{i+1}|_{X_{i,j}}$ is not a proper local homeomorphism onto $X_{i+1,j'}$ then $X_{i,j}$ is an open subset of $p_{i+1}^{-1}(X_{i+1,j'})$, and the union of $X_{i,j}$ such that $p_{i+1}|_{X_{i,j}}$ are proper local homeomorphisms to their images is closed in U_i.

(2) For each $x \in \operatorname{graph} f$ there exists $A \in \mathcal{A}_{0,x}$ such that the germ of graph f at x is some union of elements of A.

(3) For any germs A and B in $\mathcal{A}_{i,x}$ there exists a germ in $\mathcal{A}_{i,x}$ which is a common refinement of A and B. (4) Let $\{X_{i,j}\}_j$ be a C^0 $\mathfrak{X}$-stratification of an open set in $\mathbb{R}^{n_i} \times \mathbb{R}^m$ whose germ at a point x is an element of $\mathcal{A}_{i,x}$. Then the germ of $\{X_{i,j}\}_j$ at any point y near x is an element of $\mathcal{A}_{i,y}$.

(2.14) Triangulation theorem of $\mathfrak{X}$-maps. *Let $X \subset \mathbb{R}^n$ and $Y \subset \mathbb{R}^m$ be locally closed $\mathfrak{X}$-sets and let $f : X \to Y$ be an $\mathfrak{X}$-map.*

(1) *Assume Axiom* (v) *unless X and Y are compact. Then the proper f is $\mathfrak{X}$-triangulable if after a change of the coordinate system of $\mathbb{R}^n$ by an $\mathfrak{X}$-homeomorphism of $\mathbb{R}^n$ there exists a local resolution $\{\mathcal{A}_{i,x}\}_{i=0,\dots,\ell,\, x\in\mathbb{R}^{n_i}\times\mathbb{R}^m}$ of f with respect to some projections $\mathbb{R}^{n_0}\times\mathbb{R}^m \xrightarrow{p_1} \cdots \xrightarrow{p_\ell} \mathbb{R}^{n_\ell}\times\mathbb{R}^m$.*

(2) *Assume Axiom* (v) *unless X is compact. The f which is not necessarily proper is R-$\mathfrak{X}$-triangulable if, in addition, Y is an $\mathfrak{X}$-polyhedron and for any $\{X_{\ell,j}\}_j \in \mathcal{A}_{\ell,x}$, each $X_{\ell,j}$ is the germ of a semilinear set.*

(3) *If X is closed in $\mathbb{R}^n$ then these conditions are necessary for the corresponding properties.*

As a corollary we can prove

(2.15) Triangulation theorem of complex analytic functions. *A complex analytic function on a complex analytic set in $\mathbb{C}^n$ admits a subanalytic R-triangulation.*

Before we prove (2.14) and (2.15), we give a conjecture and remarks on $\mathfrak{X}$-triangulations of $\mathfrak{X}$-maps.

Let $f : X \to Y$ be a proper $\mathfrak{X}$-map between locally closed $\mathfrak{X}$-sets. Assume Axiom (v). If f is C^0 $\mathfrak{X}$-triangulable then there exists a C^0 $\mathfrak{X}$-stratification $\{X_i\} \to \{Y_j\}$ of f such that the function d_f on X, defined by

$$d_f(x) = \dim X_i - \dim f(X_i) \quad \text{for } x \in X_i,$$

is lower semicontinuous. But this condition is not sufficient. For example,

$$X = [0,\infty[\times\mathbb{R}\times[0,1], \quad Y = \mathbb{R}^2,$$

$$f(x,y,z) = (x, y^2 - xz) \quad \text{for } (x,y,z) \in [0,\infty[\times\mathbb{R}\times[0,1].$$

If f, X and Y are complex analytic then the condition that f is flat is necessary for f to be locally C^0 triangulable. However, the condition is not sufficient. For example, the map $f : \mathbb{C}^4 \to \mathbb{C}^3$, which is defined by

$$f(x_1,\dots,x_4) = (x_1, x_2, ((x_1 - x_2x_3)^2 - x_4^2)^n) \quad \text{for } (x_1,\dots,x_4) \in \mathbb{C}^4,$$

is not C^0 triangulable, where n is a sufficiently large integer.

A C^1 Whitney stratified map $f : \{X_i\} \to \{Y_j\}$ is called a *stratified map sans éclatement* if the following condition is satisfied. Let $\cup_i X_i = X \subset \mathbb{R}^n$. Let X_i and $X_{i'}$ be distinct strata with $\overline{X_i}\cap X_{i'} \neq \varnothing$, and $\{a_k\}$ a sequence of points in X_i such that $\{a_k\}$ converges to a point a of $X_{i'}$ and the sequence of the tangent spaces $\{T_{a_k}(f|_{X_i})^{-1}f(a_k)\}_k$ converges to a linear subspace $L \subset \mathbb{R}^n$. Then L includes $T_a(f|_{X_{i'}})^{-1}f(a)$.

Conjecture 3. *Let $f : X \to Y$ be a proper $\mathfrak{X}$-map between locally closed $\mathfrak{X}$-sets. Assume Axiom* (v). *If f admits a C^1 Whitney $\mathfrak{X}$-stratification sans éclatement, then f is $\mathfrak{X}$-triangulable.*

We can prove Conjecture 3 under the condition of $d_f \leq 1$ or dim $X \leq 3$.

Proof of (2.14). (3) in (2.14) (necessity of the conditions) is easy to prove, and (1) and (2) follow from the following assertion.

Assertion 1. Let $X \subset \mathbb{R}^n \times \mathbb{R}^m$ be a closed $\mathfrak{X}$-set, $p : \mathbb{R}^n \times \mathbb{R}^m \to \mathbb{R}^m$ the projection, $P \subset \mathbb{R}^m$ a closed polyhedron which includes $p(X)$, and $\mathbb{R}^n \times \mathbb{R}^m = \mathbb{R}^{n_0} \times \mathbb{R}^m \xrightarrow{p_1} \cdots \xrightarrow{p_\ell} \mathbb{R}^{n_\ell} \times \mathbb{R}^m = \mathbb{R}^m$ projections which forget the respective first factors. Assume Axiom (v) and a local resolution $\{\mathcal{A}_{i,x}\}_{i=0,\ldots,\ell,\, x \in \mathbb{R}^{n_i} \times \mathbb{R}^m}$ of X with respect to $p_1, \ldots, p_\ell$ (which is naturally defined). If $p|_X$ is proper, there exists an $\mathfrak{X}$-homeomorphism π of $\mathbb{R}^n \times \mathbb{R}^m$ of the form

$$\pi(x,y) = (\pi'(x,y), \pi''(y)) \quad \text{for } (x,y) \in \mathbb{R}^n \times \mathbb{R}^m$$

such that $\pi(X)$ is a polyhedron and π'' is invariant on P. Even in the case where $p|_X$ is not proper, if for any $\{X_{\ell,j}\}_j \in \mathcal{A}_{\ell,x}$, each $X_{\ell,j}$ is the germ of a semilinear set then we can choose the above π so that $\pi'' =$id.

By induction on ℓ, we can reduce this to

Assertion 2. Assume Axiom (v). Let $Y \supset Y_1 \supset Y_2$ be closed $\mathfrak{X}$-sets in $\mathbb{R}^n \times \mathbb{R}^m$, and let $p : \mathbb{R}^n \times \mathbb{R}^m \to \mathbb{R}^m$ and $q : Y \to \mathbb{R}^m$ denote the projection and its restriction to Y, respectively. Let L be a simplicial complex with underlying polyhedron $= q(Y)$, and let Q be a closed polyhedron in $\mathbb{R}^n \times \mathbb{R}^m$ which does not intersect with Y. Assume that q is proper, Y_1 and Y_2 are polyhedra, Y_1 is a neighborhood of Y_2 in Y, $q^{-1}(q(Y_1)) = Y_1$, and for each simplex $\sigma \in L$, the map

$$q|_{q^{-1}(\sigma^\circ)} : q^{-1}(\sigma^\circ) \to \sigma^\circ$$

is a proper local homeomorphism. Then there exists an $\mathfrak{X}$-homeomorphism π of $\mathbb{R}^n \times \mathbb{R}^m$ of the form

$$\pi(x,y) = (\pi'(x,y), y) \quad \text{for } (x,y) \in \mathbb{R}^n \times \mathbb{R}^m$$

such that $\pi(Y)$ is a polyhedron and $\pi =$ id on $Q \cup p^{-1}(q(Y_2))$.

It is not easy to construct directly π in Assertion 2. But in the same way as the proof of (2.8) we obtain easily an $\mathfrak{X}$-homeomorphism τ of $\mathbb{R}^n \times \mathbb{R}^m$ of the form

$$\tau(x,y) = (\tau'(x,y), \tau''(y)) \quad \text{for } (x,y) \in \mathbb{R}^n \times \mathbb{R}^m$$

such that $\tau(Y)$ is a polyhedron, τ'' is invariant on each simplex of a subdivision L' of L,

$$\tau'(x,y) = x \quad \text{for } (x,y) \in Q,$$

$$\tau = \text{id} \quad \text{on } p^{-1}(q(Y_2)).$$

Hence we reduce Assertion 2 to

Assertion 3. Let $Y \supset Y_2$, $p : \mathbb{R}^n \times \mathbb{R}^m \to \mathbb{R}^m$, $q : Y \to \mathbb{R}^m$, L, L' and Q be the same as above. Let π'' be an $\mathfrak{X}$-homeomorphism of $\mathbb{R}^m$ which is invariant on each simplex of L' and the identity on $q(Y_2)$. Then there exists an $\mathfrak{X}$-homeomorphism π of $\mathbb{R}^n \times \mathbb{R}^m$ of the form

$$\pi(x,y) = (\pi'(x,y), \pi''^{-1}(y)) \quad \text{for } (x,y) \in \mathbb{R}^n \times \mathbb{R}^m$$

such that π is invariant on Y,

$$\pi'(x,y) = x \quad \text{for } (x,y) \in Q,$$

$$\pi = \text{id} \quad \text{on } p^{-1}(q(Y_2)).$$

In order to apply an induction method we first generalize Assertion 3 to the case where π'' and π are parameterized as Lemma 5 in [S$_2$]. Then we can prove it in the same way as the lemma. □

Proof of (2.15). Consider the case $\mathfrak{X}$ = the family of subanalytic sets. Let f be a complex analytic function on a complex analytic set X in $\mathbb{C}^n$. Let $\mathbb{C}^n \times \mathbb{C} \xrightarrow{p_1} \mathbb{C}^{n-1} \times \mathbb{C} \xrightarrow{p_2} \cdots \xrightarrow{p_n} \mathbb{C}$ denote the projections which forget the respective first factors. By (2.14) it suffices to find a complex analytic diffeomorphism π of $\mathbb{C}^n$ and a local resolution $\{\mathcal{A}_{i,x}\}_{i=0,\dots,n,\, x\in\mathbb{C}^{n-i}\times\mathbb{C}}$ of $f \circ \pi$ with respect to $p_1, \dots, p_n$ such that each $\mathcal{A}_{i,x}$ is the germ of a complex analytic subanalytic stratification. (The other conditions in (2) of (2.14) are automatically satisfied because $Y = \mathbb{C}$ is an $\mathfrak{X}$-polyhedron and for any $\{X_{n,j}\}_j \in \mathcal{A}_{n,x}$, each $X_{n,j}$ is the germ of $\mathbb{C}$ or of x.

In the proof also we generalize the problem so that an induction method works. Let O_α and Y_α, $\alpha = 1, 2, \dots$, be open sets in $\mathbb{C}^n \times \mathbb{C}$ and complex analytic sets in O_α. We define a local resolution $\{\mathcal{A}_{i,x}\}_{i=0,\dots n,\, x\in\mathbb{C}^{n-i}\times\mathbb{C}}$ of $\{Y_\alpha\}_\alpha$ with respect to $p_1, \dots, p_n$ by replacing (2) of the definition of a local resolution with (2)′ for any Y_α and any $x \in Y_\alpha$ there exists $A \in \mathcal{A}_{0,x}$ such that the germ of Y_α at x is some union of elements of A. Then what we do is to find a complex analytic diffeomorphism π of $\mathbb{C}^n$ and a local resolution $\{\mathcal{A}_{i,x}\}_{i=0,\dots,n,\, x\in\mathbb{C}^{n-i}\times\mathbb{C}}$ of $\{\tilde{\pi}(Y_\alpha)\}_\alpha$ with respect to $p_1, \dots, p_n$ such that each $\mathcal{A}_{i,x}$ is the germ of a complex analytic stratification, where $\tilde{\pi}$ is the diffeomorphism of $\mathbb{C}^n \times \mathbb{C}$ defined by

$$\tilde{\pi}(x,t) = (\pi(x), t) \quad \text{for } (x,t) \in \mathbb{C}^n \times \mathbb{C}.$$

Since the case of $n = 0$ is trivial we proceed by induction on n. We can assume that each Y_α is irreducible and of complex dimension $\le n$, for odd α Y_α is an open set of $\mathbb{C}^n \times c$ for some $c \in \mathbb{C}$, and for even α Y_α is not so. There exists a complex bipolynomial diffeomorphism τ of $\mathbb{C}^n \times \mathbb{C}$ of the same form as the above $\tilde{\pi}$ such that for every $y \in \mathbb{C}^{n-1} \times \mathbb{C}$ and for even α, $\tau(Y_\alpha) \cap (\mathbb{C} \times y)$ is of dimension 0. Replace $\{Y_\alpha\}_\alpha$ with $\{\tau(Y_\alpha)\}_\alpha$. Then for every $y \in \mathbb{C}^{n-1} \times \mathbb{C}$ and even α, $Y_\alpha \cap (\mathbb{C} \times y)$ is of dimension 0. For each α there exists a countable covering $\{P_\beta\}_\beta$ of O_α by open subanalytic sets of the form $Q_\beta \times R_\beta \subset \mathbb{C} \times (\mathbb{C}^{n-1} \times \mathbb{C})$ such that for odd α

$$P_\beta \cap Y_\alpha = Q_\beta \times p_1(P_\beta \cap Y_\alpha),$$

and for even α $p_1|_{Y_\alpha} : Y_\alpha \to R_\alpha$ is proper. For each α let $(\{X_{0,j}(\alpha)\}_j, \{X_{1,j}(\alpha)\}_j)$ be a finite subanalytic analytic stratification of $p_1|_{Q_\alpha}$ such that Y_α is some union of $X_{0,j}(\alpha)$, the union of $X_{0,j}(\alpha)$ such that $p_1|_{X_{0,j}(\alpha)}$ are proper local homeomorphisms to their images is closed in O_α, and for each integer ℓ, $\cup_{\dim\leq\ell} X_{0,j}(\alpha)$ and $\cup_{\dim\leq\ell} X_{1,j}(\alpha)$ are complex analytic sets in O_α and R_α, respectively. Here we add all finite intersections $O_\alpha \cap \cdots \cap O_{\alpha'}$ and $Y_\alpha \cap \cdots \cap Y_{\alpha'}$ to the family $\{O_\alpha, Y_\alpha\}$. Then we can assume that for any α and α' there exist a countable number of α'' such that

$$O_\alpha \cap O_{\alpha'} = \bigcup_{\alpha''} O_{\alpha''}, \quad Y_{\alpha''} = O_{\alpha''} \cap Y_\alpha \cap Y_{\alpha'},$$

and for each $i = 0, 1$ and each α'', $\{X_{i,j}(\alpha'')\}_j$ is a refinement of $\{X_{i,j}(\alpha) \cap X_{i,j'}(\alpha') \cap O_{\alpha''}\}_{j,j'}$.

Consider $(R_\alpha, p_1(Y_\alpha))$ for all odd α and $(R_\alpha, \cup_{\dim\leq\ell} X_{1,j}(\alpha))$ for all even α and $\ell = 0, \dots, n-1$. By induction hypothesis, changing the coordinate system of $\mathbb{C}^{n-1}$ in $\mathbb{C}^{n-1} \times \mathbb{C}$ we can assume a local resolution $\{\mathcal{A}_{i,x}\}_{i=1,\dots,n,\, x\in\mathbb{C}^{n-i}\times\mathbb{C}}$ of $\{p_1(Y_\alpha)\}_{\alpha:\text{odd}} \cup \{\cup_{\dim\leq\ell} X_{1,j}(\alpha)\}_{\alpha:\text{even},\, \ell=0,\dots,n-1}$ with respect to $p_2, \dots, p_n$. For $x \in \mathbb{C}^n \times \mathbb{C}$ let $\mathcal{A}_{0,x}$ denote the germs of $\{X_{0,j}(\alpha)\}_j$ at x for all α with $x \in O_\alpha$. Then $\{\mathcal{A}_{i,x}\}_{i=0,\dots,n,\, x\in\mathbb{C}^{n-i}\times\mathbb{C}}$ satisfies the conditions of a local resolution of $\{Y_\alpha\}_\alpha$ except that for each j

$$p_1(X_{0,j}) = X_{1,j'} \quad \text{for some } j'.$$

We modify $\{\mathcal{A}_{0,j}\}_j$ as follows. Let $A \in \mathcal{A}_{0,x}$. Then there exist $B \in \mathcal{A}_{1,p_1(x)}$ and complex analytic stratifications $\{X_{0,j}\}_j$ in $\mathbb{C}^n \times \mathbb{C}$ and $\{X_{1,j}\}_j$ in $\mathbb{C}^{n-1} \times \mathbb{C}$ such that the germs of $\{X_{0,j}\}_j$ at x and of $\{X_{1,j}\}_j$ at $p_1(x)$ are A and B, respectively, and $\{X_{1,j}\}_j$ is a refinement of $\{p_1(X_{0,j})\}_j$. Replace A with the germs of $\{X_{0,j} \cap p_1^{-1}(X_{1,j'})\}_{j,j'}$ at x for all such B, $\{X_{0,j}\}_j$ and $\{X_{0,j}\}_j$. Then $\{\mathcal{A}_{i,x}\}_{i=0,\dots,n,\, x\in\mathbb{C}^{n-i}\times\mathbb{C}}$ satisfies the remaining condition. □

Next we see an $\mathfrak{X}$-version of Thom's transversality theorem [L] and an $\mathfrak{X}$-approximation theorem, which are used to prove the triviality theorem (2.18) but interesting by themselves. Let $r' < r$ be a nonnegative integer, and let X and Y be C^r $\mathfrak{X}$-submanifolds of $\mathbb{R}^n$ and $\mathbb{R}^m$, respectively, with $\dim X = n'$. Let $C^r_{\mathfrak{X}}(X,Y)$ and $B(X,Y)$ denote the families of all C^r $\mathfrak{X}$-maps from X to Y and of those which carry bounded sets in $\mathbb{R}^n$ to bounded sets in $\mathbb{R}^m$, respectively. We give naturally a topology to $C^r_{\mathfrak{X}}(X,Y)$ and $B^r(X,Y)$ in the same way as the Nash case. (See [S$_1$].) As usual, for each $(x,y) \in X \times Y$ let $J^{r'}_{x,y}(X,Y)$ denote the set of all r'-jets of germs $(X,x) \to (Y,y)$, set

$$J^{r'}(X,Y) = \bigcup_{(x,y)\in X\times Y} J^{r'}_{x,y}(X,Y),$$

and let the map

$$C^r_{\mathfrak{X}}(X,Y) \times X \ni (f,x) \to J^{r'} f(x) \in J^{r'}(X,Y)$$

be defined. Then we can imbed naturally $J^{r'}(X,Y)$ in some Euclidean space so that $J^{r'}(X,Y)$ is a $C^{r-r'}$ $\mathfrak{X}$-manifolds, and for each $f \in C^r_{\mathfrak{X}}(X,Y)$ the map

$$J^{r'}f : X \to J^{r'}(X,Y)$$

is of class $C^{r-r'}$ $\mathfrak{X}$. Then

(2.16) Transversality theorem. *Let $\{X_i\}_i$ be a finite $C^{r-r'}$ Whitney $\mathfrak{X}$-stratification in $J^{r'}(X,Y)$ with* codim $\cup_i X_i = n''$. *Assume that $\cup_i X_i$ is closed in $J^{r'}(X,Y)$ and*

$$r - r' > \max\,(n' - n'', 0).$$

Then the subset of $B^r(X,Y)$ consisting of f such that $J^{r'}f$ are transversal to each X_i is open and dense in $B^r(X,Y)$.

(2.17) Approximation theorem. *We can approximate any map in $B^{r'}(X,Y)$ by a C^r $\mathfrak{X}$-map.*

(2.18) Triviality theorem. *Let M be a bounded C^r $\mathfrak{X}$-manifold, set*

$$U = \{y \in \mathbb{R}^m : |y| < 1\}$$

and let $f : M \to U$ be a proper C^r $\mathfrak{X}$-submersion. Then f is C^r $\mathfrak{X}$-trivial, i.e. there exists a C^r $\mathfrak{X}$-diffeomorphism π from M to $f^{-1}(0) \times U$ of the form $\pi = (\pi', f)$.

Note that C^r triviality of f is trivial and local C^r $\mathfrak{X}$-triviality is easy to prove.

Proof. We proceed through two stages. First we prove that there exists a finite $C^r\mathfrak{X}$-stratification $\{U_i\}$ of U such that for each i,

$$f|_{f^{-1}(U_i)} : f^{-1}(U_i) \to U_i$$

is C^r $\mathfrak{X}$-trivial. Next we prove (2.18) in this situation. In the Nash case [C-S$_2$] carried out the first stage, and then [C-S$_2$] did the second. [C-S$_2$] did not use essentially the theory of real spectrum in the proof, and the proof proceeds in the case of any $\mathfrak{X}$. But this is not the case for the first stage. Hence we sketch the proof of the first stage. We generalize the problem for an induction procedure as follows.

Statement. Let X, N and S be bounded $\mathfrak{X}$-sets in $\mathbb{R}^n$, let U be a bounded C^r $\mathfrak{X}$-submanifold of $\mathbb{R}^{m'}$ of dimension m, let $f : \mathbb{R}^n \to U$ be a C^r $\mathfrak{X}$-map, and set

$$f_X = f|_X,\ f_N = f|_N,\ f_S = f|_S,\ f_{X-S} = f|_{X-S},$$

$$X_y = f_X^{-1}(y),\ N_y = f_N^{-1}(y),\ S_y = f_S^{-1}(y) \quad \text{for each } y \in U.$$

Assume that S is a subset of X and of N, $X - S$ is a C^r $\mathfrak{X}$-manifold of dimension $k+m$, N is a C^r $\mathfrak{X}$-manifold of dimension $k+m+1$, N includes a neighborhood of S in X, $f|_X$ is proper, $f|_{X-S}$ is a C^r submersion, for each $y \in U$, N_y is C^r smooth, S_y is of dimension 0, and there exists a C^r $\mathfrak{X}$-function φ on N such that

$\varphi^{-1}(0) = X \cap N$ and for $y \in U$ $\varphi|_{N_y}$ takes the Morse type at each point of S_y. Shrink N. Then there exist a finite C^r $\mathfrak{X}$-stratification $\{U_i\}_i$ and for each U_i and $y_i \in U_i$, a C^r $\mathfrak{X}$-imbedding

$$\pi_i : f_X^{-1}(U_i) \cup f_N^{-1}(U_i) \to (X_{y_i} \cup N_{y_i}) \times U_i$$

of the form $\pi_i = (\pi_i', f)$ such that

$$\pi_i(f_X^{-1}(U_i)) = X_{y_i} \times U_i, \quad \pi_i(f_N^{-1}(U_i)) \subset N_{y_i} \times U_i,$$

$$\varphi \circ \pi_i' = \varphi \quad \text{on } f_N^{-1}(U_i).$$

If this statement holds then we apply it to $X = M$ and $N = S = \varnothing$ in (2.18). Then (2.18) follows. Hence it suffices to prove the statement. Since the case of $k = 0$ is trivial, we prove the statement by induction on k. For simplicity of notation we assume $0 \in U$. By stratifying U we can suppose that S is C^r smooth and f_S is a proper local homeomorphism. Shrink N. Then f_N is a C^r submersion. Hence, shrinking N once more, we can define a C^r $\mathfrak{X}$-imbedding $\pi : N \to N_0 \times U$ of the form $\pi = (\pi', f)$ such that

$$\pi(N \cap X) = (N_0 \cap X_0) \times U, \quad \varphi \circ \pi' = \varphi.$$

It suffices to stratify U and to extend π to X. Hence we reduce the problem to the case of $N = S = \varnothing$. (Here f_X is no longer proper, and we need some modification of the induction procedure.) Assume a bounded C^r $\mathfrak{X}$-function ψ on X such that for each $y \in U$, $\psi|_{X_y}$ takes the Morse type at each critical point. For each $y \in U$, let Z_y denote the critical point set of $\psi|_{X_y}$. Set $Z = \cup_{y \in U} Z_y$. Then we can assume that ψ is constant on each connected component of Z. Apply the induction hypothesis to $\psi^{-1}(\psi(Z))$, X, Z, f and ψ. Then we obtain a C^r $\mathfrak{X}$-trivialization of the restriction of f to an $\mathfrak{X}$-neighborhood V of $\psi^{-1}(\psi(Z))$ in X. Next we can extend the trivialization to $X - V$ if we apply once more the induction hypothesis to

$$(f, \psi)|_{X-V} : X - V \to \mathbb{R}^{m+1}.$$

It remains to construct ψ. Let a subset Q of $J^2(X, \mathbb{R})$ be given so that

$$Q \cap J^2_{x,t}(X, \mathbb{R}) = \{J^2\lambda(x) : \lambda \in C^r_{\mathfrak{X}}(X, \mathbb{R}),\ \lambda(x) = t,\ x \text{ is a degenerate critical point of } \lambda|_{X_{f(x)}}\} \quad \text{for } (x,t) \in X \times \mathbb{R},$$

and let $\{X_i\}$ be a finite C^{r-2} Whitney $\mathfrak{X}$-stratification of Q. Using (2.17) we modify r so that $r > d + 1$. Then there exists ψ such that $(J^2\psi)^{-1}Q$ is of dimension $<$ dim U. Hence stratifying U we suppose that $(J^2\psi)^{-1}Q$ is empty, which implies that for each $y \in U$, $\psi|_{X_y}$ takes the Morse type at each critical point. □

(2.19) First isotopy lemma. *Let X be a bounded $\mathfrak{X}$-subset of $\mathbb{R}^n$, set*

$$U = \{y \in \mathbb{R}^m : \ |y| < 1\},$$

let $\{X_i\}_i$ be a C^r Whitney $\mathfrak{X}$-stratification of X, and let $f : X \to U$ be a proper $\mathfrak{X}$-map such that f is extenddable to a C^r $\mathfrak{X}$-map from an open $\mathfrak{X}$-neighborhood of

X in $\mathbb{R}^n$ to U, and the restriction of f to each stratum is a C^r submersion. Then f is $\mathfrak{X}$-trivial, i.e. there exists an $\mathfrak{X}$-homeomorphism π from X to $f^{-1}(0) \times U$ of the form $\pi = (\pi', f)$ such that for each X_i, $\pi|_{X_i}$ is a C^r $\mathfrak{X}$-diffeomorphism onto $(f^{-1}(0) \cap X_i) \times U$.

Using (2.18) in the same way as in [C-S$_1$], we can prove the following two theorems (the latter half of (2.20) is proved by (2.11)).

(2.20) Compactification theorem. *Let M be a noncompact bounded C^r $\mathfrak{X}$-submanifold of $\mathbb{R}^n$. Then there exists a unique compact C^r $\mathfrak{X}$-submanifold of $\mathbb{R}^n$ with boundary whose interior is C^r $\mathfrak{X}$-diffeomorphic to M. Here uniqueness means that if we have two compactifications of M then they are C^r $\mathfrak{X}$-diffeomorphic.*

Let N be a bounded C^0 $\mathfrak{X}$-submanifold of $\mathbb{R}^n$. Then there exists a unique compact PL manifold possibly with boundary whose interior is $\mathfrak{X}$-homeomorphic to N.

Note that there are two compact C^r manifolds with boundary which are not C^r diffeomorphic but whose interiors are C^r diffeomorphic.

(2.21) Algebraic model theorem. *A bounded C^r $\mathfrak{X}$-submanifold of $\mathbb{R}^n$ is C^r $\mathfrak{X}$-diffeomorphic to some affine nonsingular algebraic variety.*

Note that (2.21) does not necessarily hold in the C^0 case because there exists a compact PL manifold which is not smoothable.

We call two subsets X and Y of $\mathbb{R}^n$ *C^r diffeomorphic* if there exists a C^r diffeomorphism from a neighborhood of X in $\mathbb{R}^n$ to one of Y which carries X to Y.

(2.22) Diffeomorphism implies $\mathfrak{X}$-homeomorphism theorem. *Let X and Y be $\mathfrak{X}$-subsets of $\mathbb{R}^n$. Assume that X and Y are bounded unless Axiom* (v) *is satisfied. If X and Y are C^1 diffeomorphic then they are $\mathfrak{X}$-homeomorphic.*

Proof. Let $\dim X = k$, and let $f : U \to V$ be a C^1 diffeomorphism between neighborhoods of X and Y such that $f(X) = Y$. Assume that X and Y are bounded for simplicity of notation. Let $\{X_i\}_i$ and $\{Y_i\}_i$ denote the canonical Whitney C^1 stratifications of X and Y, respectively, such that for each i the dimensions of X_i and Y_i are equal to i. Then by definition of a canonical Whitney C^1 stratification, f carries each X_i onto Y_i, and by (2.6) $\{X_i\}_i$ and $\{Y_i\}_i$ are of class $\mathfrak{X}$. Let $\{T_i^X = (|T_i^X|, \pi_i^X, \rho_i^X)\}_i$ and $\{T_i^Y = (|T_i^Y, \pi_i^Y, \rho_i^Y)\}_i$ be controlled $\mathfrak{X}$-tube systems for $\{X_i\}_i$ and $\{Y_i\}_i$, respectively. (Here we extend the foregoing definition of a tube system to the C^1 case.) Let ε be a sequence of sufficiently small numbers

$$\infty \gg \varepsilon_0 \gg \cdots \gg \varepsilon_k > 0.$$

Set

$$U_{i,\varepsilon} = X \cap \rho_i^{X-1}([0, \varepsilon_i]) - \bigcup_{\ell<i} \rho_\ell^{X-1}([0, \varepsilon_\ell[),$$

$$X_{i,\varepsilon} = X_i \cap U_{i,\varepsilon} \quad \text{for } i = 0, \dots, k,$$

and define similarly $V_{i,\varepsilon}$ and $Y_{i,\varepsilon}$ by Y and $\{Y_i\}$. Then each $X_{i,\varepsilon}$ and $Y_{i,\varepsilon}$ are compact C^1 $\mathfrak{X}$-submanifolds of R^n possibly with boundary and corners, and (*)

$$\rho_i^X|_{U_{i,\varepsilon}} : U_{i,\varepsilon} \to X_{i,\varepsilon} \quad \text{and} \quad \rho_i^Y|_{V_{i,\varepsilon}} : V_{i,\varepsilon} \to Y_{i,\varepsilon}$$

are $\mathfrak{X}$-retractions. We can modify the restriction of f to each $X_{i,\varepsilon}$ so that it is a C^1 $\mathfrak{X}$-diffeomorphism onto $Y_{i,\varepsilon}$ and each face of $X_{i,\varepsilon}$ is carried to the naturally corresponding face of $Y_{i,\varepsilon}$. Let f_i denote the modified diffeomorphism. We define an $\mathfrak{X}$-diffeomorphism g from X to Y by a downward induction method as follows. On $X_{k,\varepsilon} = U_{i,\varepsilon}$ we set $g = f_k$. Let $0 \le i < k$ be an integer, and assume that we have defined

$$g : U_{i+1,\varepsilon} \cup \cdots \cup U_{k,\varepsilon} \to V_{i+1,\varepsilon} \cup \cdots \cup V_{k,\varepsilon}$$

so that

$$g = f_j \quad \text{on } X_{j,\varepsilon},\ j = i+1, \dots, k.$$

Define g to be f_i on $X_{i,\varepsilon}$. Then by (*) we can extend g to $U_{i,\varepsilon}$. Thus we obtain g. □

(2.23) Remark. (1) We conjecture the following. Let f_1 and f_2 be $\mathfrak{X}$-functions on an $\mathfrak{X}$-subset X of $\mathbb{R}^n$. Assume C^r diffeomorphisms π_1 of X and π_2 of $\mathbb{R}$ such that

$$f_1 \circ \pi_1 = \pi_2 \circ f_2 \quad \text{on } X.$$

Assume Axiom (v) or that f_1, f_2 and X are bounded. Then there exist C^r $\mathfrak{X}$-homeomorphisms τ_1 of X and τ_2 of $\mathbb{R}$ such that

$$f_1 \circ \tau_1 = \tau_2 \circ f_2 \quad \text{on } X.$$

(2) In (1) if f_1, f_2 and X are C^1 smooth then we find $\mathfrak{X}$-homeomorphisms τ_1 of X and τ_2 of $\mathbb{R}$ with the above equality. We can prove this by (2.22) because the critical value sets of f_1 and f_2 are finite locally at each point of X.

(3) If Axiom (v) is not satisfied then the boundedness condition in (2.22) and the above (1) is necessary. Indeed there exist an affine nonsingular real algebraic variety X and smooth rational functions f_1 and f_2 on X such that $f_1 \circ \pi = f_2$ for some analytic diffeomorphism π of X,

$$\dim f_1^{-1}(0) = \dim f_2^{-1}(0) < \dim X - 1,$$

$$\dim f_1^{-1}(t) = \dim f_2^{-1}(t) = \dim X - 1, \quad t \in \mathbb{R} - 0,$$

and no semialgebraic homeomorphism of X carries $f_1^{-1}(0)$ to $f_2^{-1}(0)$. For example, let X_1 and X_2 be compact affine nonsingular real algebraic varieties which are h-cobordant but not diffeomorphic. Let

$$\varphi : X_1 \times \mathbb{R} \to X_2 \times \mathbb{R}, \quad g_i : X_i \times \mathbb{R} \to \mathbb{R},\ i = 1, 2,$$

be a C^∞ diffeomorphism and polynomial functions, respectively, such that

$$\varphi(x_0, s_0) = (y_0, t_0) \quad \text{for some } (x_0, s_0, y_0, t_0) \in X_1 \times \mathbb{R} \times X_2 \times \mathbb{R},$$

$$g_1^{-1}(0) = \{(x_0, s_0)\}, \quad g_2^{-1}(0) = \{(y_0, t_0)\}, \quad g_1 = g_2 \circ \varphi \quad \text{near } (x_0, s_0).$$

Set

$$X = X_1 \times \mathbb{R} \times X_2 \times \mathbb{R} \times \mathbb{R},$$

$$f_1(x, s, y, t, r) = rg_1(x, s), \quad f_2(x, s, y, t, r) = rg_2(y, t).$$

Then X, f_1 and f_2 fulfill the requirements.

(4) There exist a compact polyhedron X and PL functions f_1 and f_2 on X such that $0 \in X$, $f_1 \circ \pi = f_2$ as germs at 0 for some homeomorphism germ $\pi : (X, 0) \to (X, 0)$ but $f_1 \circ \tau = f_2$ does not hold for any $\mathfrak{X}$-homeomorphism germ $\tau : (X, 0) \to (X, 0)$. By (2.11) we can construct such an example using a pair of compact polyhedra homeomorphic but not PL homeomorphic of Milnor.

(5) There exist analytic functions f_1 and f_2 on $\mathbb{R}^n$, $n \geq 6$, such that $f_1 \circ \pi = f_2$ for some homeomorphism π of $\mathbb{R}^n$, $f_1^{-1}(0)$ and $f_2^{-1}(0)$ are not subanalytically homeomorphic nor C^0 manifolds, and for each $t \in \mathbb{R} - 0$, $f_1^{-1}(t)$ and $f_2^{-1}(t)$ are analytic manifolds (in particular, $f_1 \circ \tau_1 = \tau_2 \circ f_2$ does not hold for any subanalytic homeomorphisms τ_1 of $\mathbb{R}^n$ and τ_2 of $\mathbb{R}$). To construct this example we use a counter-example by Siebenmann to the s-cobordism conjecture [Si]. □

3. $\mathfrak{Y}$-sets, $\mathfrak{X}^0$-sets and $\mathfrak{Y}^0$-sets

With a mind to the family $\mathfrak{S}_3$ of semilinear sets, we introduce a family $\mathfrak{Y}$ of subsets of Euclidean spaces, which satisfies the axioms:

Axiom (i)$_\mathrm{Y}$ Every semilinear set in a Euclidean space is an element of $\mathfrak{Y}$.

Axiom (ii), (iii) and **(iv)**.

Axiom (vi) For an element $Y \subset 1 \times \mathbb{R}^{n-1}$ of $\mathfrak{Y}$, the cone $0 * Y \subset \mathbb{R}^n$ is an element of $\mathfrak{Y}$.

Axiom (vii) For an element $Y \subset \mathbb{R}^n$ of $\mathfrak{Y}$, the contangent CTY of Y is an element of Y, where

$$CTY = \{(y, z) \in Y \times \mathbb{R}^n : \text{there exists a sequence } \{y_i\}_{i=1,2,\dots} \text{ in } Y \text{ convergent to } y \text{ such that } |z|(y_i - y)/|y_i - y| \text{ converges to } z\}.$$

Note that (1) for a C^1 submanifold Y of $\mathbb{R}^n$, CTY coincides with the tangent space TY, (2) for a closed polyhedron Y in $\mathbb{R}^n$, CTY is a polyhedron in $\mathbb{R}^n \times \mathbb{R}^n$ (consequently, $\mathfrak{S}_3$ is $\mathfrak{Y}$), and (3) it follows from (2.1) that a family $\mathfrak{X}$ is $\mathfrak{Y}$. We define a $\mathfrak{Y}$-*set*, a $\mathfrak{Y}$-*map*, a $\mathfrak{Y}$-*manifold*, etc. in the same way as in the $\mathfrak{X}$ case.

(3.i)$_{i=1,\dots,7}$ Theorem. *For each* $i = 1, \dots, 7$, *the* $\mathfrak{Y}$ *case of* (2.i) *holds true.*

We prove this by a way similar to the $\mathfrak{X}$ case. But we need to be careful about the following. The Grassmannian manifolds, which are convenient tools for problems of subanalytic and semialgebraic geometry, are not semilinear. Hence we avoid using them. Neither is the distance function from a $\mathfrak{Y}$-set $Y \subset \mathbb{R}^n$, $n > 1$, if we define it as usually. We replace it by the $\mathfrak{Y}$-function

$$\mathbb{R}^n \ni x = (x_1, \dots, x_n) \to \inf_{(y_1,\dots,y_n)\in Y} \max_{i=1,\dots,n} |x_i - y_i| \in \mathbb{R}.$$

We conjecture the $\mathfrak{Y}$-cases of (2.8), (2.8)′, (2.11), (2.11)′, (2.12), (2.13) and (2.14) and can prove them under a condition. Let us consider another axiom.

Axiom (vii)$_s$ For an element $Y \subset \mathbb{R}^n$ of $\mathfrak{Y}$, the strong contangent CT^sY of Y is an element of $\mathfrak{Y}$, where

$$CT^sY = \{(y,z) \in Y \times \mathbb{R}^n : y + tz \in Y \text{ for } t \in \text{(an open interval in } \mathbb{R} \text{ containing } 0)\}.$$

Note that for a closed polyhedron Y in $\mathbb{R}^n$, CT^sY coincides with CTY, and any family $\mathfrak{X}$ satisfies (vii)$_s$.

(3.i)$_{i=8,11,12,13}$ and (3.i)$'_{i=8,11}$ Theorem. *If Axiom* (vii)$_s$ *is satisfied then the* $\mathfrak{Y}$*-cases of* (2.8), (2.8)′, (2.11) *without uniqueness,* (2.11)′ *without uniqueness,* (2.12) *and* (2.13) *hold true.*

We use Axiom (vii)$_s$ to obtain only the following assertion. (We conjecture it without (vii)$_s$.)

Let $Y \subset \mathbb{R}^n$ be a bounded $\mathfrak{Y}$-set of dimension $<$ n and let b be a point of $\mathbb{R}^n$. Then there exists a point b' arbitrarily close to b such that for any $a \in \mathbb{R}^n$, $Y \cap (a + \mathbb{R}b)$ has no inner points in the line $a + \mathbb{R}b$.

(3.9) and (3.9)′ Theorem. *Assume that any* C^r ($r \geq 3$) *Whitney* $\mathfrak{Y}$*-stratification has a controlled* $\mathfrak{Y}$*-tube system. Then the* $\mathfrak{Y}$*-Hauptvermutung theorem holds (i.e. two compact* $\mathfrak{Y}$*-homeomorphic polyhedra are PL homeomorphic, and if Axiom* (v) *is satisfied then this holds for noncompact polyhedra).*

Applying (2.10) we can prove this in the same way as (2.9). But the condition on a controlled $\mathfrak{Y}$-tube system is too strong because $\mathfrak{S}_3$ does not satisfy the condition. Indeed we conjecture that any family $\mathfrak{Y}$ is $\mathfrak{X}$ if there exists at least one $\mathfrak{Y}$-tube at 0 in $\mathbb{R}^2$. We should prove the $\mathfrak{Y}$-Hauptvermutung theorem without the condition.

Under the same condition we can prove also uniqueness of $\mathfrak{Y}$-triangulations of $\mathfrak{Y}$-functions, which corresponds to uniqueness in (2.11) and (2.11)′, and the $\mathfrak{Y}$ case of (2.22).

We can not prove the $\mathfrak{Y}$-cases of (2.16), (2.17), (2.18) and (2.19) as in the $\mathfrak{X}$-cases because the proofs of (2.16), ... ,(2.19) use essentially polynomial functions.

(3.20) and (3.21) Remark. The $\mathfrak{Y}$-cases of (2.20) and (2.21) are not correct. Indeed there does not exist a compact C^1 $\mathfrak{S}_3$-submanifold of $\mathbb{R}^n$ with boundary of dimension > 1, and an affine irreducible nonsingular algebraic $\mathfrak{S}_3$-variety is a linear subspace.

But the following seems possible. For a noncompact bounded C^r $\mathfrak{Y}$-submanifold M of $\mathbb{R}^n$ there exists a unique compact C^r $\mathfrak{Y}$-submanifold with corners whose interior is C^r $\mathfrak{Y}$-diffeomorphic to M.

By (3.11) we can prove the $\mathfrak{Y}$ case of the latter half of (2.20) under Axiom (vii)$_s$, i.e. for a bounded C^0 $\mathfrak{Y}$-submanifold N of $\mathbb{R}^n$ there exists a compact PL manifold possibly with boundary whose interior is $\mathfrak{Y}$-homeomorphic to N. Moreover, this compactification is unique if the condition in (3.9) and (3.9)′ is satisfied.

Remark. The following axiom may seem more natural than Axiom (vii):

Axiom (vii)* For a $\mathfrak{Y}$-set $Y \subset \mathbb{R}^n$ the infinite cone CY of Y is a $\mathfrak{Y}$-set, where

$$\begin{aligned} CY = \{(y,z) \in Y \times \mathbb{R}^n : &\text{there exists } y' \neq y \in Y \\ &\text{such that } |z|(y'-y) = |y'-y|z\}. \end{aligned}$$

But if we replace Axiom (vii) with this axiom then $\mathfrak{Y}$ coincide with $\mathfrak{X}$ because the set $\{(z_1,z_2,z_3) \in \mathbb{R}^3 : z_1 z_2 = z_3\}$ is an element of any $\mathfrak{Y}$.

Proof. Set

$$\begin{aligned} Y_1 &= \{(y_1,y_2,y_3) \in \mathbb{R}^3 : \ y_1 = y_3 = 0\}, \\ Y_2 &= \{(y_1,y_2,y_3) \in \mathbb{R}^3 : \ y_1 = 1, \ y_2 = y_3\}, \\ Y &= Y_1 \cup Y_2. \end{aligned}$$

Then

$$\begin{aligned} &CY \cap \{(y_1,y_2,y_3,z_1,z_2,z_3) \in \mathbb{R}^6 : \ y_2 = z_2, \ z_1 > 0\} \\ &\quad = \{z_3 = z_1 y_2, \ z_2 = y_2, \ z_1 \neq 0, \ y_1 = y_3 = 0\}. \end{aligned}$$

Hence the image of this set under the projection $\mathbb{R}^6 \to \mathbb{R}^3$, which forgets the first three factors, is

$$\{(z_1,z_2,z_3) \in \mathbb{R}^3 : \ z_3 = z_1 z_2, \ z_1 \neq 0\}.$$

□

The families $\mathfrak{S}_1$ of subanalytic sets nor $\mathfrak{X}$ do not have interesting global properties. On the other hand, the families $\mathfrak{S}_2$ of semialgebraic sets and $\mathfrak{S}_3$ of semilinear sets have some global finiteness properties. If we strengthen the axioms of an $\mathfrak{X}$-family then the family becomes to keep the properties. Let $\mathfrak{X}^0$ be a family of subsets of Euclidean spaces defined by Axioms (i), (ii) and the following axioms:

Axiom (iii)^0 Let $X \subset \mathbb{R}^n$ be an element of $\mathfrak{X}^0$ and let $p : \mathbb{R}^n \to \mathbb{R}^m$ be a linear map. Then $p(X)$ is an element of $\mathfrak{X}^0$.

Axiom (iv)^0 An element of $\mathfrak{X}^0$ included in $\mathbb{R}$ is a finite union of points and intervals.

We define an $\mathfrak{X}^0$-*set*, an $\mathfrak{X}^0$-*function*, an $\mathfrak{X}^0$-*map*, etc. Note that $\mathfrak{S}_2$ is an example of $\mathfrak{X}^0$, and an $\mathfrak{X}^0$-family does not satisfy the axiom (v). Since $\mathfrak{X}^0$ is $\mathfrak{X}$, an $\mathfrak{X}^0$-family keeps the properties (2.i), $i = 1, \dots, \overset{\vee}{10}, \dots, \overset{\vee}{15}, \dots, 21$, in §2. Moreover, we have the following. (The proofs are easy because by a semialgebraic bounded imbedding of $\mathbb{R}^n$ into $\mathbb{R}^n$ we can reduce to the bounded cases.)

(3.2)^0 *For an $\mathfrak{X}^0$-set $X \subset \mathbb{R}^n$, the number of connected components of X is finite, and each connected component is an $\mathfrak{X}^0$-set.*

(3.8)^0 Semilinearization theorem of $\mathfrak{X}^0$-sets. *For an $\mathfrak{X}^0$-set $X \subset \mathbb{R}^n$ there exists an $\mathfrak{X}^0$-homeomorphism π of $\mathbb{R}^n$ such that $\pi(X)$ is semilinear.*

In (3.8)^0 we call $(\pi(X), \pi^{-1}|_{\pi(X)})$ an $\mathfrak{X}^0$-*semilinearization* of X.

(3.9)^0 Remark. In (3.8)^0, even if X is closed in $\mathbb{R}^n$, $\pi(X)$ is not necessarily unique up to semilinear homeomorphism.

For example, set

$$C_1 = \{(x_1, x_2) \in \mathbb{R}^2 : 0 \le x_1,\ 0 \le x_2 \le 1\},$$
$$C_2 = \{(x_1, x_2) \in \mathbb{R}^2 : 0 \le x_1,\ 0 \le x_2 \le 1 + x_1\},$$

and define a semilinear homeomorphism τ of $\mathbb{R}^2$ by

$$\tau(x_1, x_2) = (x_1, x_2/(1 + |x_1|)) \quad \text{for } (x_1, x_2) \in \mathbb{R}^2.$$

Then $\tau(C_2) = C_1$, but C_1 and C_2 are semilinearly distinct.

(3.11)^0 Semilinearization theorem of $\mathfrak{X}^0$-functions. *Let f be an $\mathfrak{X}^0$-function on an $\mathfrak{X}^0$-set $X \subset \mathbb{R}^n$. Then there exist a semilinear set $Y \subset \mathbb{R}^{n+1}$ and an $\mathfrak{X}^0$-homeomorphism $\pi : Y \to X$ such that $f \circ \pi$ is semilinear.*

If X is closed in $\mathbb{R}^n$ and f is proper then we can choose Y in $\mathbb{R}^n$ and extend π to an $\mathfrak{X}^0$-homeomorphism of $\mathbb{R}^n$.

(3.12)^0 Equivalence theorem of $\mathfrak{X}^0$-functions *Let f_1 and f_2 be $\mathfrak{X}^0$-functions on an $\mathfrak{X}^0$-set $X \subset \mathbb{R}^n$ such that the zero point sets coincide and the positive point sets also coincide. Then there exists an $\mathfrak{X}^0$-homeomorphism π of X such that $f_1 \circ \pi$ equals f_2 on an $\mathfrak{X}^0$-neighborhood of the zero set.*

(3.13)^0 Equivalence class theorem of $\mathfrak{X}^0$-functions. *Define an equivalence relation on all (not necessarily bounded) $\mathfrak{X}^0$-functions on all (not necessarily bounded) $\mathfrak{X}^0$-sets as in* (2.13). *Then the cardinal number of equivalence classes is countable.*

Consider (R-)$\mathfrak{X}^0$-semilinearizations of $\mathfrak{X}^0$-maps. Let $X \subset \mathbb{R}^n$ and $Y \subset \mathbb{R}^m$ be locally closed $\mathfrak{X}^0$-sets, and let $f : X \to Y$ be an $\mathfrak{X}^0$-map. An *$\mathfrak{X}^0$-semilinearization* of f is a pair of $\mathfrak{X}^0$-semilinearizations (X_1, π) of X and (Y_1, τ) of Y such that $\tau^{-1} \circ f \circ \pi : X_1 \to Y_1$ is semilinear. If $(Y_1, \tau) = (Y, \mathrm{id})$, we call (X_1, π) an *R-$\mathfrak{X}^0$-semilinearization* of f. We define a *finite local $\mathfrak{X}^0$-resolution* in the same way as in the $\mathfrak{X}$ case, adding the fifth condition that for each i there exist a finite number of C^0 $\mathfrak{X}^0$-stratifications of open $\mathfrak{X}^0$-sets in $\mathbb{R}^{n_i} \times \mathbb{R}^m$ such that each element of $\mathcal{A}_{i,x}$, $x \in \mathbb{R}^{n_i} \times \mathbb{R}^m$, is the germ of some of these stratifications.

(3.14)^0 Semilinearization theorem of $\mathfrak{X}^0$-maps. *Let $X \subset \mathbb{R}^n$ and $Y \subset \mathbb{R}^m$ be locally closed $\mathfrak{X}^0$-sets and let $f : X \to Y$ be an $\mathfrak{X}^0$-map.*

(1) *The map f is $\mathfrak{X}^0$-semilinearizable if after a change of the coordinate system of $\mathbb{R}^n$ by an $\mathfrak{X}^0$-homeomorphism of $\mathbb{R}^n$ there exists a finite local $\mathfrak{X}^0$-resolution $\{\mathcal{A}_{i,x}\}_{i=0,\dots,\ell,\, x\in\mathbb{R}^{n_i}\times\mathbb{R}^m}$ of f with respect to some projections $\mathbb{R}^{n_0} \times \mathbb{R}^m \xrightarrow{p_1} \cdots \xrightarrow{p_\ell} \mathbb{R}^{n_\ell} \times \mathbb{R}^m$.*

(2) *The map f is R-$\mathfrak{X}^0$-semilinearizable if, in addition, Y is semilinear and for any $\{X_{\ell,j}\}_j \in \mathcal{A}_{\ell,x}$, each $X_{\ell,j}$ is the germ of a semilinear set.*

(3) *If X is closed in $\mathbb{R}^n$ then these conditions are necessary for the corresponding properties.*

As a corollary we can prove

(3.15)^0 Semilinearization theorem of complex polynomial functions. *A complex polynomial function on an affine complex algebraic set is semialgebraically semilinearizable.*

Let $r' < r$ be a nonnegative integer, and let X and Y be C^r $\mathfrak{X}^0$-submanifolds of $\mathbb{R}^n$ and $\mathbb{R}^m$, respectively, with $\dim X = n'$.

(3.16)^0 Transversality theorem. *Let $\{X_i\}_i$ be a finite $C^{r-r'}$ Whitney $\mathfrak{X}^0$-stratification in $J^{r'}(X, Y)$ with* codim $\cup_i X_i = n''$. *Assume that $\cup_i X_i$ is closed in $J^{r'}(X, Y)$ and*

$$r - r' > \max\,(n' - n'', 0).$$

Then the subset of $C^r_{\mathfrak{X}^0}(X, Y)$ consisting of f such that $J^{r'} f$ are transversal to each X_i is open and dense in $C^r_{\mathfrak{X}^0}(X, Y)$.

(3.17)^0 Approximation theorem. *We can approximate any map in $C^{r'}_{\mathfrak{X}^0}(X, Y)$ by a C^r $\mathfrak{X}^0$-map.*

(3.18)^0 Triviality theorem. *Let M be a C^r $\mathfrak{X}^0$-manifold. Then a proper C^r $\mathfrak{X}^0$-submersion from M to $\mathbb{R}^m$ is C^r $\mathfrak{X}^0$-trivial.*

(3.19)^0 First isotopy lemma. *Let X be an $\mathfrak{X}^0$-subset of $\mathbb{R}^n$, let $\{X_i\}_i$ be a C^r Whitney $\mathfrak{X}^0$-stratification of X, and let $f : X \to \mathbb{R}^m$ be a proper $\mathfrak{X}^0$-map such that the restriction of f to each stratum X_i is a C^r submersion. Then f is $\mathfrak{X}^0$-trivial.*

(3.20)^0 Compactification theorem. *Let M be a noncompact C^r $\mathfrak{X}^0$-submanifold of $\mathbb{R}^n$. Then there exists a unique compact C^r $\mathfrak{X}^0$-submanifold of $\mathbb{R}^n$ with boundary whose interior is C^r $\mathfrak{X}^0$-diffeomorphic to M.*

Let N be a C^0 $\mathfrak{X}^0$-submanifold of $\mathbb{R}^n$. Then there exists a unique compact PL manifold possibly with boundary whose interior is $\mathfrak{X}^0$-homeomorphic to N.

(3.21)^0 Algebraic model theorem. *A C^r $\mathfrak{X}^0$-submanifold of $\mathbb{R}^n$ is C^r $\mathfrak{X}^0$-diffeomorphic to some affine nonsingular algebraic variety.*

(3.22)^0 Remark. The $\mathfrak{X}^0$ case of (2.22) does not hold. There exist two C^1 diffeomorphic $\mathfrak{X}^0$-subsets of $\mathbb{R}^n$ which are $\mathfrak{X}^0$-distinct (see (2.23.3)).

Strengthening the axioms of a $\mathfrak{Y}$-family in the same way as above we define a family $\mathfrak{Y}^0$ of subsets of Euclidean spaces. Then $\mathfrak{Y}^0$-sets, $\mathfrak{Y}^0$-functions and $\mathfrak{Y}^0$-maps have finite properties such as in the $\mathfrak{X}^0$ cases, which are clear by the above arguments. We omit the details.

References

[C-S$_1$] M. Coste–M. Shiota, Nash triviality in families of Nash manifolds. Invent. Math. 108 (1992), 349–368.

[C-S$_2$] M. Coste–M. Shiota, Thom's first isotopy lemma: a semialgebraic version, with uniform bound. This volume.

[L] H.T. Levine, Singularities of differentiable mappings. Proceedings of Liverpool Singularities - Symposium I, Liverpool 1971, Lecture Notes in Math. 192, 1–89, Springer 1971.

[M] J.R. Munkres, Elementary differential topology. Ann. of Math. Studies 54, Princeton Univ. Press, Princeton 1963.

[S$_1$] M. Shiota, Nash manifolds. Lecture Notes in Math. 1269, Springer 1987.

[S$_2$] M. Shiota, Piecewise linearization of subanalytic functions II. Lecture Notes in Math. 1420, 247–307, Springer 1990.

[S$_3$] M. Shiota, Geometry of subanalytic and semialgebraic sets. To appear.

[S-Y] M. Shiota–M. Yokoi, Triangulations of subanalytic sets and locally subanalytic manifolds. Trans. Amer. Math. Soc. 286 (1984), 727–750.

[Si] L.C. Siebenmann, Disruption of low-dimensional handlebody theory by Rohlin's theorem. Topology of manifolds, 57–76, Markham, Chicago, 1970.

[Ta] M. Tamm, Subanalytic sets in the calculus of variation. Acta Math. 146 (1981), 167–199.

[Th] R. Thom, Ensembles et morphismes stratifiés. Bull. Amer. Math. Soc. 75 (1969), 240–284.

[V] L. van den Dries, Tarski problem and Pfaffian functions. Logic colloquium 1984, Elsevier Science Publishers B.V. (North-Holland), 1986, 59–90.

[W_1] A.J. Wilkie, Model completeness results for expansions of the real field I: restricted Pfaffian functions. Preprint.

[W_2] A.J. Wilkie, Model completeness results for expansions of the real field II: the exponential function. Preprint.

On good points and a new canonical algorithm of resolution of singularities

Orlando Villamayor U. *

Abstract. We present a new canonical algorithm of resolution of singularities over fields of characteristic zero.

The invariants attached to the singularities are (essentially) characteristic free and grow from Abhyankar's good point theory.

Introduction

The aim of this presentation is to outline a new canonical algorithm of resolution of singularities over fields of characteristic zero.

Let W be a smooth variety, X a subvariety of W. By an algorithm of resolution of singularities of the pair (X, W) we mean a canonical assignment of:

i) A numerical invariant of the pair (X, W).

ii) A smooth closed subvariety Y of X.

These data enjoy the following property. Take the monoidal transform of W at Y:

$$\pi: W^{(1)} \to W.$$

$W^{(1)}$ is smooth because Y is smooth. Take $X^{(1)}$ as the strict transform of X. Then apply the algorithm to the pair $(X^{(1)}, W^{(1)})$.

The fundamental property of our algorithms is that the numerical invariants attached to $(X^{(1)}, W^{(1)})$ is "better" (in a suitable sence) to that assigned to (X, W). Suitably better means that repeating this procedure again and again, we arrive after a finite number of steps to $(X^{(n)}, W^{(n)})$ where $X^{(n)}$ is finally a regular subvariety of $W^{(n)}$.

The new algorithm we present here improves that of [7] because the number of steps required to transform an arbitrary subvariety X into a regular one is in general strikingly smaller.

We illustrate and compare with examples the efficiency of our new algorithm over our old one in Section 4.

---

* Partially supported by the DGICYT PB 91-0370-C0202

Furthermore, following the line of Abhyankar's good points [1], we prove in [9] that essentially the only invariant for resolution of singularities is the codimension of the Hilbert Samuel strata.

This characteristic free invariant already opens some interesting questions.

While preparing this work, I benefited from conversations with S.S. Abhyankar and also with H. Hauser.

1. Algorithmic and inductive nature of the problem of resolution of singularities

1.1. In essence, the problem of resolution of singularities of varieties over fields of characteristic zero is an algorithmic problem. We shall devote all Section 1 to clarify this point.

In order to enlight directly the simple algorithmic nature of the problem let us add some (of course non-essential) assumptions.

So let f be a polynomial in $k[z,x_1,\dots,x_{n-1}] = k[y_1,\dots,y_n]$, and let $b(\in \mathbb{N})$ denote the order (or multiplicity) of f at the origin of $\mathbb{A}^n_k$.

Assume furthermore that f has an expression of the form

$$f = z^b + a_1 z^{b-1} + \cdots + a_k \tag{1.1.1}$$

where $a_i \in k[x_1,\dots,x_{n-1}]$.

It is clear that in this case b is the highest possible multiplicity of the hypersurface defined by f in $\mathbb{A}^n_k$. We shall first focus at the set of points of $\mathbb{A}^n_k$ where this hypersurface has maximal multiplicity (multiplicity b).

So set

$$\mathrm{Sing}(< f >, b) = \{x \in \mathbb{A}^n_k / \nu_x(f) \geq b\};$$

here $\nu_x(\,)$ denotes the order defined by the local regular ring of $\mathbb{A}^n_k$ at the point x.

$\mathrm{Sing}(< f >, b)$ is a closed subset of $\mathbb{A}^n_k$ which can be expressed by the vanishing of partial derivatives

$$\mathrm{Sing}\,(< f >, b) = V(\frac{\partial^\alpha f}{\partial y^\alpha} / 0 \leq |\alpha| < b)$$

$$\alpha = (\alpha(1),\dots,\alpha(n))$$

and

$$\frac{\partial^\alpha}{\partial y^\alpha} = \frac{\partial^{\alpha(n)}}{\partial^{\alpha(n)} y_n} \cdot \frac{\partial^{\alpha(1)}}{\partial^{\alpha(1)} y_1}.$$

If we furthermore assume that $a_1 = a_1(x_1,\dots,x_n) = 0$ (1.1.1), then

$$\frac{\partial^{b-1}}{\partial^{b-1} y_1} f = \frac{\partial^{b-1} f}{\partial^{b-1} z} = b!z.$$

Since k is a field of characteristic zero

$$\mathrm{Sing}(f,b) \subset V = \{z=0\}. \tag{1.1.2}$$

where V is a regular subvariety of $\mathbb{A}^n_k$ of dimension $n-1$.

Moreover, from 1.1.2 we conclude that $x \in \mathrm{Sing}(< f >, b)$ if and only if $x \in V$ and $\nu_x(a_i) \geq i$ or equivalently,

$$x \in \mathrm{Sing}(< f >, b) \text{ if and only if } x \in V \text{ and } \nu_x(a_i^{\frac{b!}{i}}) \geq b!. \tag{1.1.3}$$

Define now the "coefficient ideal" to be

$$C(f) =< a_2^{\frac{b!}{2}}, \ldots, a_b^{\frac{b!}{b}} > \tag{1.1.4}$$

in $k[x_1, \ldots, x_{n-1}]$ and reformulate 1.1.3. as

$$x \in \mathrm{Sing}(< f >, b) \Leftrightarrow x \in V \text{ and } \nu_x(C(f)) \geq b!. \tag{1.1.5}$$

Let us now introduce some notation required to clarify the further development.

1.2. Basic objects and transformations. A basic object will be a datum of the form $(W,(J,b),E)$ where:

i) W is a smooth variety over a field of characteristic zero.
ii) $J(\subset O_W)$ is an ideal (or a sheaf of ideals) of W.
iii) $b \in \mathbb{N}$.
iv) $E = \{H_1, \ldots, H_r\}$ are smooth hypersuperfaces in W having only normal crossings.

We are mainly interested in the case that J is not the zero ideal at any point and we shall define

$$\mathrm{Sing}(J,b) = \{x \in W / \nu_x(J) \geq b\} \tag{1.2.1}$$

which is Zariski closed in W.

Let $Y \subset \mathrm{Sing}(J,b)$ be a smooth closed and (for simplicity) irreducible subvariety of W, and set $W \xleftarrow{\pi} W'$ to be the monoidal transformation with center Y. So that W' is smooth and $\pi^{-1}(Y)$ is a smooth hypersuperface say H in W'.

Since the order of J at any point in Y is at least $b(Y \subset \mathrm{Sing}(J,b))$, then

$$JO_{W'} = I(H)^b J_1 \tag{1.2.2}$$

where $I(H)$ is the ideal of functions vanishing on H and J_1 is some ideal of W_1.

We will furthermore assume that Y has only normal crossings in W with the hypersurfaces in E. In such case we define the *transform* of $(W,(J,b),E)$ to be $(W',(J_1,b),E_1)$ where J_1 is as in 1.2.2 and E_1 consists of the strict transform of hypersurfaces of E together with the hypersurface $H = \pi^{-1}(Y)$.

Usually this transform will be denoted as

$$(W,(J,b),E) \leftarrow (W_1,(J_1,b),E_1),$$

and $Y(\subset W)$ is called the center of the transformation.

1.2.3 Definition. If $\mathrm{Sing}(J_1, b) = \emptyset$ then we say that the transformation is a resolution. In general a *resolution* of a basic objet will be a sequence or concatenation of these transformations, say

$$(W, (J, b), E) \leftarrow (W_1, (J_1, b), E_1) \leftarrow \cdots \leftarrow (W_n, (J_n, b), E_n),$$

so that $\mathrm{Sing}(J_n, b) = \emptyset$.

Example 1. If $J =< f >$ (f as in 1.1.1) then $\mathrm{Sing}(J.b)$ denotes the set of points of maximal order. If

$$(W, (J.b), E) \leftarrow (W_1, (J_1, b), E_1)$$

denotes a transformation, then J_1 is a principal ideal defining in W_1 the strict transform of the hypersurface $\{f = 0\}$ (1.2.2).

The transformation is a resolution (1.2.3) if and only if the strict transform of the hypersurface has maximal order strictly smaller than b.

Example 2. In 1.1.3 the set $\mathrm{Sing}(< f >, b) \subset W$ is described as a closed subset of the smooth subvariety V of codimension 1 in W. ($V = \{z = 0\}$).

Furthermore if $J = C(f)$ (1.1.4) then we can identify J with an ideal in O_V and

$$\mathrm{Sing}(< f >, b) = \mathrm{Sing}(C(f), b!)(\subset V).$$

1.3. Let f, b be as before and set $E = \emptyset$ for simplicity. Now we have two basic objects $(W, (f, b), \emptyset)$ and $(V, (C(f), b!), \emptyset)$ where V is a smooth hypersurface in W and $\mathrm{Sing}(f, b) = \mathrm{Sing}(C(f), b)$.

Let Y be a closed and smooth subscheme of W; the monoidal transformation of W with center Y defines a transformation (1.2)

$$(W, (f, b), E = \emptyset) \xleftarrow{\pi} (W_1, (f', b), E_1), \tag{1.3.1}$$

and $Y \subset \mathrm{Sing}(f, b) = \mathrm{Sing}(C(f), b!)$ also defines a transformation

$$(V, (C(f), b!), E = \emptyset) \leftarrow (V_1, (C(f)_1, b!, \bar{E}_1). \tag{1.3.2}$$

V_1 obtained by blowing up Y in V can be identified with a smooth hypersurface of W_1: the strict transform of the smooth hypersurface $V = \{z = 0\}$.

Now let $y \in W_1$ be a rational point at which f' has order b ($x \in \mathrm{Sing}(f', b)$).

Then locally at y, f' admits an expression of the form

$$f' = (z')^b + a'_2 (z')^{b-2} + \cdots + a'_b \in k[z'_1, x'_1, \ldots, x'_{n-1}]$$

where now $a'_i \in k[x'_1, \ldots, x'_{n-1}]$ and $\{z' = 0\}$ defines V'.

Clearly the setup at y is that of 1.1. So again

$$\mathrm{Sing}(f', b) \subset V' = \{z' = 0\}$$

and furthermore

$$\mathrm{Sing}(f', b) = \mathrm{Sing}(C(f'), b!)$$

where now $C(f')$ is the coefficient ideal as in 1.1.4 defined now for f'.

The point is that the ideal $(C(f'))_1$ of 1.3.2 is exactly the ideal $C(f')$.

In other words, $\mathrm{Sing}(C(f), b!)$ describes $\mathrm{Sing}(f, b)$ at W. And if the transform of $(V, (C(f), b!), E)$ is $(V_1, (C(f)_1, b!), E_1)$, then again

$$\mathrm{Sing}((C(f))_1, b!) = \mathrm{Sing}(f', b)$$

no matter which center Y is chosen to blow-up. Furthermore since $C(f)_1$ is exactly $C(f')$, the setup is as that of the very beginning. So for any concatenation of transformation:

$$(W, (f, b), E) \leftarrow (W_1, (F', b), E_1) \leftarrow \cdots \leftarrow (W_n, (f^{(n)}, b), E_n) \tag{1.3.3.a}$$

$$(V, (C(f), b!), \bar{E}) \leftarrow (V_1, (C(f)_1, b!), \bar{E}_1) \leftarrow \cdots \leftarrow (V_n, (C(f)_n, b!), \bar{E}_n) \tag{1.3.3.b}$$

$$\mathrm{Sing}(f^{(n)}, b) = \mathrm{Sing}(C(f)_n, b!).$$

The problem, to begin with, is that of defining a suitable sequence of transformations as in 1.3.3.a such that $\mathrm{Sing}(f^{(n)}, b) = \emptyset$. Now we know that this is equivalent to defining the corresponding sequence 1.3.3.b with

$$\mathrm{Sing}(C(f)_n, b!) = \emptyset.$$

In other words, the problem of lowering the highest multiplicity of the hypersurface defined by f (i.e. the problem of finding a resolution of the basic object $(W, (f, b), E)$) is equivalent to that of finding a resolution of $(V, (C(f), b!), \bar{E})$. But $\dim V = \dim W - 1$, so $C(f)$ involves one variable less than f. This is the heart of induction in the algorithmic (or constructive) proof of resolution of singularities.

2. On the monomial case

2.1.1. Let the notation be that of 1.3.3.a and 1.3.3.b. Assume that $C(F)_n$ (ideal in V_n) is monomial in the following sense.

If $\bar{E}_n = \{H_1, \dots, H_p\}$ are the hypersurface of V_n with normal crossings as in 1.2 then we require that

$$C(f)_n = I(H_1)^{a(1)} \cdots I(H_p)^{a(p)}$$

where $I(H_i)$ is the ideal defining H_i and $a(1), \dots, a(p)$ are non negative integers.

In this case

a) $$x \in \mathrm{Sing}(C(f)_n, b!) \text{ if and only if } \sum_{x \in H_i} a(i) \geq b!.$$

If $Y_n \subset \mathrm{Sing}(C(f)_n, b!) (= \mathrm{Sing}(f^{(n)}, b))$ is the center of a transformation

$$(V_n, (C(f)_n, b!) \bar{E}_n) \leftarrow (V_{n+1}, (C(f)_{n+1}, b!), \bar{E}_{n+1})$$

then

b) $$C(f)_{n+1} = I(H_1)^{a(1)} \cdots I(H_p)^{a(p)} \cdots I(H)^{b(p+1)},$$

where H_i denotes again the strict transform of the hypersurface $H_i(\subset V_n)$ of $\bar{E}_n \cdot I(H_i) = O_{V_{n+1}}$ if $H_i(\subset V_{n+1})$ is empty and

$$b(p+1) = \sum_{Y_n \subset H_i} a(i) - b!.$$

One can check that the formula in b) is that of the transformation as in 1.2 of our $C(f)_n$.

2.1.2 Remark. Suppose that $a(1) \geq b!$, then:

i) $H_1 \subset \mathrm{Sing}(C(f)_n, b!)$.

ii) H_1 is a smooth and has normal crossing with hypersurfaces of $\bar{E}_n$.

iii) If $(V_n, (C(f)_n, b!), \bar{E}_n) \leftarrow (V_{n+1}, (C(f)_n, b!), \bar{E}_{n+1})$ is the transformation defined by blowing up $Y = H_1$ (1.2) then $V_{n+1} = V_n$ (the blow up of a hypersurface is the identity),

$$C(f)_{n+1} = I(H_2)^{a(2)} \cdots I(H_p)^{a(p)} I(H_{p+1})^{a(1)-b!}$$

and $H_1 = \emptyset$ at V_{n+1}.

2.1.3 Example. Suppose that locally at $x \in \mathrm{Sing}(f^{(n)}, b)$ $f^{(n)}$ is defined by $z^2 + x^2y^2$ where $\{z = 0\}$ defines (locally) V_n, $\{z = 0; x = 0\}$ defines H_1; $\{z = 0; y = 0\}$ defines H_2 and $\bar{E}_n = \{H_1, H_2\}$.

So $C(f^{(n)}) = C(f) = x^2y^2 = I(H_1)^2 I(H_2)^2$; $b = b! = 2$.

Blowing up $H_1 \subset V_n \subset W_n$ we obtain at $V_{n+1}(\subset W_{n+1})$

$$C(f)_{n+1} = I(H_2)^2.$$

Now $\mathrm{Sing}(C(f)_{n+1}, b!)$ is H_2 and blowing up H_2

$$C(f)_{n+2} = O_{V_{n+2}},$$

so $\mathrm{Sing}(f^{(n+2)}, b) = (\mathrm{Sing}(C(f)_{n+2}, b!)) = \emptyset$.

2.1.4 Example. With the notation as before, but now $f^{(n)} = z^3 + x^2y^2$.

Here, $b = 3$, $b! = 6$, $C(f^{(n)}) = (x^2y^2)^2$ and $\mathrm{Sing}(C(F^{(n)}, b!)$ is the point $H_1 \cap H_2$.

Blowing up this point we obtain

$$C(f^{(n+1)}) = I(H_1)^4 I(H_2)^4 I(H_3)^2.$$

Now $H_1 \cap H_2 = \emptyset$ and

$$\mathrm{Sing}(f^{(n+1)}, b) = (H_1 \cap H_3) \cup (H_2 \cap H_3) \subset V_{n+1}.$$

2.1.5 Remark. In the case that $C(f)$ is monomial then a resolution is simple to obtain by blowing up convenient intersections of hypersurfaces H_i in V_n. Within this case the simplest subcase is the following.

2.1.6 Definition. As before, let the notation be as in 1.3.3.a) and 1.3.3.b).

We shall say that $(W_n, (f^{(n)}, b), E_n)$ is in the *simple monomial case* if

a) $C(f)_n$ is monomial (2.1.1), so

$$C(f)_n = I(H_1)^{a(1)} \cdots I(H_p)^{a(p)}$$

where $a(i)$ is a nonnegative integer ($\bar{E}_n = \{H_1, \dots, H_p\}$).

b) $\sum_{i=1}^{p} \bar{a}(i) < b!$ where $\bar{a(i)}$ denotes the remainder of $a(i)$ divided by $b!$.

Observation. The simple monomial case is the case in which $C(f)_n$ is monomial (2.1.1) and furthermore, can be solved by transformations as in 2.1.2 (i.e. can be extended to a resolution by blowing up only at hypersurfaces H_i).

So 2.1.3 is an example of a simple monomial case, while Example 2.1.4 is in the monomial but not in the simple monomial case.

2.1.7 Remark. Let the notation be that of 1.3.3.a and 1.3.3.b.

2.1.5 says that it is not difficult to extend 1.3.3.a to a resolution if $C(f)_n$ is monomial, i.e., if

$$C(f_n) = I(H_1)^{a(1)} \cdots I(H_p)^{a(p)},$$

$H_i \in \bar{E}_n$.

Within this case the observation in 2.1.6 says that the simple monomial case is that in which a resolution is obtained by blowing up the H_i corresponding to $a(i) \geq b!$.

The strategy to follow in the general case will be the following.

a) For any sequence 1.3.3.a, let

$$C(f)_n = I(H_1)^{a(1)} \cdots I(H_p)^{a(p)} \mathcal{A}_n$$

be the unique expression of $C(f)_n$ as product of ideals where $\mathcal{A}_n$ does not vanish generically at any $H_i \in \bar{E}_n$. Of course $\mathcal{A}_n = O_{V_n}$ is the monomial case, so:

b) If $C(f)_n$ is not monomial we will enlarge the sequence so that after adding s steps more

$$\mathcal{A}_{n+s} = O_{Vn+s}.$$

2.1.8 Exceptionally good points. Exceptionally good points are points which are good in the sense of [1] but always exceptional, in the sense that they show up at the exceptional locus.

The algorithm tells you which center to choose, after applying the transformation (1.2) at such center new singular points can arise at the exceptional locus. The algorithm itself will be able to decide if such exceptional singular point is an *exceptionally good point.*

Exceptionally good points will be good points. But the notion of good points presented in [1] and [2] is quite more ample.

Definition. With notation as in 1.3.3.a and 1.3.3.b, assume that

$$C(f)_n = I(H_1)^{a(1)} \cdots I(H_p)^{a(p)} \mathcal{A}_n$$

($\bar{E}_n = \{H_1, \ldots, H_p\}$) is as in 2.1.7 b).

We shall say that a point $x \in \mathrm{Sing}(C(f)_n, b!) = \mathrm{Sing}(f^{(n)}, b)$ is exceptionally good if

$$\nu_x(I(H_1)^{\bar{a(1)}} \cdots I(H_p)^{\bar{a(p)}} \mathcal{A}_n) < b!$$

where $\bar{a(i)}$ denotes the remainder of $a(i)$ modulo $b!$.

Example 1. $W = \mathbb{A}_k^3$, $E = \emptyset$, $f = z^2 + x^6 + y^7$, $V = \{z = 0\}, \bar{E} = \emptyset$.

Here $b = 2, b! = 2, C(f) = < x^6 + y^7 >$.

After blowing up the origin

$$(W, (f, 2), E = \emptyset) \leftarrow (W_1, (f', 2), E_1 = \{H\})$$
$$(V, (C(f), 2), \bar{E} = \emptyset) \leftarrow (V_1, (C(f)_1, 2), \bar{E}_1 = \{\bar{H}\})$$

and $C(f) = I(H)^4 \mathcal{A}_1$. There is only one point y_0 where $C(f)_1$ is not monomial: the origin of the chart at which

$$C(f)_1 = I(H)^4 \mathcal{A}_1 = y^4(\left(\frac{x}{y}\right)^6 + y).$$

The remainder of 4 modulo 2 being zero and order of $\mathcal{A}_1$ at y_0 being one asserts that y_0 is an exceptionally good point.

Example 2. In the simple monomial case (2.1.6) any singular point is exceptionally good.

3. The numerical invariants

3.1. Consider the equation

$$f = x^2 + y^3 + z^5$$

in $\mathbb{A}_k^3$. Here we can define V by

$$V = \{z = 0\}.$$

Consider now the equation

$$f = x^2 + y^2 + z^5.$$

We can first define $V_1 = \{x = 0\}$, and by iterating the procedure developed in §1 and §2 we can define

$$V_2 = \{x = 0 \wedge y = 0\}$$

and $C_2(f)$ and $b!$ in accordance with the coefficients, involving only the variable z.

The distinction between the first example and the second is that in the first one the initial form can be expressed as polynomial in one variable, while in the second case it requires two.

For a fixed basic object $(W,(J,b),E)$ (1.2) and for any $x \in \mathrm{Sing}(J,b)$ (1.2.1) consider

$$\mathrm{In}_b(J) \subset [\mathrm{gr}(O_{W,x})]_b$$

(i.e., $\frac{J+M^{b+1}}{M^{b+1}} \subset \frac{M^b}{M^{b+1}}$ where M denotes the maximal ideal of the local regular ring $O_{W,x}$). Now define $COD(x)$ to be the least number of variables in the polynomial ring $\mathrm{gr}(O_{W,x})$ required to express all equations of $\mathrm{In}_b(J)$. With this definition set

$$COD: \ \mathrm{Sing}(J,b) \to \mathbb{Z}.$$

Finally put

$$COD(W,(J,b),E) = \inf\{COD(x)\} \ x \in \mathrm{Sing}(J,b),$$

which is called the *codimension of the basic object.*

3.2 Remark. If $COD(W,(J,b),E) = e$, for any $x \in \mathrm{Sing}(J,b)$ there is a neighborhood, say W', a smooth subvariety $V_e \subset W'$ of codimension e, and a basic object

$$(V_e,(C(J),b!),\bar{E})$$

such that $\mathrm{Sing}(C(J),b!) = \mathrm{Sing}(J,b)/W'$ (restricted to W') and the same holds by applying any transformation. In other words the case treated in 1.3 is the particular case of $COD(W,(f,b),E) = 1$ where $W' = W$.

So take for simplicity $W' = W$. Now any sequence

$$(W,(J,b),E) \leftarrow \cdots \leftarrow (W_n,(J_n,b),E_n) \tag{3.2.1.a}$$

is equivalent to

$$(V_e,(C(J),b!),\bar{E}) \leftarrow \cdots \leftarrow (V_e^n,(C(J)_n,b!),\bar{E}_n) \tag{3.2.1.b}$$

and

$$\mathrm{Sing}(C(J)_k,b!) = \mathrm{Sing}(J_k,b) \qquad k = 0,1,\ldots,n.$$

Furthermore, $C(J)_n$ admits a unique expression

$$C(J)_n = I(H_1)^{a(1)} \cdots I(H_p)^{a(p)} \mathcal{A}_n \tag{3.2.2}$$

where $\bar{E}_n = \{H_1,\ldots,H_p\}$, and $\mathcal{A}_n$ does not vanish generically along any H_i (see 2.1.7.a)).

3.2.3 Definition. With notation as in 3.2.1.a, 3.2.1.b and 3.2.2 where

$$COD(W,(J,b),E) = e,$$

a point $x \in \mathrm{Sing}(J_n, b)$ is said to be e-exceptionally good if

$$\nu_x(I(H_1)^{\overline{a(1)}} \cdots I(H_p)^{\overline{a(p)}} \mathcal{A}_n) < b!$$

where $\overline{a(i)}$ denotes the remainder of $a(i)$ divided by $b!$.

3.2.3.1.a) Definition. Let the *reduction* of $C(J_n)$ (3.2.2) be $\overline{C(J_n)}$:

$$\overline{C(J_n)} = I(H_1)^{\overline{a(1)}}, \ldots, I(H_p)^{\overline{a(p)}} \mathcal{A}_n,$$

$0 \leq \overline{a(i)} < b!$ and $\overline{a(i)} \equiv a(i)(b!)$.

Clearly $\mathrm{Sing}(\overline{C(J_n)}, b!) \subset \mathrm{Sing}(C(J_n), b!)$ and we can reformulate Definition 3.2.3 by saying that $x \in \mathrm{Sing}(J_n, b)$ is e-exceptionally good if

$$x \in \mathrm{Sing}(C(J_n), b!) \text{ and } x \notin \mathrm{Sing}(\overline{C(J_n)}, b!).$$

3.2.3.1.b) Definition. A point $x \in \mathrm{Sing}(J_n, b)$ is said to be (e, k)-*exceptionally good* for some index $k \leq n$ if

b.i) The sequence of transformations

$$(V_e^k, (C(J)_k, b!), \bar{E}_k) \longleftarrow \cdots \longleftarrow (V_e^k, (C(J)_n, b!), \bar{E}_n)$$

in 3.2.1.b) induces a sequence of transformations of $(V_e^k, \overline{C(J)}_k, b!), \bar{E}_k)$, say

$$(V_e^k, (\overline{C(J)}_k, b!), \bar{E}_k) \longleftarrow \cdots \longleftarrow (V_e^k, ((\overline{C(J)_k})_{n-k}, b!), \bar{E}_n), \text{ and}$$

b.ii) $$x \in \mathrm{Sing}(C(J_n), b!) \text{ but } x \notin \mathrm{Sing}((\overline{C(J_k)})_{n-k}, b!).$$

One can check that a point which is (e, k)-exceptionally good is also an e-exceptionally good point.

3.2.4 Definition. With the notation as before (3.2.2), $(W_n, (J_n, b), E_n)$ is said to be e-monomial (is said to be locally e-monomial at $x \in \mathrm{Sing}(J_n, b)$) if $\mathcal{A}_n = \mathcal{O}_{W_n}$ (if $(\mathcal{A}_n)_x = \mathcal{O}_{W,x}$).

3.2.5 Remark. If any point $x \in \mathrm{Sing}(J_n, b)$ is e-exceptionally good (3.2.3) then a resolution can be achieved by choosing smooth centers of codimension $e + 1$ in W_n. The argument being exactly as that of the observation of 2.1.6 for the simple monomial case.

3.3. Let $(I, \geq)$ denote a set with a total order and $f : S \to I$ a function which takes only finitely different values. We define

$$\mathrm{Max}\ f = \text{ maximal value achieved by } f, \text{ and}$$

$$\underline{\mathrm{Max}}\, f = \{x \in S / f(x) = \mathrm{Max}\ f\}.$$

Clearly $\mathrm{Max}\ f \in I$ and $\underline{\mathrm{Max}}\, f \subset S$.

In our context S will be a topological space and f will be upper-semicontinuous in the sense that

$$\forall \alpha \in I,\ F_\alpha = \{x \in S / f(x) \geq \alpha\}$$

is a closed subset of S.

3.3.1 Lemma. *For each $e \in \mathbb{N}$ there is a totally ordered set $(I_e, \geq)$ and, attached to any basic object $(W, (J, b), E)$ for which $COD(W, (J, b), E) \geq e$ (3.1), an upper-semi-continuous function*

$$f_e : \operatorname{Sing}(J, b) \to I_e.$$

Moreover, for any sequence

$$(3.3.1.1) \qquad (W, (J, b), E) \leftarrow \cdots \leftarrow (W_k, (J_k, b), E_k),$$

there are upper-semi-continuous functions $f_e^j : \operatorname{Sing}(J_j, b) \to I_e$ so that

$$\operatorname{Max} f_e \geq \cdots \geq \operatorname{Max} f_e^k$$

if all centers of transformations are included in $\underline{\operatorname{Max}}\, f_e^i$.

I) *If the codimension of the closed set $\underline{\operatorname{Max}}\, f_e^k$ in W_k is e, then Y = the union of all components of codimension e of $\underline{\operatorname{Max}}\, f_e^k$ is a center for a transformation:*

$$(W_k, (J_k, b), E_k) \leftarrow (W_{k+1}, (J_{k+1}, b), E_{k+1})$$

and

i) $\operatorname{Max} f_e^{k+1} < \operatorname{Max} f_e^k$ *if $\underline{\operatorname{Max}}\, f_e^k$ has codimension e at any point.*

ii) *If* $\operatorname{Max} f_e^{k+1} = \operatorname{Max} f_e^k$ *then $\underline{\operatorname{Max}}\, f_e^{k+1}$ has no point of codimension e, and* II) *and* III) *below will also hold.*

II) *If the codimension of $\underline{\operatorname{Max}}\, f_e^k$ in W_k is bigger than e at any point then there is a basic object $(W_k, (J'_k, b'), E_k)$ such that*

i) $COD(W_k, (J'_k, b'), E_k) > e.$

ii) $\operatorname{Sing}(J'_k, b) = \underline{\operatorname{Max}}\, f_k^e.$

iii) *For any sequence of transformations*

$$(*) \qquad (W_k, (J_k, b), E_k) \leftarrow \cdots \leftarrow (W_{k+s}, (J_{k+s}, b), E_{k+s})$$

with centers in $\underline{\operatorname{Max}}\, f_{k+i}^e$, either

a) $\operatorname{Max} f_k^e > \operatorname{Max} f_{k+s}^e$, *or*

b) $\operatorname{Max} f_k^e = \operatorname{Max} f_{k+s}^e$, *in which case (*) induces a sequence of transformations*

$$(W_k, (J'_k, b'), E_k) \leftarrow \cdots \leftarrow (W_{k+s}, (J'_{k+s}, b'), E_{k+s})$$

and $\operatorname{Sing}(J'_{k+i}, b') = \underline{\operatorname{Max}}\, f_{k+i}^e$, $i = 0, \ldots, s$.

Remark A. If we accept the existence of resolutions of basic objects (1.2.2) of codimension bigger than e (3.1), then I) and II) are already saying that by enlarging the sequence of transformation (from k to $k + s$) one can force

$$\operatorname{Max} f_k^e > \operatorname{Max} f_{k+s}^e.$$

Remark B. An algorithm should indicate at each step exactly which center to choose, this is the strength of I. In case II) this will follow by induction on e. If

$$COD(W_k,(J_k,b),E_k) = \dim W_k$$

then only case I) will occur, and by blowing up $\underline{\mathrm{Max}} f_e^k$ we will obtain a resolution (1.2.3).

III) *If one can enlarge the sequence* 3.3.1 *from length* k *to* $k+s$ *such that* Max $f_k^e >$ Max f_{k+s}^e, *and after repeating this procedure finitely many times (say* $r(\geq s)$*) then every point of* $\mathrm{Sing}(J_{k+r},b)$ *will be either e-exceptionally good* (3.2.3) *or e-monomial* (3.2.4).

Remark C. Let the setup at $k+r$ be as in III). Then, by blowing up centers of codimension exactly $e+1$, the basic object $(W_{k+r},(J_{k+r},b),E_{k+r})$ will be successively transformed

$$(W_{k+r},(J_{k+r},b),E_{k+r}) \longleftarrow \cdots \longleftarrow (W_{k+t},(J_{k+t},b),E_{k+t})$$

in such a way that either $\mathrm{Sing}(J_{k+t},b)=\emptyset$ or at any point

$$x \in \mathrm{Sing}(J_{k+t},b),$$

J_{k+t} is locally e-monomial (3.2.4). Then it will be a simple combinatorial game to find a resolution of $(W_{k+t},(J_{k+t},b),E_{k+t})$.

3.4 The function f^e. With notation as in 3.2.1.a, 3.2.1.b and 3.2.2,

$$C(J)_n = I(H_1)^{a(1)}\cdots I(H_p)^{a(p)}\mathcal{A}_n$$

for a point $x \in Sin(J_n,b)$ let $x\_k \in \mathrm{Sing}(J_k,b)$ denote the image of x at $W_k (k \leq n)$.
Define

1) $$w\text{-}\mathrm{ord}_e : \mathrm{Sing}(J_n,b) \to \mathbb{Q}.$$

$$w\text{-}\mathrm{ord}_e(x) = \frac{\nu_x(\mathcal{A}_n)}{b!} \in \mathbb{Q} \geq 0.$$

2) $$n_e : \mathrm{Sing}(J_n,b) \to \mathbb{Z}.$$

2i) $$n_e(x) = 0 \text{ if } w\text{-}\mathrm{ord}_e(x) = 0.$$

2ii) If $w\text{-}\mathrm{ord}_e(x) > 0$ set k to be the smallest index for which

$$w\text{-}\mathrm{ord}_e(x) = w\text{-}\mathrm{ord}_e(x_k).$$

Set $E_{k,x_k} = \{H_i \in E_k / x_k \in H_i\}$ and define

$$n_e(x) = \#\{H_i'/H_i \in E_{k,x_k} \text{ and } x \in H_i'\}$$

where H_i' denotes the strict transform at W_n of the hypersurface H_i of W_k.

3) $$\mathrm{Ab}_e : \mathrm{Sing}(J_n,b) \to \{M,G,B\}$$

where $\{M, G, B\}$ is totally ordered by $M < G < B$.

Let k be as in 2ii) and define:

3i) $\mathrm{Ab}_e(x) = M$ if $w\text{-}\mathrm{ord}(x_k) = 0$ and x_k is not e-exceptionally good (3.2.3).

3ii) $\mathrm{Ab}_e(x) = G$ if x_k is e-exceptionally good (3.2.3).

3iii) $\mathrm{Ab}_e(x) = G$ if x_k is not e-exceptionally good (3.2.3), but x is (e, k) exceptionally good (3.2.3.1, b)).

3iv) $\mathrm{Ab}_e(x) = B$ in any other case.

Finally set

$$I_e = \{M, G, B\} \times \mathbb{Q} \times \mathbb{Z}$$

with lexicographic order, and

$$f_e : \ \mathrm{Sing}(J_n, b) \to I_e$$

$$f_e(x) = (\mathrm{Ab}_e(x), w\text{-}\mathrm{ord}_e(x), n_e(x)).$$

Notice that $w\text{-}\mathrm{ord}_e(x) \in \mathbb{Q} \geq 0$ and $n_e(x) \in \mathbb{Z} \geq 0$. If Max $f_e = (\alpha, \beta, \gamma)$ then $\beta \geq 0$ and $\gamma \geq 0$. Condition II) of the lemma in 3.3 will hold whenever $\alpha = B$ and $\beta > 0$. This will be good enough for the purpose of resolutions.

In fact the case $\alpha = B$ and $\gamma = 0$ will never occur because 3i) or 3ii) will hold if $w\text{-}\mathrm{ord}(x) = 0$.

If $\alpha = G$ then any point $x \in \mathrm{Sing}(J_n, b)$ will be either e-exceptionally good or monomial (3i) or (3ii). But then, after blowing up centers of codimension $e + 1$ we will come to the monomial case (Remark C of 3.3.1). That is, to the case $\alpha = M$.

The resolution in this last case is a simple combinatorial problem.

4. Examples

1) $W = \mathbb{A}^3_k, E = \emptyset, f = z^2 + xy^2$ so $b = 2; b! = 2$, we can define

$$V = \{z = 0\}$$

and

(1.1) $$C(f) = < xy^2 > .$$

To simplify notation, let $L_1 = \{x = 0\} \subset V$ and $L_2 = \{y = 0\} \subset V$ so that

$$I(L_1) = < x > \text{ and } I(L_2) = < y >,$$

and we express $C(f) = < I(L_1) \cdot I(L_2)^2 >$.

$$\mathrm{Sing}(C(f), 2) = L_2,$$

$w\text{-}\mathrm{ord}(x) = 1 \ \forall x \in L_2$, except for $x_0 = L_1 \cap L_2$ (the origin):

$$w\text{-}\mathrm{ord}(x_0) = \frac{3}{2}.$$

Since $E = \emptyset$, $n(x) = 0$ $\forall x \in \mathrm{Sing}(C(f), 2)$ and $\mathrm{Ab}(x) = B$, $\forall x \in \mathrm{Sing}(C(f), 2)$.

So $\underline{\mathrm{Max}}\, f = x_0$. After blowing up x_0

$$C(f)_1 = I(H_1)I(L_1)I(L_2)^2$$

where L_i denotes again the strict transform of L_i ($i = 1, 2$) and H_1 the exceptional hypersurface.

Now $L_1 \cap L_2 = \emptyset$ and $\mathrm{Sing}(C(f)_1, 2) = L_2 \cup \{H_1 \cap L_1\}$.

As for the function w-ord:

$w\text{-}\mathrm{ord}(H_1 \cap L_1) = \frac{1}{2}$.

$w\text{-}\mathrm{ord}(H_1 \cap L_2) = 1$.

$w\text{-}\mathrm{ord}(y) = 1$ $\forall y \in L_2$.

$n(x)$ will be zero except for $x_0 = H_1 \cap L_2 :$ $n(x_0) = 1$ (3.4., 2)).

Since $\mathrm{Ab}(x) = B$ $\forall x \in \mathrm{Sing}(C(f)_1, 2)$ it follows that $\underline{\mathrm{Max}}\, f = x_0 = H_1 \cap L_2$.

After blowing up this point one can check that $\underline{\mathrm{Max}}\, f$ will be L_2 (the strict transform of L_2), this is of codimension 1 in V. So condition I) of 3.3.1 says that $\underline{\mathrm{Max}}\, f$ should be smooth (which is clear in the example).

Following this condition I) we will blow-up L_2 and still one more quadratic transformation will be required to obtain a resolution.

Remark. In this first example the invariants Ab plays no role, actually the canonical algorithm of resolution of singularities introduced in [7] to prove the constructiveness of the theorem of resolution of singularities, is an algorithm which relies entirely on the functions $w\text{-}\mathrm{ord}_e$ and n_e. The next example will show why in general this new algorithm can be much faster and efficient.

2) $W = \mathbb{A}^3_k$, $E = \emptyset$.

$$f = z^2 + (x^3 + y^4)(y^3 + x^4)(x + 2y)(y + 2x)$$

so $b = 2$; $b! = 2$.

Set $V = \{z = 0\}$

$$C_1 = \{x^3 + y^4 = 0\}; C_2 = \{y^3 + x^4 = 0\}; L_1 = \{x + 2y = 0\};$$

$$L_2 = \{y + 2x = 0\}$$

and $C(f) = I(C_1)I(C_2)I(L_1)I(L_2)$.

So $\mathrm{Sing}(f, 2) = \mathrm{Sing}(C(f), 2)$ is the origin. After blowing up that point:

$$C(f)_1 = I(H_1)^6 I(C_1)I(C_2)I(L_1)I(L_2)$$

where now C_1, C_2, L_1, L_2 (denote the strict transform of the old ones) are all smooth and there is no intersection among them. One can check that

$$\mathrm{Sing}(C(f)_1, 2) = H_1$$

and

$$\mathrm{Ab}(x) = G, \quad \forall x \in H_1.$$

In the new algorithm the center will be H_1, and such transformation repeated three times, is a resolution.

Altogether four steps are required for a resolution of our original problem with the new algorithm. The resolution of the same singularity requires over 20 transformations with the old algorithm.

References

[1] Abhyankar, S.S. Good points of a Hypersurface. Adv. Math. 68, 87–256 (1988).

[2] Abhyankar, S.S. *Weighted Expansions for Canonical Desingularization.* Lecture Notes in Math. 910, Springer-Verlag, Berlin, Heidelberg, New York, 1982.

[3] Cossart, V., Giraud, J., Orbanz, U., *Resolution of Surface Singularities.* Lecture Notes in Math. 1101, Springer-Verlag, Berlin, Heidelberg, New York, 1984.

[4] Hironaka, H., Idealistic exponent of singularity. In: *Algebraic Geometry.* The John Hopkins centennial lectures, 52–125. Baltimore; John Hopkins. University Press 1977.

[5] Lipman, J., Introduction to resolution of singularities. In: *Algebraic Geometry Arcata* 1974. Amer. Math. Soc. Proc. Symp. Pure Math. 29, 187–230.

[6] Moh, T.T., Canonical Uniformization of hypersurface singularities of characteristic zero. J. Pure Appl. Algebra (to appear).

[7] Villamayor, O., Constructiveness of Hironaka's resolution. Ann. Sci. École Norm. Sup. (4) 22, 1–32, 1989.

[8] Villamayor, O., Patching local uniformizations. Ann. Sci. École Norm. Sup. (4) 25, 629–677, 1992.

[9] Villamayor, O., Good points and algorithmic resolution of singularities. To appear.

[10] Villamayor, O., Introduction to the algorithm of resolution. In: *Proceedings of La Rabida III* (to appear).

List of Contributors

Carlos Andradas, Departamento de Algebra, Facultad de Matemáticas, Universidad Complutense de Madrid, 28040 Madrid, Spain

Eberhard Becker, Fachbereich Mathematik, Universität Dortmund, 44221 Dortmund, Germany

Ralph Berr, Fachbereich Mathematik, Universität Dortmund, 44221 Dortmund, Germany

John F. Canny, Computer Sciences Division, University of California at Berkeley, Berkeley, CA 94720-0001, USA

Michel Coste, Institut de Recherche Mathématique de Rennes (CNRS, URA 305), Université de Rennes 1, Campus de Beaulieu, 35042 Rennes Cedex, France

Charles N. Delzell, Department of Mathematics, Louisiana State University, Baton Rouge, Louisiana 7080-4918, USA

Jean-Pierre Françoise, UFR 920, URA 213, Université de Paris VI, 45-46, 5ème, 4 place Jussieu, 75252 Paris Cedex 05, France

Danielle Gondard, Université Paris VI, Institut de Mathématiques Pures et Appliquées, UER 47, 4 Place Jussieu, 75252 Paris Cedex 05, France

Maria Jose González-López, Departamento de Matemáticas, Facultad de Ciencias, Universidad de Cantabria, 39071 Santander, Spain

Zbigniew Hajto, Departamento Algebra, Geometría y Topología, Facultad de Ciencias, Prado de la Magdalena s/n, 47005 Valladolid, Spain

Ilia V. Itenberg, DMI, Ecole Normale Supérieure, 45 rue d'Ulm, 75005 Paris, France

V.M. Kharlamov, UER Mathématique, Université Louis Pasteur, rue R. Descartes 7, 67084 Strasbourg Cedex, France

Krzysztof Kurdyka, U.E.R. de Mathématiques, Université de Savoie, Campus Scientifique, 73376 Le Bourget-du-Lac Cedex, France

James J. Madden, Department of Mathematics, Louisiana State University, Baton Rouge, Louisiana 7080-4918, USA

Tomas Recio, Departamento de Matemáticas, Facultad de Ciencias, Universidad de Cantabria, 39071 Santander, Spain

Jean-Jacques Risler, DMI, ENS, 45 Rue d'Ulm, 75005 Paris Cedex 05, France

Felice Ronga, Section de Mathématiques, 2-4 Rue du Lièvre, Case Postale 240, 1211 Genève 24, Switzerland

Jesús M. Ruiz, Departamento de Geometría y Topología, Facultad de Matemáticas, Universidad Complutense de Madrid, 28040 Madrid, Spain

Francisco Santos, Departamento de Matematica, Faculdad de Ciencias, Universidad de Cantabria, 39071 Santander, Spain

Claus Scheiderer, Fakultät für Mathematik, Universität Regensburg, Universitätsstrasse 31, 93040 Regensburg, Germany

Masahiro Shiota, Department of Mathematics, College of General Education, Nagoya University, Chikusa-ku, Nagoya 464-01, Japan

Orlando Villamayor, Departamento de Matematica, Universidad Autonoma de Madrid, Ciudad Universitaria de Canto Blanco, 28049 Madrid, Spain

Thierry Vust, Section de Mathématiques, 2-4 Rue du Lièvre, Case Postale 240, 1211 Genève 24, Switzerland

List of Participants

Francesca Acquistapace (Università di Pisa, Italy)
Carlos Andradas (Universidad Complutense de Madrid, Spain)
Edoardo Ballico (Università di Trento, Italy)
Ralph Berr (Universität Dortmund, Germany)
Jacek Bochnak (Vrije Universiteit, Amsterdam, The Netherlands)
Jean-Paul Brasselet (Faculté des Sciences Luminy, Marseille, France)
Ludwig Bröcker (Westf. Wilhelms Universität, Münster, Germany)
Fabrizio Broglia (Università di Pisa, Italy)
Mike Buchner (University of New Mexico, Albuquerque, USA)
John F. Canny (University of California at Berkeley, USA)
Ana Castilla Carbajo (Universidad Complutense de Madrid, Spain)
Michel Coste (Université de Rennes I, France)
Marie Francoise Coste-Roy (Université de Rennes I, France)
Maria Jesús De La Puente Muñoz (Universidad Complutense de Madrid, Spain)
Charles N. Delzell (Louisiana State University, Baton Rouge, USA)
Massimo Ferrarotti (Università di Pisa, Italy)
Elisabetta Fortuna (Università di Pisa, Italy)
Jean-Pierre Françoise (Université de Paris VI, France)
Margherita Galbiati (Università di Pisa, Italy)
José Manuel Gamboa (Universidad Complutense de Madrid, Spain)
Danielle Gondard (Université de Paris VI, France)
Francesco Guaraldo (Università "La Sapienza", Roma, Italy)
Zbigniew Hajto (Facultad de Ciencias, Valladolid, Spain)
Jean-Pierre Henry (Ecole Polytechnique, Palaiseau, France)
Heisuke Hironaka (Japan Association for Mathematical Sciences, Tokyo, Japan)
Ilia V. Itenberg (Ecole Normale Supérieure, Paris, France)
Alain Jacquemard (Laboratoire de Topologie, Dijon, France)
V. Kharlamov (Université Louis Pasteur, Strasbourg, France)
Krzysztof Kurdyka (Université de Savoie, Le Bourget-du-Lac, France)
Francesca Li Volsi (Università di Pisa, Italy)
Domenico Luminati (Università di Pisa, Italy)
James J. Madden (Louisiana State University, Baton Rouge, USA)
Maria Grazia Marinari (Università di Genova, Italy)
Tadeusz Mostowski (Warsaw University, Poland)
Barbara Nelli (Università di Pisa, Italy)
Laurent Noirel (Université de Provence, Marseille, France)
Alexander Prestel (Universität Konstanz, Germany)
Tomas Recio (Universidad de Cantabria, Santander, Spain)
Jean Jacques Risler (Ecole Normale Superieure, Paris, France)
Felice Ronga (Université de Genève, Switzerland)
Mario Salvetti (Università di Pisa, Italy)

Francisco Santos (Universidad de Cantabria, Santander, Spain)
Claus Scheiderer (Universität Regensburg, Germany)
Niels C. Schwartz (Universität Passau, Germany)
Mika Seppälä (University of Helsinki, Finland)
Ivonne Sgura (Università di Pisa, Italy)
Masahiro Shiota (Nagoya University, Japan)
Robert Silhol (Université Montpellier II, France)
Zbigniew Szafraniec (University of Gdansk, Poland)
Alessandro Tancredi (Università di Perugia, Italy)
Alberto Tognoli (Università di Trento, Italy)
Jean Claude Tougeron (Université de Rennes I, France)
David Trotman (Université de Provence, Marseille)
Orlando Villamayor (Universidad Autonoma de Madrid, Spain)
Oleg Viro (Steklov Math. Institute, St. Petersburg, Russia)

Bei Fragen zur Produktsicherheit wenden Sie sich bitte an:
If you have any questions regarding product safety,
please contact:

Walter de Gruyter GmbH
Genthiner Straße 13
10785 Berlin
productsafety@degruyterbrill.com